Victims of Fashion

Animal products were used extensively in nineteenth-century Britain. A middle-class Victorian woman might wear a dress made of alpaca wool, drape herself in a sealskin jacket, brush her hair with a tortoiseshell comb and sport feathers in her hat. She might entertain her friends by playing a piano with ivory keys or own a parrot or monkey as a living fashion accessory. In this innovative study, Helen Cowie examines the role of these animal-based commodities in Britain in the long nineteenth century and traces their rise and fall in popularity in response to changing tastes, availability and ethical concerns. Focusing on six popular animal products – feathers, sealskin, ivory, alpaca wool, perfumes and exotic pets – she considers how animal commodities were sourced and processed, how they were marketed and how they were consumed. She also assesses the ecological impact of nineteenth-century fashion.

Helen Louise Cowie is Professor of History at the University of York, where she researches the history of animals and the history of science. She is the author of *Conquering Nature in Spain and Its Empire, 1750–1850* (2011), *Exhibiting Animals in Nineteenth-Century Britain: Empathy, Education, Entertainment* (2014) and *Llama* (2017).

SCIENCE IN HISTORY

Series Editors

Simon J. Schaffer, University of Cambridge

James A. Secord, University of Cambridge

Science in History is a major series of ambitious books on the history of the sciences from the mid-eighteenth century through the mid-twentieth century, highlighting work that interprets the sciences from perspectives drawn from across the discipline of history. The focus on the major epoch of global economic, industrial and social transformations is intended to encourage the use of sophisticated historical models to make sense of the ways in which the sciences have developed and changed. The series encourages the exploration of a wide range of scientific traditions and the interrelations between them. It particularly welcomes work that takes seriously the material practices of the sciences and is broad in geographical scope.

Victims of Fashion

Animal Commodities in Victorian Britain

Helen Louise Cowie

University of York

CAMBRIDGE
UNIVERSITY PRESS

CAMBRIDGE
UNIVERSITY PRESS

University Printing House, Cambridge CB2 8BS, United Kingdom

One Liberty Plaza, 20th Floor, New York, NY 10006, USA

477 Williamstown Road, Port Melbourne, VIC 3207, Australia

314–321, 3rd Floor, Plot 3, Splendor Forum, Jasola District Centre, New Delhi – 110025, India

103 Penang Road, #05–06/07, Visioncrest Commercial, Singapore 238467

Cambridge University Press is part of the University of Cambridge.
It furthers the University's mission by disseminating knowledge in the pursuit of education, learning, and research at the highest international levels of excellence.

www.cambridge.org
Information on this title: www.cambridge.org/9781108495172
DOI: 10.1017/9781108861267

First published 2022

Printed in the United Kingdom by TJ Books Limited, Padstow Cornwall

A catalogue record for this publication is available from the British Library.

Library of Congress Cataloging-in-Publication Data
Names: Cowie, Helen (Helen Louise), author.
Title: Victims of fashion : animal commodities in Victorian Britain / Helen Louise Cowie, University of York.
Description: Cambridge, United Kingdsom ; New York, NY : Cambridge University Press, 2022. | Series: Science in history | Includes bibliographical references and index.
Identifiers: LCCN 2021031766 (print) | LCCN 2021031767 (ebook) | ISBN 9781108495172 (hardback) | ISBN 9781108817080 (paperback) | ISBN 9781108861267 (epub)
Subjects: LCSH: Fashion–Great Britain–History–19th century. | Clothing and dress–Great Britain–History–19th century. | Animal products–Great Britain–History–19th century. | Great Britain–Social life and customs–19th century. | BISAC: TECHNOLOGY & ENGINEERING / History
Classification: LCC GT737 .C75 2022 (print) | LCC GT737 (ebook) | DDC 391.00941/09034–dc23
LC record available at https://lccn.loc.gov/2021031766
LC ebook record available at https://lccn.loc.gov/2021031767

ISBN 978-1-108-49517-2 Hardback

Contents

Figures

Acknowledgements

Victims of Fashion began its slow journey from fledgling idea to book around seven years ago, when I stumbled across a newspaper article about alpaca smuggling in nineteenth-century Peru. Since then, it has been refashioned multiple times, taking me to topics and places I would not have anticipated and encompassing a range of coveted – and sadly persecuted – animal species, from fur seals to parrots. In the process, I have incurred many debts and learned a great deal about human–animal relations in the past and the present.

The research for this book was undertaken at the University of York, where I benefited from the support and friendship of my colleagues in the history department. Tara Alberts, Henrice Altink, Oleg Benesch, Sanjoy Bhattacharya, Martha Cattell, Sabine Clarke, David Clayton, Hanne Cottyn, Joanna de Groot, Jasper Heinzen, Stephanie Howard-Smith, Mark Jenner, Catriona Kennedy and Sam Wetherell all pointed me in the direction of valuable primary and secondary sources, while Amanda Behm, Lawrence Black, Simon Ditchfield, Natasha Glaisyer, Hannah Grieg, Jon Howlett, David Huyssen, Tom Johnson, Gerard McCann, Shaul Mittelpunkt, Harry Munt, Emilie Murphy, Mark Roodhouse, Lucy Sackville, Hugo Service, Laura Stewart, Craig Taylor, Miles Taylor, Pragya Vohra and Sethina Watson contributed much appreciated support and friendship throughout the period in which this book was in gestation. Members of the Centre for Eighteenth-Century Studies provided further inspiration. I would also like to extend my thanks to several cohorts of students on my various animal history modules whose seminar contributions helped shape my work.

Beyond York, I owe thanks to a variety of institutions and individuals. First, I would like to thank the librarians and archivists who assisted in locating material and securing image permissions, in particular staff from the British Library (especially Boston Spa Reading Rooms), the London Metropolitan Archives, the Victoria and Albert Museum, the State Library New South Wales, the Archivo de Indias and the Smithsonian Archives. I would also like to extend a special thanks to Sky Duthie for allowing me to browse his fantastic personal collection of Humanitarian League magazines.

Second, I am extremely grateful for financial assistance from the Eccles Centre and the Leverhulme Trust, without which the book could not have been written – or at least would have taken much longer to finish. I am also grateful to the University of York for granting me departmental research leave in which to pursue this project.

Third, I owe thanks to a range of scholars from around the world who read portions of the book, invited me to speak at conferences, shared their own research with me or gave feedback on seminar presentations, all of which made me rethink aspects of my research. These include (though are not limited to) John McAleer, Isabelle Charmantier, William Gervase Clarence-Smith, McKenzie Cooley, John Curry-Machado, Rebecca Earle, Alison FitzGerald, Oliver Hochadel, Dominik Hünniger, Elle Larson, Claudia Leal, Anthony McFarlane, Rory Miller, Kaori Nagai, Robin Peel, Neil Pemberton, Graciela Iglesias-Rogers, Sally Shuttleworth, Tess Somervell and Ryan Sweet. I would also like to thank members of the LA Global research network, which ran throughout much of the writing of this book, especially Mark Thurner and Juan Pimentel, for organising several stimulating workshops, and Adrian Masters and Elisa Sevilla, for taking the time to send me valuable references on vicuñas.

I am grateful to Lucy Rhymer, Rachel Blaifeder and Emily Sharp at Cambridge University Press for their encouragement throughout the project, and to the series editors, Jim Secord and Simon Schaffer, who provided useful advice on reworking it. I am also very grateful to the two anonymous peer reviewers, both of whom made valuable suggestions for improvements to the text.

Last but not least, my thanks go to my parents, Peter and Susan Cowie, my sister, Alice Cowie, who shared her expertise on animal behaviour, and, above all, my husband, Paul Williams, who has stoically lived with this project through its many ups and downs and has uncomplainingly accompanied me on many a quest in search of archival and living animals. I would also like to pay tribute to my cat, Daisy Cowie II, who has been a source of feline inspiration during the book's final stages and has reminded me loudly, charmingly and persistently why animals matter.

Introduction

In 1894, a journalist published an article in *The Standard* advocating the establishment of a kangaroo farm in England. Over the course of several pages he enumerated the multiple benefits to be gained from acclimatising this antipodean marsupial, whose hide produced 'excellent leather' for making boots and gloves, whose thighs 'taste much like those of the reindeer', and whose tail made 'a rich and most delicious soup'. Despite understandable fears to the contrary, the journalist insisted that kangaroos would do well in similar terrain to sheep and were 'sufficiently hardy animals to withstand even the trying variability of our English winters'. He also claimed that they would require comparatively little looking after, '[a]ccommodation in the shape of open shedding' providing 'sufficient shelter to keep the animals in health' and a seasonal supply of 'hay and other food' meeting their dietary needs. Although its proponent's primary concern was profitability, the scheme was potentially timely, as kangaroo numbers were rapidly decreasing in Australia due to overhunting and competition with sheep.[1] Writing just five years earlier, another journalist had reported: 'Large quantities of kangaroos are killed for the sake of the skin, which has become fashionable as a material for the making of boots, shoes and other articles, and unless this indiscriminate slaughter is stopped the kangaroo will soon have shared the fate of the dodo.'[2]

The article on kangaroo farming elicited an extensive, if mixed, response. One reader, a glove manufacturer, disputed the idea that kangaroo hide might be used in his business 'to supplement the rather short supply of buck- and doe-skin', his own earlier experiments having proven that it was the wrong texture – too 'close, hard and thick' – and 'very awkward in shape'. A second respondent, A.A.H., challenged the belief that the kangaroo could survive the British winter, since it might 'stand cold to a certain extent, but not wet'. He remarked, furthermore,

---

[1] 'Kangaroo Farming in England', *The Standard*, 11 January 1894.
[2] 'Extinction of the Kangaroo', *Northampton Mercury*, 7 September 1889.

1

that '[d]uring a residence of nearly two years in Australia, I never once saw kangaroo meat placed on the table', although it was true that 'the tail makes very fair soup'.[3] A third writer, James Troubridge Critchell, reacting to A.A.H.'s letter, surmised that the Australians' disdain for kangaroo meat was not proof of its inedibility, but rather of the 'most limited and unenterprising character' of Australian cuisine, which consisted of a 'never-varying' diet of 'beef and mutton, washed down with copious libations of tea'.[4] A fourth, proud Victoria native S. H. Palmer, rallied in turn to defend the cuisine of his homeland, insisting, 'If ever Mr Critchell visits Victoria he will find that "Australian wines", "salads" and "market gardens" are far from being "curiosities" or "conspicuous by their absence".'[5] Drawing things back to the matter at hand, a fifth respondent, G. A. Haig, suggested: 'Before we start farming kangaroos in England, would it not be advisable for some of the Australian meat companies to bring home some kangaroo meat, refrigerated, and let us see how we like it?'[6] A sixth, R. A. Swayne, recollected fondly that 'I once bought a tin of kangaroo tails' meat many years ago, when tinned meats were first introduced, and found it excellent, equal to a good game pie'.[7]

In the event, the proposed kangaroo farm never materialised and the prospect of domesticating the species faded from public debate. Individual kangaroos did make it to Britain in the nineteenth century, proving that the animals could survive the rigours of an intemperate climate; one man from Beaumaris in north Wales owned a 'pretty small bush kangaroo, from Tasmania, tame and healthy', which had 'spent two winters in an open garden'.[8] Large-scale kangaroo farming, however, did not take off in Britain – or, indeed, Australia – and kangaroo meat remained an occasional curiosity rather than a regular element of the British diet. In 1897, when a consignment of kangaroo tails was delivered to Leadenhall Market in London, '[t]he public discovered that the tail of a kangaroo made into soup was a very succulent dish, rich and highly nutritious, and like Oliver Twist, they asked for more, but the stock of 500 caudal appendages was soon exhausted'.[9]

[3] 'Kangaroo Farming in England', *The Standard*, 13 January 1894.
[4] 'Kangaroo Farming', *The Standard*, 17 January 1894. Critchell was a reporter for the *Pastoral Review* and later co-authored *A History of the Frozen Meat Trade* (London: Constable, 1912) with Joseph Raymond.
[5] 'Kangaroo Farming in England', *The Standard*, 20 January 1894.
[6] 'Kangaroo Farming', *The Standard*, 17 January 1894.
[7] 'Kangaroo Farming in England', *The Standard*, 20 January 1894.
[8] 'Country House', *The Bazaar, the Exchange and Mart and Journal of the Household*, 28 August 1872, p. 158.
[9] 'Kangaroo Tails', *The Standard*, 15 September 1898.

Despite its failure, Britain's brief flirtation with the idea of kangaroo farming was a significant episode that encapsulates a much wider pattern of engagement with animal commodities in the Victorian era. First, it highlights the desire for new luxuries, either for the table or the dressing room – or, in this case, both – and the degree to which changing tastes shaped emerging fashions. Second, it illustrates the ecological effects of these fashions, which often resulted in the over-exploitation of animal species to meet the growing market demand. Third, it demonstrates one of the most frequently proposed solutions to this ecological problem: the domestication and/or relocation of coveted animals as a way of either halting or managing the slaughter. Fourth, the responses generated by the kangaroo farming proposal reveal the role of new technologies – canning, refrigeration, steamship transportation – in driving demand for new commodities, and the ways in which professional knowledge, individual experience and imperial/colonial pride could shape discussions over where, how, and indeed whether, certain species should be exploited. The great kangaroo debate of 1894 thus represents in microcosm some of the broader questions manufacturers, farmers and conservationists would have to tackle as they extracted the varied products of the animal world.

★ ★ ★

A year after the kangaroo farm hit the headlines, another bizarre animal project gripped the nation's attention. In this instance, the scheme in question was a proposed cat farm, which was reportedly about to be established in the Netherlands. According to an article in the *Sun*, the farm would consist of several hundred cats, all black, and would breed and raise the animals solely 'for the sake of their skins'. Where the original felines would come from was not stated. Perhaps they would be pedigree specimens, donated by their owners; more likely, they would be feral cats, rounded up on the streets.

Like the kangaroo farm, the 'projected cat farm' generated a significant amount of comment, in this case most of it negative. Unlike kangaroos, cats were, by the late nineteenth century, popular household pets and not routinely subject to slaughter for their fur.[10] Cared for by loving owners, photographed in touching pictures (Figure 0.1) and even, on occasion,

---

[10] Earlier in the century the status of the cat had been less assured. In 1854, a 'ruffianly-looking fellow named Richard Calvert' was brought before magistrates in Clerkenwell charged with 'skinning cats while they were alive'. The pelts were reportedly sold to manufacturers and used as 'linings of gentlemen's winter clothing'. 'Clerkenwell', *The Morning Chronicle*, 28 February 1854.

Figure 0.1 'A Tête-à-Tête', *The Animals' Friend*, 1911, p. 36.

rescued by the fire brigade (Figure 0.2), cats were common animal companions in the Victorian era and often formed a close bond with their human carers. The idea that they might be reduced to the status of cattle and reared solely for their skins thus shocked cat lovers across Britain, prompting vocal protests against the concept. How could beloved pets be treated in this way? How could anyone justify the 'barbarity of such [a] projected industry'?[11]

Among the gasps of horror, one more nuanced critique of the proposed cat farm appeared in a somewhat unexpected place: the Royal Society for the Prevention of Cruelty to Animals' (RSPCA's) monthly magazine, *The Animal World*. In a surprisingly measured response to the *Sun*'s original report, the author reflected on the complex ethics surrounding cat farming and conceded that, while '[w]e are cat lovers, and we would rather the company should not succeed', if 'the cats are destroyed humanely, we fail to see the illegality of the practice, in law or in morals'. Meditating on whether it was inherently wrong to raise cats for their skins, the writer concluded that it was not, or at least no more so than farming sheep in Australia 'solely for the purpose of the market volume of their fat', killing Alaskan seals 'for the purpose of producing skins for jackets' or creating 'ostrich farms … in Africa … only for the

---

[11] 'A Projected Cat Farm', *The Animal World*, March 1895, p. 38.

Figure 0.2 'Humane Rescue of a Cat by Fireman Cave', *The Animal World*, June 1893, p. 92.

market of ostrich feathers'. Considering, similarly, whether it was intrinsically worse to kill animals for clothing than for food, he suggested that using the skins of creatures originally killed for their meat might be more acceptable, although in the specific case of cats, the reverse appeared to be true, since behind the 'outcry against the new company' may have been a fear 'that the bodies of the cats would be used for food, without certainty of detection'. Whether cat meat entered the food chain or not, the acceptability of the cat farm ultimately came down to two issues, both directly connected to the cats' well-being: 'the happiness of the animals when they arrive here [at the cat farm] and the method by which their lives will be terminated'. On the second issue, the author dismissed the concern that cats might be skinned alive; since '[t]he glint or gloss of fur is not lost if the skin be removed from the body after death but before the carcase becomes cold, and as the mutilation of an animal endowed with means of defence like a cat is easier after death than during life, surely such dreaded evil has no foundation'. On the first issue, however, he was less certain, for '[i]t is doubtful whether cats can be reared in communities without deterioration of health, or the spread of disease, or the transmission of infirmities'. Moreover, 'a cat without human affection is a wild animal, which without wild life must be unhappy'. On these grounds, and not on the grounds of killing cats per se, the writer ultimately opposed the proposed cat farm, suggesting that no cat would be happy there while alive.[12]

The reflections inspired by the Dutch cat farm – like those elicited by kangaroo farming – touched on some of the key moral questions surrounding any animal-based commodity. Was it ethical to rear animals simply in order to kill them for their flesh or coats? Was the method of killing the key determinant of this, or was it the treatment of the animals while they were alive that mattered most? If the latter, were there some species that could not, and should not, be kept in captivity, because their natures and habits militated against their happiness in this state? And did it make a difference whether the species in question was long-domesticated, essentially wild or, in the case of the cat, hovering somewhere between the two categories? These were questions that nineteenth-century commentators repeatedly grappled with, and which remain highly pertinent in the modern world. As we shall see, they would surface again and again in relation to different animal commodities, as manufacturers, humanitarians and conservationists pondered the morality of shooting elephants, clubbing seals, farming civets and keeping parrots as pets.

* * *

Animal products were used extensively in nineteenth-century Britain. A middle- or upper-class Victorian woman might wear a dress made of fine alpaca wool, drape herself in a sealskin jacket, strap herself into a whalebone corset, brush her hair with a tortoiseshell comb and sport the feathers or sometimes the entire bodies of wild birds in her hat or on her earrings. She might entertain her friends and family by playing a piano

---

[12] Ibid.

with ivory keys, own a parrot or monkey as a living fashion accessory and, at the dinner table, feast on Argentine beef, New Zealand lamb, or, if she were feeling more adventurous, kangaroo tail. She might chew these delicacies with dentures made from the teeth of a hippopotamus. The use and consumption of such luxuries attested to her place in high society and her attention to the latest sartorial and culinary fashions. They also had momentous and often dire environmental implications for the animal sources of these products, which perished in their thousands to satisfy the latest trends and caprices of the 'civilised' world.

Victims of Fashion examines the role of animal-based commodities in Britain in the period c.1800–1914. Focusing on six animal products employed for fashionable living – birds' feathers, sealskin, ivory, alpaca wool, perfumes (civet, musk, ambergris and bear's grease) and exotic pets – the book highlights the pervasive nature of animal-based consumables in the Victorian and Edwardian eras and traces their rise and fall in popularity in response to changing tastes, availability and ethical concerns. *Victims of Fashion* considers how animal commodities were sourced and processed, how they were marketed and how they were consumed. It also assesses the humanitarian and ecological issues raised by the consumption of exotic luxuries and the moral dilemmas these posed for consumers.

Animal commodities were not, of course, a novelty in the nineteenth century. Animal products had been used extensively in pre-industrial societies and some were traded over long distances. Beaver pelts were procured in North America and sent to Europe to make felt hats.[13] Pearls were sourced from the East and West Indies to adorn the bodies of European elites.[14] Silk was transported from China to Europe across Central Asia, while sable fur reached Western Europe from Russia.[15] What was new in the Victorian era, however, was the range and volume of products in circulation. Faster transportation in the form of railways and steamships brought large quantities of feathers, wool and fur over long distances, while imperial penetration opened up once remote regions to commercial exploitation. Shorter journey times and new preservation techniques also allowed perishable goods to traverse continents and oceans for the first time, bringing New Zealand lamb and Argentine

[13] Shepard Krech III, *The Ecological Indian* (New York and London: Norton, 1999), pp. 173–209.
[14] Molly Warsh, *American Baroque: Pearls and the Nature of Empire, 1492–1700* (Chapel Hill: University of North Carolina Press, 2018).
[15] Susan Whitfield, *Life along the Silk Road* (Berkeley: University of California Press, 2015); Janet Martin, *Treasure of the Land of Darkness: The Fur Trade and Its Significance for Medieval Russia* (Cambridge: Cambridge University Press, 1986).

beef to British consumers – initially as live cargo, later as dried, salted, tinned, frozen or chilled flesh.[16] This had implications, too, for the trade in live animals, which entered Britain in increasing numbers for sale as menagerie inmates or exotic pets.

Animal luxuries also became more widely accessible in the nineteenth century, giving rise to increased demand. In the early modern period, exotic animal commodities were confined to a privileged elite, their use sometimes regulated by sumptuary laws or limited, in practice, by high prices. In the wake of the Industrial Revolution, however, new manufacturing processes made goods cheaper, while more and more people had the disposable income necessary to spend on non-essential items.[17] Products that had once been the preserve of the nobility came increasingly within reach of the expanding middle classes, eager to keep up with the latest fashions. Even servants and artisans could afford to own feather bonnets or parrots, sometimes procured second-hand. The nineteenth century thus saw a greater range of animal goods being traded in greater numbers and over longer distances than ever before to satisfy the whims of a much wider spectrum of consumers.

Victims of Fashion examines this new influx of animal commodities and considers its social and ecological impact. The book centres on four key themes. First, I explore how each animal product was sourced and processed. I chart the commodity chains that brought elephants' tusks and egrets' feathers from Africa and South America to Europe and I unpick the global networks that facilitated them. Sealskin, for example, was procured by Aleut hunters on the Pribilof Islands off the coast of Alaska, shipped to San Francisco and forwarded to London for processing and sale. Alpaca wool was collected by Peruvian shepherds, passed to British merchants in the city of Arequipa and shipped to Liverpool, where much of it was bought by Bradford industrialist Titus Salt. The finished products were then re-exported to the rest of the world. The successful procurement and manufacture of luxury animal products required the knowledge and expertise of a wide range of people and operated, in many cases, on a truly global level. I explore how this expertise was generated and how it was transmitted between different cultures and nations. I trace animal commodities from the sierra to the salon, the jungle to the dressing table.

---

[16] Jack Goody, 'Industrial Food' in Carole Counihan and Penny Van Esterik (eds), *Food and Culture: A Reader* (London: Routledge, 1997), pp. 338–56; Rebecca J. Woods, 'From Colonial Animal to Imperial Edible: Building an Empire of Sheep in New Zealand, ca.1880–1900', *Comparative Studies of South Asia, Africa and the Middle East* 35:1 (2015), pp. 117–36.

[17] Peter N. Stearns, *The Industrial Turn in World History* (London: Routledge, 2017), p. 88.

Closely related to the sourcing of exotic commodities was the manner in which they were marketed. The nineteenth century witnessed important innovations in the way in which goods were bought and sold, as well as new opportunities for advertising. Cheaper manufacturing and increased choice brought what had once been elite products to a wider range of middle- and even working-class consumers, while the advent of mass media, monthly fashion magazines, catalogues and, towards the end of the nineteenth century, the department store forged a new niche for the female consumer, whose changing tastes in hats, coats and accessories shaped demand for a plethora of animal commodities.[18] From the1870s, moreover, seasonal fashion and a desire for novelty exerted a growing influence on the demand for different animal items, often taking precedence over utility and quality. As one commentator remarked in 1912 in relation to the sealskin industry, 'people do not pay for what is best; they pay for what fashion demands. If the fashion should demand baby seals, they would have to be taken.'[19] Situating animal commodities within this broader economic context, *Fashion Victims* considers the appeal and affordability of furs, feathers and fragrances and examines how they were advertised to the purchasing public. I explore, in particular, whether publicity for animal products suppressed, emphasised or fabricated their exotic origin, and how this reflected broader concerns about authenticity and adulteration.

A third element of the book focuses on attempts to appropriate, acclimatise and 'improve' useful animals by domesticating them, selectively breeding them or introducing them to new territories. Agriculturalists, for example, attempted to naturalise the alpaca on British soil, first in Scotland and Ireland, later in Australia, with the aim of both increasing wool production and gaining direct control over this valuable commodity. There were similar schemes in South Africa and the Belgian Congo to domesticate the ostrich and the African elephant respectively, and, as we have seen, even plans to acclimatise the kangaroo in Britain to farm it for its meat. Historians have become increasingly interested in the field of 'economic botany', charting efforts

---

[18] On the rise of the fashion magazine, see Margaret Beetham and Kay Boardman (eds), *Victorian Women's Magazines: An Anthology* (Manchester: Manchester University Press, 2001), pp. 10–20. On the advent of the department store, see Erika Diane Rappaport, *Shopping for Pleasure: Women and the Making of London's West End* (Princeton: Princeton University Press, 2000), pp. 142–77.

[19] 'Hearings Before the Committee of Expenditures in the Department of Commerce and Labour House of Representatives on House Resolution No. 73, To Investigate the Fur-Seal Industry of Alaska', 8, 18 and 20 June 1912, p. 1011, cited in William T. Hornaday, *Scrapbook Collection on the History of Wild Life Protection and Extermination*, Vol. 5, Wildlife Conservation Society Archives Collection, 1007-04-05-000-a.

to cultivate valuable commodities such as rubber, quinine and tea in different parts of the British Empire.[20] Animal acclimatisation has received less attention – in part, no doubt, because it was less successful – but it was promoted repeatedly as a means of improvement and even a form of conservation. Drawing on archival records, scientific treatises and articles in the contemporary press, I situate animal acclimatisation within a wider programme of biopiracy and livestock improvement, which sought to appropriate commercially valuable species and 'improve' them through careful husbandry.[21]

The consumption of animal products entailed a significant degree of suffering and pain for the animals used to manufacture them and ultimately threatened the existence of some species. This raised important moral and economic questions, and these form the final strand of this book. First, there were issues of sustainability. Overhunting, particularly the indiscriminate killing of females and young animals, drastically reduced the numbers of certain species and raised the spectre of their extinction. The Pacific fur seal, already exterminated in the southern hemisphere, was believed to be in danger in Alaska by the 1880s due to unregulated 'pelagic' sealing (catching seals in the sea rather than on land). Many bird species and the alpaca's wild relative the vicuña were also pushed to the verge of extinction. So devastating was the trade in ivory to African elephant populations that an article in *The Review of Reviews* in 1899 asked 'Is the elephant following the dodo?'[22] These severe environmental issues led to efforts to control the trade in over-exploited species by creating reservations, imposing a close season on hunting and setting quotas for the number of animals that could be killed. *Victims of Fashion* examines the genesis and enforcement of these regulations and emphasises the importance of international collaboration in ensuring their effectiveness. It also points to the complications

---

[20] Important studies of 'economic botany' include Lucille Brockway, *Science and Colonial Expansion: The Role of the British Royal Botanic Gardens* (New Haven: Yale University Press, 2002 [1979]); Richard Grove, *Green Imperialism: Colonial Expansion, Tropical Island Edens and the Origins of Environmentalism, 1600–1860* (Cambridge: Cambridge University Press, 1995); Richard Drayton, *Nature's Government: Science, Imperial Britain and the 'Improvement' of the World* (New Haven: Yale University Press, 2000); Emma Spary, *Utopia's Garden: French Natural History from Old Regime to Revolution* (Chicago: University of Chicago Press, 2000); Londa Schiebinger, *Plants and Empire: Colonial Bio-Prospecting in the Atlantic World* (Cambridge: Harvard University Press, 2004).

[21] On breeding and animal improvement, see 'Sex and the Single Animal' in Harriet Ritvo, *Noble Cows and Hybrid Zebras: Essays on Animals and History* (Charlottesville: University of Virginia Press, 2010), pp. 13–28; Rebecca J. Woods, *The Herds Shot around the World: Native Breeds and the British Empire, 1800–1900* (Chapel Hill: University of North Carolina Press, 2017).

[22] 'Is the Elephant Following the Dodo?', *The Review of Reviews*, September 1899, p. 287.

surrounding animal protection legislation, particularly when the animal in question lived in the sea or the air and crossed territorial boundaries.

The concerns expressed about the survival – or possible disappearance – of certain species need to be understood within the context of a rising awareness of extinction as both a conceivable phenomenon and one that could be brought about by human action. Until the early nineteenth century, the idea that God's perfectly designed creations could become extinct was simply unthinkable and contradicted entrenched biblical understandings of the cosmos. The discovery of the bones of giant ground sloths, mammoths, moa and other creatures for which no living counterpart could be found, however, gradually forced naturalists to accept (often reluctantly) the possibility that species could die out, radically changing man's perception of nature.[23] More pressingly, examples from recent times of animals that had been pushed to the brink of extinction (or beyond) by overhunting proved that it was possible for even the most numerous species to become extinct. This was illustrated most graphically by the plight of the bison, reduced from millions to just a few hundred animals in less than a century of incessant slaughter, and the passenger pigeon, gone from the wild by 1900 and completely extinct by 1914.[24] The realisation that even creatures that once existed in huge numbers could be exterminated stimulated a growing sense of nostalgia, sometimes tinged with shame and guilt, and triggered fears that many other species might go the same way if nothing were done to curb their exploitation. Writing in 1914 – the year when the last passenger pigeon died in Cincinnati Zoo – a contributor to *The Animal World* listed several animals believed to be 'verging on extinction', among them the Siberian sable and the Andean chinchilla, slaughtered in large numbers for their fur, the Tasmanian tiger or thylacine, killed by farmers for its alleged depredations on sheep and poultry; the flightless kakapo from New Zealand, under threat from habitat loss and introduced predators; and the 'inoffensive' koala, 'in much demand as a pet, owing to its chubby appearance and amiable disposition'.[25] Whether all of these animals were worth saving was another question, as we shall

---

[23] On the genesis of the concept of extinction, see 'The Mastodon's Molars' in Elizabeth Kolbert, *The Sixth Extinction: An Unnatural History* (London: Bloomsbury, 2014), pp. 23–46.

[24] Andrew Isenberg, *The Destruction of the American Bison* (Cambridge: Cambridge University Press, 2001); Mark V. Barrow, *Nature's Ghosts: Confronting Extinction from the Age of Jefferson to the Age of Ecology* (Chicago: University of Chicago Press, 2009); Ursula K. Heise, *Imagining Extinction: The Cultural Meanings of Endangered Species* (Chicago: University of Chicago Press, 2016), pp. 43–4.

[25] 'Animals Verging on Extinction', *The Animal World*, February 1913, pp. 27–30. For a detailed study of the extinction of the thylacine, see Robert Paddle, *The Last Tasmanian*

see. For at least some contemporaries, however, the extinction of any of God's animals was cause for regret, and, in some cases, remedial action.

Related to, but distinct from, these broader environmental concerns, animal rights activists also began to question animal exploitation from a humanitarian perspective. As several historians have shown, concern for animal suffering became increasingly vocal from the early nineteenth century, and the first legislation was passed in Britain to criminalise particular forms of abuse.[26] In the 1820s and 1830s, the focus was on brutal blood sports, such as bull and bear baiting (outlawed in 1835), and on the cruel treatment of horses and cattle (protected by law from 1822). By the 1870s and 1880s, however, concern was shifting to other forms of abuse, most notably the cruelties inflicted by vivisectionists in the name of science.[27] While domestic animals were at the centre of most of these campaigns, captive exotic animals also inspired sympathy and (occasionally) legal intervention, particularly if their abuse happened in public. In 1891, for instance, the RSPCA successfully prosecuted Italian organ grinder Loni Verrichia for 'thrash[ing]' a monkey 'with a stick' and nearly 'dash[ing]' out its brains against the woodwork of the organ'.[28] In 1901, the Society secured another conviction against 'Frank Delawar, a lion tamer', for beating a 'sulky' lion 'about the head with the stock of a whip'.[29]

Animals used as fashion accessories were, in general, obtained overseas, so the pain inflicted on them was less visible. As the nineteenth century progressed, however, awareness grew in Britain that the acquisition of several cherished luxuries entailed some form of cruelty, whether this was through skinning seals, removing shells from live turtles, shooting elephants for their ivory or slaughtering birds of paradise for their feathers. Humanitarians increasingly invoked this cruelty in their propaganda, often issuing stomach-churning descriptions of how animal products were procured. In 1897, D. Harrison accused the Hudson's Bay Company of poisoning bears with strychnine to make 'the fur glossy',

---

Tiger: The History and Extinction of the Thylacine (Cambridge: Cambridge University Press, 2000).

[26] See, for example, 'A Measure of Compassion' in Harriet Ritvo, *The Animal Estate: The English and Other Creatures in the Victorian Age* (Cambridge: Harvard University Press, 1987), pp. 125–66; Hilda Kean, *Animal Rights: Political and Social Change in Britain since 1800* (London: Reaktion Books, 1998), pp. 30–1.

[27] On the rise of anti-vivisection, see Coral Lansbury, *The Old Brown Dog: Women, Workers and Vivisection in Edwardian England* (Madison: University of Wisconsin Press, 1985); 'Cruelty and Kindness' in Anita Guerrini, *Experimenting with Humans and Animals: From Galen to Animal Rights* (Baltimore: Johns Hopkins University Press, 2003), pp. 70–92.

[28] 'Crystal Palace Flower Show', *Lloyd's Illustrated Newspaper*, 22 March 1891.

[29] 'Cruelty to a Lion', *Sunderland Daily Echo*, 27 May 1901.

leaving the animals 'doubled up' in 'agony'.[30] Six years later another member of the Humanitarian League, Joseph Collinson, denounced the cruelty perpetrated against turtles, which were hung alive over burning leaves to remove their precious shells and then, if they survived the ordeal, 'turned back into the sea to re-plate [themselves]'.[31] Both men attributed the ultimate blame for these horrors to the wearers of fur coats and tortoiseshell hair combs (presented – not always accurately – as mainly female), urging British and American consumers to boycott animal products and thereby spare animal lives. Both also emphasised the protracted nature of this animal suffering, drawing parallels with the 'torture' of vivisection. *Fashion Victims* considers what role cruelty played in discussions of animal commodities and assesses the role of gender in the animal welfare debate, noting both the growing consumer power of women as the prime buyers of fashionable animal products and their prominent role in animal protection societies.[32] I also examine some of the human abuses connected to the trade in animal commodities, which included the exploitation of indigenous peoples in Peru and the Pribilof Islands in pursuit of alpaca wool and sealskins and a much debated link between the ivory trade and the slave trade in colonial Africa; General Charles Gordon described the ivory trade as 'only the slave trade under another name' because it involved the impressment of African porters.[33]

Two further issues underlie all of these themes and are analysed throughout the book. One is the role of scientists and engineers. Scientists intervened in all aspects of the exotic animal trade, from supervising the breeding of animals to classifying species and devising tests to determine the authenticity of finished products. They also acted as vital consultants in debates about animal conservation, providing the evidence and expertise needed to formulate protection measures. Engineers were equally important, designing and building the machines used to convert exotic materials into finished products and the railways and steamships that transported them around the world, not to mention the firearms that would have such a devastating effect on so many species. *Victims of Fashion* explores the contribution of science and technology to the production of animal commodities and the subsequent role of science in the preservation of wildlife.

---

[30] 'The Fur Industry', *The Animals' Friend*, September 1897, p. 244.

[31] Joseph Collinson, 'What Tortoiseshell Is', *Humanity: The Journal of the Humanitarian League*, February 1900, p. 14.

[32] On the connection between women and animal protection, see Diana Donald, *Women against Cruelty: Protection of Animals in Nineteenth-Century Britain* (Manchester: Manchester University Press, 2019).

[33] 'The African Elephant', *The Anti-Slavery Reporter*, January 1879, p. 139.

A second important issue is the international dimension of the trade in exotic animal products. Animal commodities came to Britain from all over the world. Acquiring them necessitated multiple exchanges and cross-cultural encounters. Preserving them demanded international agreement and diplomacy. While focusing predominantly on Britain, *Victims of Fashion* emphasises this global context and explores the relationships and tensions created by the trade in animals. I examine species that originated from several distinct regions of the globe, including egrets from Florida and Venezuela, ostriches from South Africa, fur seals from the Bering Sea, elephants from East Africa and the Belgian Congo, alpacas from Peru and Bolivia, bears from Russia, musk deer from Tibet, civets from Abyssinia and grey parrots from West Africa (Figure 0.3).

The book encompasses a range of different voices and perspectives. Perhaps inevitably, scientists and politicians feature prominently in all of the case studies examined, not least because they wrote treatises and reports, gave speeches and organised conservation campaigns, generating an abundance of published material. At the same time, however, we also hear from less celebrated figures, whose role in the proceedings was less visible but equally important: the women who abjured the wearing of sealskin and birds' feathers; the Bolivian alpaca herders who travelled to Australia to assist with the animals' acclimatisation; the London hairdressers who kept live bears on their premises; and the Leeds housewife prosecuted for neglecting her pet parrot.[34] The inclusion of these lesser-known actors reveals the ubiquity of animal commodities in Victorian society and the complex, and sometimes conflicting, attitudes towards them.

Animal voices are obviously harder to recover, but some non-human actors did leave behind traces of their acquiescence or resistance to human projects. On a species level, animals often frustrated human plans, shaping what was and was not possible in terms of both exploitation and conservation. Slow reproduction cycles, for instance, made animals such as elephants more vulnerable to over-exploitation and harder to domesticate. Large ranges and unknown migration routes made fur seals and birds more difficult to protect when they strayed across artificial human boundaries. Centuries of adaptation to life at high altitudes made alpacas and vicuñas vulnerable to the hot, humid conditions of the tropics, thwarting multiple acclimatisation plans. On an individual level, some animals intruded more directly into the historical

[34] 'Alleged Cruelty to a Parrot', *Leeds Mercury*, 21 April 1903.

Figure 0.3 'Map Showing the Distribution of Animals'. (London: Cassell, 1900)

15

record, making a documented impact on the human world. Asian elephants Sundhar Gaj ('Beautiful Elephant'), Nadir Baksh ('Wonder Inspiring'), Phul Masla ('Flower Garland') and Susan Kali ('Budding Lily') were praised by their human handlers for their exploits during an African expedition.[35] One female alpaca imported to Adelaide in 1858 objected to being 'fondled and patted' by her Australian admirers and returned their 'caresses by ejecting in their faces a kind of steamy saliva, accompanying the impoliteness by a short peculiar noise, something between a hiss and a cough'.[36] A few animals – specifically parrots – even had their utterances recorded for posterity. Though mainly acted upon, therefore, animals did play a part in shaping their interactions with humans, exhibiting a significant, though limited, degree of agency in the commodification process.[37]

[35] 'Royal Belgian Expedition to Central Africa', *The Times of India*, 20 August 1880.
[36] 'Stock by the Orient', *Adelaide Observer*, 25 September 1858.
[37] On animal agency, and the possibility of recovering it in a historical context, see Susan Nance (ed.), *The Historical Animal* (Syracuse: Syracuse University Press, 2015); Philip Howell, 'Animals, Agency and History' in Philip Howell and Hilda Kean, *Handbook for Animal–Human History* (London: Routledge, 2018), pp. 197–221.

1 Murderous Millinery

TO THE LADIES OF ENGLAND
Why do you trifle with my skin
To decorate your bonnet?
Your friends would value you as much
Without myself upon it.   'The Ghost of a Hummingbird', *The Animal
World*, September 1876, p. 135

Among the extensive collection of jewellery in the Victoria and Albert Museum, there is a beautiful but macabre object: a pair of earrings made from the stuffed heads of two hummingbirds (Figure 1.1). Mounted on gold bases, the little birds stare blankly at the viewer through red glass eyes. Their delicate feathers glisten red, green and yellow and their tiny, gold-embossed beaks probe the air. The museum's catalogue states that the earrings were manufactured around 1865 by Harry Emanuel, a London jeweller who worked in both feathers and ivory. They belonged to Mrs Sidney Matilda Adams (1806–86) and were donated to the museum by her great-great-granddaughter, Katharine Mortimer.[1]

The sight of two dead birds on women's jewellery is shocking to a modern viewer and strikes us as bizarre, ghoulish and repulsive. In the nineteenth century, however, the practice of wearing birds' corpses on earrings, dresses and, most commonly, hats was regarded as the height of fashion. From the 1860s until the First World War, women adorned themselves with the plumage of hummingbirds, egrets and other attractive birds, competing with one another to achieve the most elaborate creations. One lady, attending a drawing room at Dublin Castle in 1878, wore a dress 'trimmed with the skins of 300 robins'.[2] Across the Atlantic, a correspondent of the American magazine *Forest and Stream* spotted 'on 700 hats 542 birds' walking down Fifth Avenue in New York in 1886, 'the common tern and quail being the most frequent'.[3] In such a

---

[1] See collections.vam.ac.uk/item/O86513/earrings-emanuel-harry/.
[2] 'Bird Murder', *The Animal World*, March 1878, p. 43.
[3] 'Birds, Butchers and Beauties', *Pall Mall Gazette*, 5 January 1886.

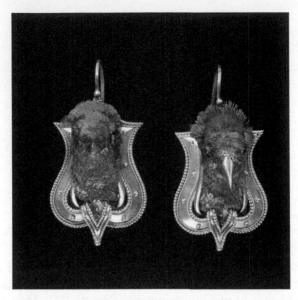

Figure 1.1 Hummingbird earrings, c.1865. Victoria and Albert
Museum M.11:1, 2-2003

setting, a pair of shimmering hummingbird earrings would certainly not
have seemed out of place.

This chapter examines the craze for birds' plumage in the late nineteenth and early twentieth centuries and assesses its severe environmental impact. A product of new manufacturing techniques, changing tastes
and expanding commercial networks, the plumage trade was big business. It provided stylish headgear for women in Europe and the USA and
drew upon a global workforce of hunters, merchants and milliners. It was
also, however, a highly controversial industry that attracted searing criticism from conservationists and humanitarians. The chapter traces the
rise and fall of the trade in feathers and explores the ethical and ecological
implications of this arresting but destructive fashion. It looks, too, at the
organisations that emerged to protect endangered birds and the challenges they faced in changing laws and attitudes.

While almost any creature with feathers was a target for the plume
hunter, two birds in particular bore the brunt of the demand: the egret
(*Ardea alba*) and the ostrich (*Struthio camelus*). The former, a type of
heron, was slaughtered on a massive scale for its handsome breeding
plumage and was all but wiped out in Florida; it became the cause
célèbre of the bird protection movement, playing a prominent role in

much of its propaganda. The latter, the largest living species of bird, was also initially hunted for its feathers, and it too faced extinction. From the 1860s, however, ostriches began to be domesticated and farmed in Cape Colony, offering what most contemporaries saw as a sustainable and humane source of plumes. This chapter explores the contrasting fates of these two birds and shows, in the case of the egret, how milliners propagated lies to defend a coveted commodity.

The Plumage Trade

In the 1860s, a fashion took hold for adorning women's hats with the plumage of dead birds. The craze began with British birds, such as robins, wrens, goldfinches and kingfishers, but quickly extended to more exotic species such as hummingbirds. Feathers – and later whole birds – appeared on bonnets, dresses, fans, earrings and even shoes.[4] In 1875, a lady appeared at a ball wearing a dress 'trimmed with the plumage of eight hundred canaries'.[5]

The birds desired by milliners came from across the globe. Hummingbirds emanated from the West Indies and America, where they were caught in nets, to prevent damage to their feathers, or 'killed by sand being blown at them by means of a tube'.[6] Birds of paradise arrived from New Guinea, shot 'with blunted arrows' by indigenous hunters, 'so that stunned, they fall to the ground'.[7] Lyre birds arrived from Australia, egrets from across the Americas, especially Florida and later Venezuela, and mirasol birds from the pampas, their feathers fetching 'at least 2,300 dollars per kilo'.[8] Colonisation and steam shipping opened up new bird populations to exploitation as the nineteenth century wore on, while better weapons accelerated the pace and volume of the killing. Changes in fashion also had a significant impact on the species of birds that were most desired; in 1908, 'there was a sudden demand for the metallic

---

[4] Fiona Clark, *Hats* (London: Batsford, 1982), p. 36.
[5] 'Revolting Cruelty to Seals', *The Animal World*, April 1875, p. 52.   [6] Ibid.
[7] 'Feathers', *North-Eastern Daily Gazette*, 13 July 1886. From around 1870, Moluccan hunters armed with guns began to take over the trade from local Papuans, increasing the slaughter. See Peter Boomgaard, 'Oriental Nature, Its Friends and Its Enemies: Conservation of Nature in Late-Colonial Indonesia, 1889–1949', *Environment and History* 5 (1999), p. 279.
[8] RSPB, *Feathers and Facts: Statement by the Royal Society for the Protection of Birds* (London: RSPB, 1911), pp. 29, 23–4; 'Tit-Bits on the Fashion Feather Mania', *The Animal World*, December 1893, p. 183.

Figure 1.2 A selection of the latest feathered bonnets in the *Paris Millinery Trade Review*, January 1897, plate 4. The lady on the bottom left is wearing egret feathers.

breast-patch of [the six-plumed bird of paradise], and great numbers were caught'.[9]

Although they were harvested from around the world, the majority of these birds ended up at dealerships in New York, Paris, and, most prominently, London, where they were sold at quarterly auctions. After passing through the auction houses, feathers were purchased by milliners and dressmakers, who transformed them into stylish adornments for female consumers. Corpses were stuffed, eyes replaced with glass beads, wings pulled off or rearranged and feathers dyed to create more striking colours. The finished articles would then appear in the windows of high-street stores, by turn tempting and shocking the passing public (Figure 1.2). Walking

[9] RSPB, *Feathers and Facts: Statement by the Royal Society for the Protection of Birds* (London: RSPB, 1911), p. 69.

along Bond Street in 1885, one horrified viewer witnessed 'a spray of five goldfinches, wired so as to be worn cross the bodice of a dress'.[10]

A journalist from the *Pall Mall Gazette* visited a dealership in birds' feathers in 1886 to get a first-hand impression of the trade. On entering the premises, he was met by one of the managers, who guided him through the store and some of the workrooms. Hummingbirds, he was told, were 'most always much worn', and were sold for '5s to 60s per dozen'. The tangara was a 'great favourite of ladies', being 'the only bird which is naturally red'. The 'white Java sparrow, which we sell for 4s each, is one of the most expensive birds … White birds are always expensive because they are rare.' After examining '[b]ox after box, chest after chest of birds of all colours', the journalist was introduced to a second employee, a workman, who described how the birds were mounted. The latter explained that corpses were first 'put out twelve hours on damp sand, and covered with a damp cloth, in order to soften them' before being 'stuffed with cotton wool' and placed over an internal frame. A wire was then 'passed through each tail feather, so that the tail may be bent to any shape' and 'balls of glass … stuck into the bird's head' to serve as eyes. Some plumes were dyed – a job done by female convicts at the dealership's warehouse in Berlin; others were combined to create composite birds never found in nature. Many of the avian cadavers rotted en route to the dealers and had to be thrown away.[11]

What was the extent of the feather trade, and what impact did it have on bird populations across the world? Surviving trade catalogues and newspaper reports give us some idea of the volume. In 1892, one London auction room sold '6,000 birds of paradise, 5,000 Impeyan pheasants, 400,000 hummingbirds, and other birds from North and South America, and 360,000 feathered skins from India' in a single week.[12] Six years later, the Society for the Protection of Birds (SPB) published a list of birds and feathers sold at the London Commercial Sales Room in Mincing Lane in one of the quarterly auctions. The list included:

Osprey feathers or aigrettes, 6,800 ounces; peacock feathers, 22,107 bundles; peacock neck feathers, 878 pounds; parrots, 35,497 skins; hummingbirds, 24,956 skins; jays, 16,107 skins; bee-eaters, 2,216 skins; Impeyan pheasants, 1,317 skins; kingfishers, 1,327 skins; trogons, 1,403 skins; argus pheasants, 122 skins; paradise birds, 15 skins; orioles, 32 skins; thrushes, 73 skins; owls, 108; toucans' breasts, 29; various birds, 7,595.[13]

---

[10] 'A Plumage League', *The Times*, 25 December 1885.
[11] 'Birds, Butchers and Beauties', *Pall Mall Gazette*, 5 January 1886.
[12] 'Tit-Bits on the Fashion Feather Mania', *The Animal World*, December 1893, p. 183.
[13] 'The Traffic in Birds', *New York Tribune*, 4 September 1898.

Naturalist and bird defender W. H. Hudson visited a feather auction in Mincing Lane on 14 December 1895 and observed '125,300 specimens' of parrots, 'mostly from India'. 'Spread out in Trafalgar Square they would have covered a large portion of that square with a gay grass-green carpet, flecked with vivid purple, rose and scarlet.'[14]

Horrifying in themselves, these figures conceal the true extent of the slaughter. First, quantities expressed in terms of weight obscure the real number of victims, since some birds produced only a few feathers each, which meant that many had to be killed to supply the requisite amount. 'To produce a kilo ... of small plumes,' for instance, '870 birds [egrets] have to be killed, an appalling fact, when one realises that, besides the mutilated and dying 870 large birds, countless numbers of small birds are left starving in the nests.'[15] Second, as this quotation indicates, the killing of nesting birds could also result in the premature deaths of their chicks, which perished from lack of food when their parents were slaughtered. Bird species with small ranges, moreover, were especially vulnerable to overhunting and quickly succumbed to the onslaught of the feather trade. Birds of paradise, for example, shaped by years of isolation and sexual selection, were often localised to a particular island chain or mountain range, making them susceptible to rapid extinction. As Margaretta Lemon explained,

The common sense of every thoughtful woman must at once tell her that no comparatively rare tropical species, such as the Bird of Paradise, can long withstand this appalling drain upon it, and that this ruthless destruction, which merely panders to the caprice of a passing fashion, will soon place one of the most beautiful denizens of our earth in the same category as the Great Auk or the Dodo.[16]

The craze for feathered millinery thus precipitated a wholesale massacre of the world's birds, which perished in their thousands to decorate hats, dresses and bonnets.

The Lady and the Law

While many women were clearly comfortable with wearing the corpses and feathers of dead birds, others found the practice repulsive. To kill a wild animal purely to decorate a bonnet appeared frivolous in the extreme. The sheer volume of the trade, moreover, raised fears of species extinction, both locally and globally. For this reason, the feather trade

[14] 'The Trade in Birds' Feathers', RSPB Pamphlet No. 28, 1895, p. 2.
[15] 'Florida', Bird Lore, Vol. XIII, 1911, p. 367.
[16] 'The Bird of Paradise', RSPB Pamphlet No. 20, 1895, p. 2.

was one of the first animal-based industries to generate widespread disapproval, and – of all the industries examined in this book – the one that spawned the most organised and sustained opposition. An examination of the campaign against 'murderous millinery' thus reveals attitudes, tactics and challenges that would surface again in critiques of fur, ivory and exotic pets.

Individuals had spoken out against the fashion for wearing feathers from its inception, but it was in the 1880s that the first formal bird protection organisations came into existence. In Britain, two women, Eliza Phillips and Margaretta Lemon, founded the SPB in 1889, 'in the hope of inducing a considerable number of women, of all ranks and ages, to unite in discouraging the enormous destruction of bird-life exacted by milliners and others for purely decorative purposes'.[17] In the USA, the first Audubon Society (named after the naturalist John James Audubon) came into existence in 1886 to prevent: '1) the killing of any wild bird not used for food, 2) the destruction of nests or eggs of any wild bird, and 3) the wearing of feathers as ornaments or trimming for dress'.[18] Both movements expanded rapidly over the following decade, attracting a growing following: the SPB boasted 22,000 members by 1899, while the Audubon Society, after a brief decline from 1888, was re-founded in 1896 and had branches in thirty-five states by 1904.[19] Women made up a considerable proportion of the movements' memberships and, in the SPB in particular, were instrumental in its leadership, heading the majority of regional branches.[20]

In attempting to preserve the avian victims of the plumage trade, campaigners essentially had two options. The first was to secure a change in the law to prevent the collection, export and sale of feathers. The second, more difficult, but ultimately more effective option was to foster a change in consumer attitudes, thus removing the demand side of the

---

[17] SPB, *First Annual Report* (London: J. Davy and Sons, 1891), p. 7.

[18] William Dutcher, 'Report: History of the Audubon Movement', *Bird Lore*, Vol. VI, 1904, p. 50.

[19] SPB, *Ninth Annual Report* (London: J. Davy and Sons, 1899), p. 7; William Dutcher, 'Report: History of the Audubon Movement', *Bird Lore*, Vol. VI, 1904, pp. 54–6, 47. The SPB also spawned branches in the colonies: an Australian branch was founded in Adelaide in 1896 by Mrs Playford while an Indian branch was founded in Lucknow in 1900 by William Jesse. See SPB, *Sixth Annual Report* (London: J. Davy and Sons, 1896), p. 3; SPB, *Tenth Annual Report* (London: J. Davy and Sons, 1900), p. 13.

[20] On female involvement and influence in the SPB, see 'Writing for the Birds' in Barbara T. Gates, *Kindred Nature: Victorian and Edwardian Women Embrace the Living World* (Chicago: University of Chicago Press, 1998), pp. 114–24. For a detailed overview of the rise of bird protection in the late nineteenth century, see Robin Doughty, *Feather Fashions and Bird Preservation: A Study in Nature Protection* (Berkeley: University of California Press, 1975).

equation. Bird protection organisations on both sides of the Atlantic experimented with each of these methods simultaneously, hoping to change public attitudes before it was too late. A closer look at their two-pronged strategy illustrates both the problems they were up against and the campaigning techniques that would become central to present-day animal conservation.

Curbing the Slaughter

When it came to legislation, bird protectors initially lobbied for measures that would end, or at least regulate, the slaughter in the field. These generally consisted of close seasons, during which specified birds could not be shot, and the creation of formal reservations or refuges, where all hunting was prohibited. The British Wild Birds Act of 1872 (amended in 1876 and 1880) established a close season for birds between 1 March and 15 August, giving vulnerable species protection during the breeding season.[21] Similar measures were enacted in British Guiana (1878), where the Court of Policy prohibited the killing, sale or export 'at any time of year' of forty native bird species, and in India, where the Wild Birds' Protection Act (1887) gave local governors the power to prohibit the hunting of selected birds during the breeding season.[22] In the USA, individual states passed laws to protect local birds, making it a crime to kill selected species at certain times of the year. Twenty-eight states had enacted some form of bird protection by 1904, and several formal bird refuges had been created across the country.[23]

While anti-hunting legislation represented an important step in bird protection, it did not, on its own, put an end to the plumage trade. For a start, game laws were hard to police and reservations worked only if they were properly guarded. The Audubon Society, recognising this, employed wardens to patrol the newly established refuges, funding their work from supporters' donations. Policing large areas was both difficult and dangerous, however, and by 1909, following the murder of Guy Bradley and two fellow wardens in Florida, the society was expressing considerable pessimism over the effectiveness of the warden system – at least in the Sunshine State. As one commentator lamented,

---

[21] 'Wild Birds Protection Act', *Western Times*, 21 August 1872; 'Protection of Wild Birds', *York Herald*, 28 April 1871; 'The Wild Birds Protection Act', *Essex Standard*, 2 November 1880.

[22] 'The Close Time for Birds', *The Animal World*, May 1878, p. 66; RSPB, *Feathers and Facts*, p. 61.

[23] William Dutcher, 'Report of the National Association of Audubon Societies', *Bird Lore*, Vol. VI, 1904, p. 50.

This Association has spent thousands of dollars in trying to preserve the birds of Florida without any seeming result, as there are far less plume birds in the state than there were when warden Guy Bradley was appointed. As we have already lost two wardens by violent deaths, it does not seem as though the Association were warranted in appointing any further wardens, especially on the west coast, for the present at least, certainly not until the citizens of Florida awake to the value of birds as an asset to the state and establish a Game Commission in order to see that the bird and game laws are enforced.[24]

Bird protection legislation, therefore, was only effective insofar as it was enforced, and bird refuges needed to be constantly monitored, not merely identified on a map.

Another major problem was, of course, that birds migrated, and even the most diligently policed sanctuary could not protect those residents that left its boundaries. Britain or Massachusetts might have stringently enforced bird laws, but these would not help the birds when they were over-wintering in Alabama or Italy, where no such legislation existed. Nor, moreover, was there anything to stop milliners buying the bodies of foreign birds from unprotected areas of the globe, or, indeed, smuggling illegally hunted plumes from one state into another. According to one Royal Society for the Protection of Birds (RSPB) report, between 20 December 1907 and 15 February 1908, 'twenty-three cases of dead bird-skins from India were imported as cowhair or horsehair', while 'osprey' feathers were sent from India to Britain 'by parcel post, declared as dress material'.[25] To make real progress, therefore, national and international legislation was essential. As William Dutcher remarked, 'The only way to stop such an unholy traffic in the lives of these beautiful and innocent creatures is to have enacted international laws, prohibiting the possession and sale of the feathers of *all* wild birds.'[26]

With this in mind, the second phase of bird protection legislation focused explicitly on the trade in coveted plumes, pushing for measures to curb their export and importation. In the USA, the first major breakthrough on this front came with the Lacey Act of 1900, which prohibited the interstate shipment of birds killed in violation of state laws. This meant that a bird killed illegally in Pennsylvania could not be transported to New York, and that any New York milliner found in possession of

---

[24] 'Third Audubon Warden Murdered', *Bird Lore*, Vol. XI, 1909, p. 52. On the death of Guy Bradley, see Stuart B. McIver, *Death in the Everglades: The Murder of Guy Bradley, America's First Martyr to Environmentalism* (Gainesville: University Press of Florida, 2003).

[25] RSPB, *Feathers and Facts*, p. 63. The SPB became the RSPB in 1904.

[26] William Dutcher, 'Plume Sales', *Bird Lore*, Vol. VIII, 1906, p. 180.

such a bird could be fined.[27] Unfortunately, the Lacey Act did not apply in states that had not enacted bird protection legislation and did not cover foreign birds such as the bird of paradise or the egret. The Audubon Society therefore continued its fight against the feather trade, urging all states to adopt its so-called Model Law and working for tighter federal legislation. An attempt in 1909 to ban the importation of egret feathers to New York ended in failure, following stiff opposition from the millinery lobby, but in 1910 William Dutcher secured a crucial piece of legislation when he persuaded the New York legislature to pass the Audubon or Plumage Act, banning the sale of birds for commercial purposes.[28] The 1913 Underwood Tariff Act finally sounded the death knell of the plumage trade, explicitly prohibiting 'the importation of aigrettes, egret plumes or so-called osprey plumes, and the feathers, quills, head, wings, skins or parts of skins of wild birds, either raw or manufactured, and not for scientific or educational purposes'.[29]

In Britain and its colonies, the struggle was more tortuous, and a ban on feather imports took longer to accomplish. India got things off to a positive start in 1902 when its government enacted a law prohibiting the export of 'the skins of all birds other than domestic birds, except ostrich feathers, and skins and feathers exported bona fide as specimens of Natural History'.[30] London, however, remained 'the head of the giant octopus of the "feather trade"', and despite repeated efforts from the SPB, a ban on the importation of plumage proved elusive.[31] In 1908, when Lord Avebury introduced into Parliament a bill to prohibit the 'sale or exchange of the plumage, skin or body of any bird', the millinery trade mounted a vocal defence and prevented it from becoming law.[32] In 1914, a second Plumage Bill met a similar fate, derailed, in this instance, by the outbreak of the First World War; in 1921, when a third Plumage Bill was introduced, opponents again threatened to block it, arguing that the fact that the bill had taken over a decade to enact was clear evidence of its flawed conception.[33] Despite the opposition, this final bill did ultimately become law, bringing to an end a cruel but lucrative trade.

---

[27] T. S. Palmer, 'Some Fundamental Principles of Bird Laws', *Bird Lore*, Vol. III, 1901, pp. 79–80.

[28] 'Interesting Items from the *Millinery Trade Review*', *Bird Lore*, Vol. XI, 1909, p. 232; 'The New York Plumage Law', *Bird Lore*, Vol. XII, 1910, pp. 128–9.

[29] Camilo Quintero Toro, *Birds of Empire, Birds of Nation* (Bogotá: Universidad de los Andes, 2012), p. 29.

[30] 'Bird Protection in India', *The Times of India*, 2 October 1902.

[31] William Temple Hornaday, *Our Vanishing Wild Life: Its Extermination and Preservation* (New York: Clark and Fitts, 1913), p. 117.

[32] RSPB, *Feathers and Facts*, p. 14.

[33] 'The Plumage Trade: An Appeal to Women', *The Spectator*, 27 March 1920, p. 5; 'Plumage Bill: Evidence against the Trade', *The Times*, 9 May 1921.

The long struggle to end the trade in bird plumage highlighted the difficulties of protecting wild animals hunted for commercial purposes. First, a major issue for bird protectors was the international dimension of the problem and the excuses and/or defeatism this sometimes induced. Critics of the 1908 Plumage Importation Bill argued that if Britain ceased to import feathers the trade would simply shift to the continent and British ladies would travel to Paris to get their hats rather than buying them in London. A feather ban would thus damage the British economy without saving any birds.[34] In the USA, meanwhile, William Dutcher complained that American birds, though protected at home, were supplying milliners in other countries. 'America cannot protect her own birds if the countries of the Old World offer a market for the plumage of American birds, as they are now doing ... How can the Americans protect their Hummingbirds if they may be killed in South America and sold in England for use wherever birds are used for millinery ornaments?'[35] The global nature of both birds and the feather trade thus complicated any protective legislation and gave ammunition to opponents of conservation, who argued that there was no point in sacrificing the economic interests of one state merely to benefit those of a neighbour and rival. Committed bird protectors, however, insisted that there was value in taking a moral lead, even if the economic benefits were not immediate. As *The Spectator* remarked in 1920, 'if this argument be accepted as valid, we should have to say that it was useless for Great Britain to put an end to the slave trade while others were still carrying it on!'[36]

A second obstacle to reform was the militancy of the milliners themselves and their refusal to countenance any legislation that might interfere with their business. This meant that whenever a bird protection bill was introduced to national or state legislatures, it faced substantial opposition, often bankrolled by representatives of the industry. In the case of the USA, the Feather Importers Association of New York reportedly paid $4,200 to fight two bills designed to prohibit the sale of aigrette and other wild bird feathers within New York State.[37] In Britain, meanwhile, milliners battled successive attempts to ban the importation of feathers from overseas, disputing the accuracy of conservationists' claims. One vocal British milliner, C. F. Downham, denounced the folly

[34] 'The Importation of Plumage Bill', *The Times*, 19 October 1908.
[35] William Dutcher, 'Some Reasons Why International Bird Protection Is a Necessity', *Bird Lore*, Vol. XII, 1910, p. 169.
[36] 'The Plumage Trade: An Appeal to Women', *The Spectator*, 27 March 1920, p. 5.
[37] 'Interesting Items from the *Millinery Trade Review*', *Bird Lore*, Vol. XI, 1909, p. 232.

of attacking a business that 'forms an important and increasing industry in this country, much in the same manner as the use of ivory, tortoiseshell, furs, etc.' and denied 'that we are as soulless destroyers of wild life as the worst collectors of rubber in the Belgian Congo have been destroyers of human life'. He also suggested that, if certain species were declining (which he considered questionable), it was down to human encroachment on previously unsettled areas and hunting for food and sport, not the plumage trade.[38] Another opponent of reform, a London feather broker, insisted that 'there were no signs of extinction of the birds mentioned in [Lord Avebury's] Bill' and that 'in the case of the Birds of Paradise the skins have for a long time been worth from 20s. to 50s. each, and yet, in spite of the tempting price, the supply is larger than ever'. A third questioned why 'the life of a bird that is a means of income through its beauty [should] be preserved more than any other'.[39] Milliners were thus willing to refute the claims of conservationists and, where necessary, to falsify or misconstrue science to defend their industry – a practice particularly notable in the case of the egret, as discussed later in this chapter. The effectiveness of their tactics highlights the ways in which a comparatively small but powerful group of lobbyists could delay important environmental legislation and mislead public opinion.

Finally, bird protectors were victims of the simple laws of supply and demand. All manner of laws could be passed, and, if enforced, they could be effective, but as long as demand remained strong the birds were still in danger. Hunting bans, bird refuges and trade prohibitions certainly made milliners' lives harder, but, as long as women were prepared to pay good money for plume-bedecked hats, someone would always be willing to provide them, even if that meant resorting to poaching and smuggling. To really put an end to the trade, therefore, something had to be done to tackle the demand side of the equation. Could ladies or milliners be persuaded not to use the feathers of dead birds on their hats? Could education supplement legislation?

The Press, the Pulpit and the Schoolmaster

Bird protectors certainly hoped so. Alongside lobbying governments to pass protective legislation into law, conservationists devoted considerable

---

[38] C. F. Downham, *The Feather Trade: The Case for the Defence* (London: London Chamber of Commerce, 1911), pp. 119, 69, 48, 71. Downham claimed, for instance, that the Australian lyre bird had been killed by newly introduced foxes, that the emu had declined as a result of the spread of 'agriculture and cattle-raising' and that birds of paradise were plentiful in the interior of New Guinea.

[39] 'The Importation of Plumage', *The Bazaar*, 3 July 1908.

energy to reshaping public opinion, conscious that only a permanent change in fashion would bring about the end of the feather trade. A steady stream of pro-bird propaganda emanated from the SPB and from the various Audubon societies, targeted primarily at female readers. This aimed, by turn, to enlighten, admonish, enlist and advise women, persuading them that a change in their buying choices could have a major impact on the survival of threatened birds.

Bird protection organisations employed a wide range of tactics to connect with a female audience. In Britain, the SPB appealed to the clergy 'to include the Protection of Birds when preaching on Man's Duty to Animals on the Fourth Sunday after the Trinity', delivered lectures on birds, illustrated with lantern slides, and took to the streets in July 1911 in the famous 'sandwich board protests', showing the public images of slaughtered egrets.[40] In the USA, the Audubon Society adopted similar tactics, creating a special lending library for schools and publishing educational leaflets alerting readers to the plight of threatened species.[41] Both British and American societies relied heavily on the press to spread the gospel of bird protection, using newspapers as key conduits for their message. Writing in 1896, the SPB explicitly thanked '*The Times* and *The Standard*, and also many other London and Provincial newspapers which have published letters on the subject of bird protection, reported meetings of the Society and published articles in support of its work'.[42] Five years later, Garrett Newkirk of the Florida Audubon Society reported that 'I have seen a number of articles written by women in such papers as the *New York Tribune* and *St Louis Globe Democrat*, pleading for the birds and remonstrating against the wicked custom of wearing them on hats. Such articles are quoted and talked about in the country, and have a great influence.'[43] Bird protectors thus mobilised the combined forces of 'the Press, the Pulpit and the Schoolmaster' to influence public opinion in their favour.[44]

The content of these articles, sermons and pamphlets fulfilled several key functions. First, it aimed to inform, making readers aware of the nesting, breeding and feeding habits of different species of birds and

---

[40] SPB, *Eighth Annual Report* (London: J. Davy and Sons, 1898), p. 5.

[41] William Dutcher's 1904 leaflet 'The Snowy Heron', for instance, sold 4,000 copies, of which 500 went to the Millinery Merchants' Protective Association for distribution among its members, 1,000 to 'a prominent wholesale millinery firm in Ohio' and a further 2,500 to the RSPB. See William Dutcher, 'National Committee News', *Bird Lore*, Vol. VI, p. 75.

[42] SPB, *Seventh Annual Report* (London: J. Davy and Sons, 1897), p. 5; SPB, *Sixth Annual Report*, p. 6; Hornaday, *Our Vanishing Wild Life*, p. 128.

[43] Garrett Newkirk, 'For Our Encouragement', *Bird Lore*, Vol. III, 1901, p. 183.

[44] SPB, *Eighth Annual Report*, p. 5.

alerting them to the dangers the birds faced from the plumage trade. William Dutcher's 1904 pamphlet on 'The Snowy Heron' (a species of egret), a particularly influential example, began with a detailed description of the heron, covering its anatomy, habits and geographical range. It then went on to explain why egrets were especially vulnerable during the breeding season, when they 'gather in colonies' and 'lose all sense of fear and wildness'. Shifting from zoology to morality, Dutcher condemned the use of egret feathers by the millinery trade, painting a highly emotive picture of starving egret chicks 'clamouring piteously for food which their dead parents could never again bring them'. This sobering verbal message was reinforced by two contrasting illustrations – the first showing happy egrets nesting peacefully with their young and the second featuring the plucked carcass of a dead bird.[45] Campaigners hoped that seeing and reading facts like these would make consumers aware of the problem and change their behaviour; at the very least they could no longer use ignorance as an excuse for their actions.

Beyond merely alerting consumers to the plight of persecuted birds, conservationists often went further, seeking to actively shame them. One especially cutting article on the subject compared feather-wearing women to savages, drawing an unflattering parallel between fashionable British ladies and the daughter of a South American dictator (probably Manuela de Rosas) who allegedly 'appeared at a ball in a magnificent dress festooned with several hundred human ears'.[46] Another text, an evocative poem published in *The Animal World*, expressly blamed women for a whole range of abuses against birds, suggesting that many put fashion above moral qualms:

> True, the sterner sex are ready of their monster bags to boast,
> Though their big battues are legal scarcely half a year at most;
> But for you, my dainty madam, with your winsome wiles and ways,
> *Your* demand is so incessant that the butcher's hand ne'er stays.
> Day by day, to deck your toilets, birds of Paradise are slain,
> Hummingbirds are shot by thousands, striving to escape in vain;
> Argus pheasants, rare kingfishers, birds of every clime and hue,
> Are destroyed in ruthless numbers, thanks, and thanks alone, to you![47]

As for William Dutcher, his 'Snowy Heron' leaflet concluded with an impassioned appeal to mothers, urging women to empathise more closely with their non-human counterparts: 'Oh human mother! Will you again wear for personal adornment a plume that is the emblem of her married

---

[45] William Dutcher, 'The Snowy Heron', *Bird Lore*, Vol. VI, 1904, pp. 37–40.
[46] 'An Old and Welcome Correspondent', *The Times*, 19 December 1885.
[47] 'Fashionable Bird Murder', *The Animal World*, November 1885, p. 170.

life as the golden circlet is of your own, the plume that was taken from her bleeding body because her motherhood was so strong that she was willing to give up life itself rather than abandon her helpless infants?'[48] According to all of these texts, feather-wearing women were barbarous, hypocritical, un-maternal and unnatural, putting personal vanity above the suffering of avian life.

If one aim of the bird protection literature was to make women feel guilty, another more constructive one was to make them aware of their power as consumers and to enlist them into the ranks of the conservationists. This meant targeting influential individuals and institutions run by and for women and persuading them to take a lead in the anti-plumage crusade. To this end, the Audubon Society concentrated on recruiting women's clubs, entreating 'the club women of America' to use 'their powerful influence' to 'take a strong stand against the use of wild birds' plumage', especially egret feathers.[49] The RSPB took an equally proactive approach, asking female members to 'refrain from wearing the feathers of any bird not killed for purposes of food, the ostrich excepted'. To make the withdrawal process easier, the organisation publicised the Millinery Depot owned by a Mrs White at '8, Lower Seymour Street, Portman Square', London, where conscience-stricken females could purchase hats manufactured 'strictly in accordance with the principles of this Society'.[50] Bird protectors believed that such actions, though small in themselves, could exert an influence on fashion and would serve both to save the birds and to redeem the female sex from accusations of unthinking cruelty. As campaigner Mabel Osgood Wright remarked: 'Every well-dressed, well-groomed woman who buys several changes of head-gear a year can exert a positive influence upon her milliner, if she is so-minded, and by appearing elegantly charming in bonnets devoid of the forbidden feathers, do more to persuade the milliner to drop them from her stock than by the most logical war of words.'[51]

Finally, if female consumers were not willing to abjure the fashion for the sake of the birds, then another group – female (and male) milliners – could also effect change by refusing to stock or sell hats featuring feathers and corpses. Although they were usually painted as the villains of the piece (and not without cause), milliners often insisted that they were merely bowing to the demands of their customers. If this were the case, they might be open to reform, providing conservationists with an

[48] William Dutcher, 'The Snowy Heron', *Bird Lore*, Vol. VI, 1904, p. 40.
[49] 'Suggestions for the Coming Year', *Bird Lore*, Vol. VII, 1905, p. 308.
[50] SPB, *First Annual Report*, p. 8; SPB, *Eighth Annual Report*, p. 10.
[51] Mabel Osgood Wright, 'Hats!', *Bird Lore*, Vol. III, 1901, p. 40.

alternative conduit for change. Bird protectors on both sides of the Atlantic grasped this opportunity, forwarding propaganda to willing milliners and making them more aware of the ecological cost of their precious feathers. In 1904, for instance, the Audubon Society sent 500 copies of William Dutcher's pamphlet on the snowy heron to the Millinery Merchants' Protective Association for distribution among its members, and a further 1,000 to 'a prominent wholesale millinery firm in Ohio'.[52] Four years earlier, the Southport branch of the SPB contacted local milliners directly, 'asking their cooperation in the preservation of our fast-decreasing foreign birds by refraining from exhibiting egrets, herons and paradise plumes in their models, and by not suggesting their use when not expressly ordered'. While much of this propaganda doubtless fell on deaf ears, there is evidence that a few milliners were open to a change in materials. Three of the milliners of Southport supported the SPB's appeal, two of them stating explicitly that 'they considered their customers were entirely to blame, and that personally they would gladly cease the supply'.[53] Another industry insider, a dressmaker named Mrs Leach, condemned the 'dreadful' fashion for 'tiny bright-plumaged birds' and advised customers to decorate their costumes instead with 'the feathers of geese, ducks, fowls, pheasants, partridges, etc., sent for the food of man', which could be 'painted with water colours' to imitate 'the wing of some tropical bird'.[54] Conservationists thus targeted both the sellers and the buyers of plumage in the hope that one group would have a redeeming influence on the other.

Ostriches and Egrets

How successful were bird protectionists in their efforts, and what impact did their activities have on the fate of the plumage trade's favourite species? To explore the consequences of the feather craze in more detail, the following two sections focus on two birds with contrasting experiences: the ostrich and the egret. The ostrich, as we have seen, was officially exempted from both Audubon and SPB pledges of plumage abstinence and was classified as a sustainable and acceptable source of feathers. Was this truly the case? And how, where and by whom were ostrich feathers harvested? The egret, on the other hand, was consistently painted as the most tragic victim of the millinery industry, slaughtered in large numbers in China, Florida and later South America. It was also,

[52] William Dutcher, 'National Committee News', *Bird Lore*, Vol. VI, p. 75.
[53] 'The Protection of Birds', *Liverpool Mercury*, 22 March 1900.
[54] 'A Dressmaker Pleading for the Birds', *The Animal World*, April 1887, p. 51.

however, a critical resource for milliners, who were not prepared to relinquish it without a fight. The final section of the chapter explores the arguments mustered by plume hunters to defend their business and the ways in which science was mobilised by both sides to substantiate or discredit rival claims.

'The Ostrich Excepted'

For women who wanted to continue wearing feathers but who opposed the wanton slaughter of small birds, ostrich plumes represented a sustainable and humane alternative. Originally worn by men, ostrich feathers began to enter female fashion in the second half of the nineteenth century and constituted a viable substitute for egret plumage and hummingbird wings. Although they were traditionally procured by hunting and killing the ostriches that bore them, ostrich feathers assumed a more benign character in the 1860s when farmers in Cape Colony started to rear the birds on their estates, plucking them alive and sending their feathers to Europe. While the humanity of the plucking process remained a matter for debate, most contemporaries saw it as preferable to indiscriminate slaughter of smaller birds and urged women to wear ostrich feathers in place of avian corpses. There ensued a dramatic growth in the ostrich farming industry as South African breeders sought to maximise their profits and agricultural entrepreneurs attempted to naturalise the species in other countries.

Until the mid-nineteenth century, ostrich feathers had been obtained by hunters in the African interior and dispatched to Marseilles, Livorno, Southampton and other major European ports by merchants in Cairo, Tripoli, the Cape, Morocco and Senegal.[55] In the 1860s, however, there was a shift from hunting to domestication, as farmers in Cape Colony experimented with raising wild ostriches on their estates. Mr L. von Maltitz, one of the first to attempt the process, purchased seventeen four-month-old ostriches in 1864 and successfully reared them on his farm, where they became 'so tame that they allow themselves to be handled and their plumage minutely examined'.[56] Messrs Heugh and Meinjes followed his example in the early 1870s; by 1875, they owned a large flock of the birds.[57] A third farmer, Mr Arthur Douglas, possessed

---

[55] Julius de Mosenthal and James Edmund Harting, *Ostriches and Ostrich Farming* (London: Trübner and Co., 1877), pp. 222–3.
[56] 'Ostrich Farming', *The Times*, 29 November 1864.
[57] 'Ostrich Farming', *The Times*, 9 September 1875.

seventy-one ostriches by 1874.[58] Though generally regarded as inferior to the plumage of their wild counterparts, the feathers produced by these ostriches were deemed to be of suitable quality for export and promised good financial returns for their owners. Von Maltitz estimated that each of his birds would earn him £12 and 10 shillings per biannual plucking – a profitable return on birds that had cost him 5 shillings each to procure.[59]

Encouraged by these early successes, ostrich farming expanded massively during the 1870s, competing with Cape Colony's other main industries of sheep farming and diamond mining. Writing in 1881, when the business was at its peak, *The Star* reported that '[f]arm after farm [in Port Elizabeth] is being cleared of sheep to make room for ostriches, now all the rage', with the result that mutton prices in the region had risen 'to 6d and 7d per lb, on account of sheep farming being pushed aside by ostriches'.[60] Four years later, *The Times* noted that, 'of the millions of goods annually exported from South Africa, 4 millions, or one half, is merely for the adornment of ladies', ostrich feathers accounting for 1 million and diamonds for the other 3 million.[61] A total of 96,582 pounds in weight of ostrich feathers was exported from Cape Colony in 1879, serving a growing demand for the product in Europe and the USA.[62]

For those thinking of investing their time and money in ostriches, the primary requirements were large tracts of land and plentiful food. Mr Douglas from Grahamstown kept 300 birds on an estate of around 1,200 acres, providing each breeding pair with a separate paddock.[63] An ostrich-farming guide published in 1877 recommended between 1,000 and 5,000 acres for a profitable ostrich farm.[64] Feeding ostriches was relatively simple, given their rather catholic tastes, although many farmers grew lucerne to sustain the birds during the dry season.[65] Anthony Trollope, visiting Mr Douglas's farm in 1877, reported that the ostriches there fed 'themselves on … grapes and shrubs unless when sick or young'.[66] An article in the *Weekly Standard and Express* claimed that the ease of feeding a 'full-grown ostrich … may be inferred from the fact that it will eat seeds, boots, insects and small reptiles as well as sand, pebbles, bones and pieces of old iron … There is even a case on record of

---

[58] Mosenthal and Harting, *Ostriches and Ostrich Farming*, p. 202.
[59] 'Ostrich Farming', *The Times*, 29 November 1864.
[60] 'Ostrich Farming at the Cape', *The Star*, 6 January 1881.
[61] 'South Africa as a Field of Commerce', *The Times*, 22 October 1885.
[62] 'Ostrich Feathers', *Manchester Courier*, 2 August 1881.
[63] 'A South African Ostrich Farm', *Hampshire Telegraph*, 7 December 1877.
[64] 'Ostriches and Ostrich Farming', *Pall Mall Gazette*, 11 December 1876.
[65] William Beinart, *The Rise of Conservation in South Africa: Settlers, Livestock and the Environment, 1770–1950* (Oxford: Oxford University Press, 2003), p. 162.
[66] 'A South African Ostrich Farm', *Hampshire Telegraph*, 7 December 1877.

Figure 1.3 'Removing the Plumes', from Harold J. Hepstone, 'Modern Ostrich Farming', *The Animal World*, November 1912, p. 207.

a colonist who was accustomed every evening to give his ostrich the newspaper for supper!'[67]

Plucking ostriches was more challenging, as the birds, if frightened, were capable of delivering a debilitating kick with their powerful legs. To counteract this problem, ostrich farmers generally herded their birds into corrals in order to more easily control them, forcing them close together to immobilise their limbs. Trollope recorded that 'the birds are enticed into a pen by a liberal display of mealies, as maize is called in South Africa'; 'when the pen is full, a moveable side, fixed on wheels, is run in, so that the birds are compressed together beyond the power of violent struggling'.[68] A later account described how 'a stocking' was 'slipped over' the birds' heads when they entered the enclosure to quieten them, and the quills 'cut very carefully with special clippers'.[69] Plucking usually occurred every six to eight months, giving the feathers time to re-grow, and involved either pulling the feathers out manually or cutting them with scissors (Figure 1.3). Whether or not the procedure was painful was

[67] 'Life on an Ostrich Farm', *Weekly Standard and Express*, 3 February 1900.
[68] 'A South African Ostrich Farm', *Hampshire Telegraph*, 7 December 1877.
[69] 'Ostrich Farming in Cape Colony', *The Times*, 5 November 1910.

a matter for debate, as we shall see, but the general consensus was that ostrich feathers represented a significant humanitarian and environmental improvement on the plumage of dead songbirds.

While South Africa took the lead in the ostrich farming business, other states also showed an interest in the industry, making the ostrich a target not only for domestication, but for naturalisation in other locales. In Egypt, an ostrich farm at Heliopolis, outside Cairo, boasted twenty-five birds by 1881, fed on 'carrots, turnips, clover, etc., prepared in a specially made "ostrich food mincing machine"'.[70] In the same year, the French consul in Tripoli sent fifty-two ostriches to Algeria for farming in his nation's North African colony, while in 1904 five pairs of ostriches were sent from South Africa to Madagascar, resulting in seventy-six birds by 1907.[71] Beyond Africa, Australia also got in on the ostrich boom, with attempts to acclimatise ostriches in both Victoria and New South Wales. One hundred ostriches lived on a farm at Murray Downs Station by 1882, giving rise to a special factory in Melbourne to wash, trim and dye the feathers they produced.[72] Twenty years later, twenty-two ostriches were successfully reared 'in a Sydney suburb' by an 'enterprising Australian'; one of the birds, nicknamed 'the Duke', sported feathers '27 inches in length, 15 inches in width and of the purest white'.[73]

A particularly successful acclimatisation operation took place in the US state of California, where climatic conditions proved favourable to the rearing of this African bird. In 1886, a Californian farmer chartered a ship to convey sixty ostriches from Durban, South Africa, to Galveston, keeping the birds in padded boxes to prevent injury during the seventy-day voyage. After a rest of several days in the Texan port, the fifty-two surviving ostriches were transported to Los Angeles by train, 'in specially prepared cars', arriving in good health on their new owner's farm. The Californian rancher could not offer his animals the same amount of space as his South African counterparts, but he found that they did well on the half-acre plots with which he supplied them. For food, he relied primarily on alfalfa, corn, barley, sugar beet and the occasional orange.[74] By 1895,

---

[70] 'Ostrich Farming in Egypt', *Glasgow Herald*, 4 January 1881.

[71] 'Ostrich Feathers', *The Star*, 22 October 1881; 'Ostrich Farming Experiments', *Manchester Courier*, 27 November 1908.

[72] 'Ostrich Farming in Australia', *Belfast News-Letter*, 22 December 1882.

[73] 'An Australian Ostrich Farm', *Devon and Exeter Gazette*, 16 July 1902.

[74] 'Ostrich-Farming in California', *Birmingham Daily Post*, 8 April 1890. The first ostriches were imported into the USA by a Dr Protheroe in 1882, although only twenty-two survived the voyage to New York. See 'The Ostrich', *Bird Lore*, Vol. VII, 1905, p. 154.

there were two ostrich-rearing establishments in California – at Coronado Beach and Norwalk – each boasting sixty-five ostriches.[75] Ten years later there were also ostrich farms in Arizona, Florida and North Carolina, with an estimated 'three thousand Ostriches' in the USA as a whole.[76] The profitability of the US ostrich industry was assisted by the US Government, which imposed a 25 per cent import duty on foreign-produced ostrich feathers to protect the nascent industry from competition.[77] In retaliation, South Africa imposed a duty of £100 on every ostrich exported, attempting to retain a monopoly on this prized commodity.[78]

Despite the less favourable climatic conditions, there were also plans to farm ostriches in Europe. In Paris, ostriches featured prominently in the Jardin d'Acclimatation in the Bois de Boulogne, and a prize was offered in 1857 for their successful introduction.[79] In Britain, an ostrich-farming syndicate was formed in 1912 in Sussex, purchasing a piece of land for the establishment of an ostrich farm and publishing an article in the *Penny Illustrated News* to raise the profile of the business. The syndicate's members claimed that ostrich farming was more profitable and 'far less laborious than ordinary farming'. They predicted that, 'in a few years', ostriches would 'become as common a feature on the British landscape as sheep and cows'.[80]

While neither the French nor the British operations moved beyond the level of exotic curiosity, an ostrich farm of significant size was established by the German animal dealer Carl Hagenbeck at his menagerie in Stellingen, Hamburg. A strong proponent of animal acclimatisation (he also promoted the domestication of the African elephant), Hagenbeck discovered that ostriches did not need a hot climate in order to survive and experimented with rearing some of his birds in an outdoor enclosure, 'with no other shelter than that afforded by an unheated open shed'.[81] Initially a small-scale breeder, Hagenbeck expanded his operation over the years, stocking his farm with 'over 100 individuals belonging to the five finest varieties of African ostrich in existence'.[82] He also espoused

[75] 'On an Ostrich Farm', *Berrow's Worcester Journal*, 31 August 1895.
[76] 'The Ostrich', *Bird Lore*, Vol. VII, 1905, p. 155. The Arizona Ostrich Company was established in 1898; in the same year, Mr A. Y. Pearson and Mr J. Taylor established the Florida Ostrich Farm at Jacksonville.
[77] 'Ostrich-Farming in California', *Birmingham Daily Post*, 8 April 1890.
[78] 'Ostrich Farming as an Industry', *Huddersfield Daily Chronicle*, 1 May 1893.
[79] 'Ostriches in France', *The Times*, 25 September 1858.
[80] 'Ostrich Farming in England', *Penny Illustrated Paper*, 5 October 1912.
[81] 'Ostriches Reared in Germany', *The Times*, 9 October 1911.
[82] Carl Hagenbeck, *Beasts and Men*, translated by Hugh S. R. Elliot and A. G. Thacker (London: Longmans, 1912), p. 260.

more ambitious acclimatisation plans, recommending the naturalisation of ostriches in South America, where they could roam 'the immense prairies' alongside existing cattle herds. While some feared that this proliferation of ostrich farms might decrease the value of ostrich feathers, Hagenbeck thought otherwise, confident that the market could sustain it. 'I have little doubt that before long laws will be passed in civilised countries for prohibiting the importation of ornamental birds for ladies' hats', and 'an immediate result of such legislation would be that the demand for and consequently the value of ostrich feathers would rapidly increase; so that there is little danger that this commodity will ever become a drug on the market'.[83]

As Hagenbeck's comment illustrates, the rise of the ostrich industry in South Africa and beyond was directly connected to the environmental and humanitarian concerns associated with the plumage trade and the changing fashions of European women. Like other animal commodities, moreover, it was facilitated by specific innovations and forms of expertise, as farmers, acclimatisers and manufacturers grappled with the challenges of a new and evolving industry. What were the market forces that stimulated the boom in ostrich farming and what further knowledge exchanges allowed the industry to flourish?

First, the shift from hunting ostriches to farming them was made possible by significant technological advances, the most important of which was the invention of a bespoke ostrich egg incubator. Credited to a South African pioneer, Mr Douglas, the incubator enabled multiple eggs to be hatched at once and relieved the parent birds of the duty of sitting on them while they developed, reducing damage to their feathers and fooling the females into laying additional eggs. Trollope, who visited Douglas's farm in 1877, described one of these early incubators in detail, emphasising the care and expertise required to operate it:

The incubator is a low, ugly piece of deal furniture, standing on four legs, perhaps eight or nine feet long. At each end are two drawers, in which the eggs are laid, with a certain apparatus of flannel, and these drawers, by means of screws beneath them, are raised or lowered to the extent of two or three inches. The drawer is lowered when it is pulled out, and is capable of hatching a fixed number of eggs. I saw, I think, fifteen in one. Over the drawers, and along the top of the whole machine, there is a tank filled with hot water, and the drawer when closed is closed up so as to bring the side of the egg in contact with the bottom of the tank. Hence comes the necessary warmth. Below the machine, and in the centre of it, a lamp, or lamps, are placed, which maintain the heat that is required. The eggs lie in the drawer for six weeks, and then the bird is brought out – not without

---

[83] Ibid., p. 266, 272.

some midwifery assistance. All this requires infinite care, and so close an observation of nature that it is not every possessor of an incubator who is able to bring out young ostriches.[84]

As time went by, the incubation process became more sophisticated, increasing in scale and efficiency. In 1908, an ostrich farm in Pasadena, California, was equipped with a large incubator house and a neighbouring 'brooder', where newly hatched chicks were fattened up on a diet of 'alfalfa, dry bone and fine ground grit' before being released into the open paddocks.[85] In 1912, Hagenbeck's ostrich farm comprised 'stables, [a] pond … feeding troughs, 10 breeding pens … and a hospital for sick and injured birds', as well as 'a chick-house', 'a show room and [a] factory for feathers'.[86] Ostrich rearing was thus becoming an industrial activity, assisted by the latest machinery.

Another important element to ostrich rearing – and specifically the naturalisation of ostriches in Europe – was the role played by zoos and menageries. Ostriches appeared regularly in zoos from the mid-nineteenth century and even featured in several international exhibitions. This generated interest in the birds and moved some to recommend their naturalisation as livestock. When twenty-three living ostriches were exhibited at the Crystal Palace in 1895, as part of a Somali village, enthusiast G. H. Lane rhapsodised over the birds' hardiness, observing that they had been 'exposed in the daytime to all variations of weather and housed at night in a shed'.[87] A journalist from the *Pall Mall Gazette*, meanwhile, took the opportunity to quiz the keeper, Mr Peace, on the practicalities of caring for the animals, receiving assurances that they 'came through the frost last winter on a diet of cabbages' and had even 'survived the Canadian winter'.[88] Captive ostriches thus tickled the entrepreneurial imagination and offered an insight into the nuances of ostrich behaviour. An *Animal Care Book* compiled by keepers at Manchester's Belle Vue Zoo noted that the ostriches in their care would 'eat a heaped bucket of lawn-mown grass', could 'peck [themselves] within 12 inches of the base of the neck', and, in one case, 'refuse[d] to eat when watched' – all useful information for anyone thinking of rearing ostriches for profit.[89]

[84] 'A South African Ostrich Farm', *Hampshire Telegraph*, 7 December 1877.
[85] 'Ostrich Incubator House', *Manchester Courier*, 3 July 1908.
[86] 'Ostrich Farming in England', *Penny Illustrated Paper*, 5 October 1912.
[87] 'Ostrich Farming in England', *The Standard*, 12 October 1895.
[88] 'Ostrich Farming in England', *Pall Mall Gazette*, 23 August 1895.
[89] *Animal Care Journal*, Belle Vue Gardens, Jennison Collection, Chetham's Library, Manchester, F.5.04, 28 May 1910, 12 August 1912 and 6 June 1912.

If keepers dispensed valuable knowledge about the handling of ostriches, the sale and marketing of ostrich feathers depended on the expertise and connections of mostly Jewish entrepreneurs. As Sarah Abreyava Stein has demonstrated, Jews were central players in the ostrich feather trade and were visible in all of its main arenas. Jewish merchants purchased feathers from Boer farmers in South Africa. Jewish immigrant workers (mostly women) trimmed, willowed and curled ostrich feathers at workshops in London and New York.[90] Jewish brokers such as the London-based Salaman family bought feathers at auction for sale to manufacturers, while Jewish financiers bankrolled ostrich farming experiments in the USA. Jewish merchants in Livorno had also played a crucial part in the Mediterranean feather trade, choreographing the movement of feathers across the Sahara by camel train. Stein argues that Jews operated as crucial conduits between farmers and milliners, capitalising on pre-existing skills and transnational contacts. Without this global Jewish diaspora, the ostrich feather trade could not have functioned effectively and would not have assumed the form it did.[91]

What about the supposed ecological and humanitarian benefits of wearing ostrich plumes? Was ostrich farming a benign enterprise, or was it, too, tainted by cruelty? Did the shift from hunting to farming mark an important moment in ostrich preservation?

The question of cruelty centred on whether or not live plucking inflicted suffering. For some commentators the answer was clear: it did not, and ostrich feathers provided a pain-free alternative to the plumage of wild birds, which could be obtained only by killing them. One proponent of this view, vegetarian and prominent anti-vivisectionist Anna Kingsford, explicitly recommended the use of ostrich feathers instead of fur for shawls and wraps, for 'while sealskins and other furs are obtained with great cruelty, and while small-bird plumage is only procured at the cost of their lives, ostrich feathers … are shorn painlessly from the bird twice a year – much as the fleece is taken from the sheep'.[92] Another campaigner, Mrs Sturge of the Bristol Branch of the SPB, insisted that self-interest ensured a level of humanity among ostrich owners, since violent and premature plucking would damage their future profits.

---

[90] Willowing entailed lengthening the shorter strands of interior feathers by tying additional strands, called flues, to them, to give them the desired volume, essentially making the feathers longer and fluffier.
[91] Sarah Abreyava Stein, *Plumes: Ostrich Feathers, Jews and a Lost World of Global Commerce* (New Haven: Yale University Press, 2008).
[92] 'Ostrich Feathers', *Pall Mall Gazette*, 14 October 1887.

There has been a good deal said lately as to how the ostrich feathers are obtained. Though in the old days needless cruelty may have been practised, there is now no excuse for obtaining them in a cruel manner. The birds are now kept on large farms and have to be well cared for, as they are delicate birds. A short time before the feathers are quite matured the birds are driven into cages one at a time and the best feathers cut off with a sharp instrument just about an inch from the skin. The bird is then let loose again. A few weeks later they are again caged and the stumps, if not already out, gently withdrawn, for if blood is drawn feathers will not grow again.[93]

While this benign view of ostrich farming predominated, other commentators expressed greater scepticism as to the humanity of the feather-plucking process, suggesting that there was at least the potential for cruelty. Writing to the RSPCA's *The Animal World* in 1891, a member of the society's Essex branch asked whether ostrich feathers were procured with cruelty, fearing that this must be the case. 'I have always had the idea that there was considerable cruelty exercised in obtaining ostrich feather plumes – in fact, I have heard they are plucked from the bird when alive. Is that so?'[94] The magazine's editor responded to this, and to subsequent queries, in reassuring tones, asserting that '[o]striches are seldom ill-treated on farms, where their feathers are cut off, not plucked'.[95] An earlier piece in *The Animal World*, however, had been less sanguine, suggesting that the profit-driven nature of the business would inevitably lead to some level of violence. 'If ripe feathers only can be "plucked" without cruelty,' the author speculated, 'is it not likely that mistakes are often made by the pluckers, or that occasional carelessness or recklessness occurs, or that the greed or need of owners induces to early plucking – as in this country it leads to too early shearing of sheep?' Some ostrich farmers seemed to give weight to these fears when they admitted to using 'a salve on the skins of the birds, kept for the purpose of healing the wounds' caused by plucking, suggesting that the procedure was not painless. One French merchant stated that his ostriches, 'after being completely plucked, have their skins rubbed over with olive oil to alleviate the irritation which naturally ensues from the operation'.[96] A later account of ostrich farming in Somaliland reported that '[e]very six months the ostriches are plucked, and a debilitated bird's strength sustained after the painful operation by feeding it with chopped mutton and fat, and rubbing ghee or clarified butter as a emollient over the lacerated follicles'.[97] Ostrich feathers were not, therefore,

[93] 'Society for the Protection of Birds', *Bristol Mercury*, 22 March 1894.
[94] 'Are Ostrich Feathers Procured by Cruel Means?', *The Animal World*, November 1891, p. 174
[95] 'Can Ostrich Feathers Be Had without Cruelty?', *The Animal World*, March 1893, p. 47.
[96] 'Ostriches and Their Feathers', *The Animal World*, July 1886, p. 100.
[97] 'Ostrich Feathers', *The Animals' Friend*, 1906, p. 148.

entirely cruelty-free, and Kingsford's analogy with sheep shearing may have been overplayed (or, as the *Animal World* article implied, it was only accurate insofar as both species were abused). An article in *The Cornishman* claimed that extracting a feather was about as painful as pulling out a human tooth, accusing 'ladies who wear ostrich feathers' of 'encouraging the most cruel and barbarous torture which man can inflict upon a bird'.[98]

If ostrich farming was a mixed blessing from a humanitarian perspective, its impact as a conservation measure seemed less contentious, at least for contemporaries. Before they were domesticated, ostriches appeared to be on the road to extinction. Writing in 1876, the *Pall Mall Gazette* claimed that 'it is inevitable that they must die out with the progress of civilisation'.[99] A decade later, the *North-Eastern Daily Gazette* expressed a similar view, noting that the earlier practice of hunting ostriches for their feathers was on the point of 'exterminating them' when 'by accident it was discovered that ostriches, notwithstanding their extreme shyness, could be domesticated and would breed in captivity'.[100] In both cases, the domestication of the ostrich was presented as the bird's salvation, and the subsequent success of ostrich farming in South Africa was seen as a major conservation success. Speaking in 1920, at a meeting of the African Society, Sir Harry Johnston claimed: 'If it had not been for the ostrich feather industry in South Africa, the ostrich would now be as extinct as the dodo.'[101] In that sense, the successful farming of the ostrich was seen as a uniformly good thing from a conservation perspective and represented a much more successful version of the parallel efforts to domesticate the African elephant (see Chapter 3). Again, though, as in the case of the elephant, the choice was very much a stark one between utility and extinction – there was no room for the wild ostrich in this scenario.

The Biography of a Lie

If the plumage trade ultimately proved a mixed blessing for the ostrich, it was an unmitigated disaster for our second species, the egret. Among the birds most heavily exploited by the millinery trade, egrets (sometimes referred to inaccurately as ospreys) were coveted for their handsome white feathers, which were shipped to fashion retailers in huge volumes.

---

[98] 'Wearing of Ostrich Feathers', *The Cornishman*, 9 February 1905.
[99] 'Ostriches and Ostrich Farming', *Pall Mall Gazette*, 11 December 1876.
[100] 'Feathers', *North-Eastern Daily Gazette*, 13 July 1886.
[101] 'Ostrich Eggs as Food', *The Times*, 29 May 1920.

Figure 1.4 'An Egret', from H. Vicars Webb, 'The Egret Plume Trade
and What It Has Accomplished', *The Animal World*, June 1910, p. 103.
The delicate feathers on the bird's back were those most desired
by milliners.

In 1905, in London alone, 'the feathers of 15,000 herons and egrets were
sold at auction'; in Paris, the following year, the figure was 230,000.[102]
Egret populations could not sustain these heavy losses and the species
soon became the cause célèbre for bird defenders, who fought hard to
save it from extinction. Their efforts elicited a storm of protest from
milliners, who denied claims of cruelty and attempted to paint the
plumage trade in a better light.

Egret feathers were particularly controversial on account of the
manner in which they were obtained. The plumes desired by milliners
came from the back of the bird and grew only during the mating season,
when the egrets were nesting (Figure 1.4). This meant that not only were
the adult birds slaughtered in large numbers to obtain a small volume of
feathers, but that their young chicks would starve to death following the
killing of their parents. An article in *The Star* (1894) recounted how
'when the killing is finished and a few handfuls of coveted feathers have
been plucked out, the slaughtered birds are left in a white heap to fester
in the sun and wind in sight of their orphaned young, that cry for food
and are not fed'.[103] A US pamphlet on 'The Horrors of the Plume Trade'
was even more emotive, featuring photographs of egret carcasses and
famished chicks; the caption for one of the images read: 'Fatherless and

---

[102] 'Murderous Millinery', *Living Age*, 8 December 1906; 'The Plume that Brings Death',
Washington Post, 3 March 1907.
[103] 'The Protection of Birds', *The Star*, 23 January 1894.

motherless – no-one to feed them – growing weaker – one already dead from starvation and exposure.'[104]

While cruelty towards wildlife was, sadly, far from unusual, the melancholy fate of the egret struck a particular chord with people at the time and became a key battleground in the fight against 'murderous millinery'. Conservationists on both sides of the Atlantic seized upon the egret's demise as an example of the heartlessness of women and the insatiable greed of the millinery trade. In response, milliners issued a series of claims designed to justify their actions and to ensure the survival of their industry. These ranged from outright denials of any kind of abuse to assertions that past abuses had now ceased, enabling women to purchase egret feathers with a clear conscience. Although these claims were ultimately exposed as untrue, their existence was highly damaging to egret conservation, sending out deceptive messages to consumers that bird defenders mobilised to rebut. A close examination of the ensuing debates reveals both the lengths to which milliners were prepared to go in order to protect a lucrative line of business and the growth in consumer demand for ethically sourced products. It also illustrates the role of science and expertise in discrediting the comforting fictions disseminated by defenders of the feather trade.

In the face of harrowing stories about mutilated adults and starving chicks, the millinery trade's first response was denial. This is most apparent in explanations for the decline of egrets in the US state of Florida, where the birds had virtually disappeared by the early twentieth century. In one controversial paper read at a meeting of the London Chamber of Commerce in November 1910, milliner Charles F. Downham argued that egret numbers in Florida had indeed diminished in recent decades, but that this was due to 'American commercial development' – by which he meant increased human settlement in the area – and definitely not the result of the plume trade, which had never been 'of importance' in that state.[105] Earlier, summoned before a House of Lords Committee, Downham had given a different spin to this argument, asserting that the egrets of Florida 'were not exterminated; they migrated'. As he elaborated, 'You might just as well say that because you do not see foxes on Hampstead Heath, foxes are exterminated.'[106] Other milliners supported Downham's view that the plume trade was not

---

[104] William Dutcher, 'The Horrors of the Plume Trade: The National Association of Audubon Societies, Special Leaflet No. 21', cited in *Bird Lore*, Vol. XI, 1909, p. 3. On the importance of photographs in publicising the egret's plight, see J. Keri Cronin, *Art for Animals: Visual Culture and Animal Advocacy, 1870–1914* (University Park: Pennsylvania State University Press, 2018), pp. 87–92.

[105] Downham, *The Feather Trade*, p. 40.   [106] RSPB, *Feathers and Facts*, p. 54.

responsible for the decrease in the egret population of Florida, although the positions they adopted were not entirely consistent. Speaking before the House of Lords Committee, one witness, Mr Weiler, conceded that some egrets may just possibly have been killed in Florida for their plumes but that 'the tale about the birds being shot at breeding time is a fairy myth'. Another milliner, Mr G. K. Dunstall, declined to say whether or when egrets were shot but attributed the extermination of the birds in Florida to the fact that 'there never were many Egrets' in the region, making them vulnerable to both natural and human hazards. As he helpfully explained, with an analogy from closer to home, 'You can soon exterminate a small number of birds in any small part of the country. If there were Egrets in the Isle of Wight they would soon be exterminated.'[107]

All of these arguments had the potential to undermine the RSPB's claims of cruelty, so bird defenders in both Europe and America moved quickly to dismiss them. Addressing one claim made by Downham – that the continued supply of egrets proved the sustainability of the industry – the president of the Pennsylvania Audubon Society, Witmer Stone, issued a vehement denial, insisting that: 'The only reason that the supply continues undiminished is that there are still wild coasts in less fre-quented parts of the world where the birds can still be procured.'[108] Responding to the specific issue of whether or not egrets were killed in Florida for their feathers, an RSPB publication, *Feathers and Facts* (1911), cited the testimony of numerous ornithologists, offering example after example of unregulated slaughter. A report by 'the late Mr W. E. D. Scott, member of the American Ornithologists' Union', described finding 'a huge pile of half-decayed birds lying on the ground', their skin and plumes torn off. A similar report from Gilbert Pearson recounted how, while visiting Horse Hummock in central Florida, he had seen 'heaps of dead Herons festering in the sun, with the back of each bird raw and bleeding', and 'young herons left in the nests to perish from exposure and starvation'.[109] Such evidence discredited the notion that egrets had declined naturally or migrated and instead corroborated William Dutcher's claim that the egret had been 'exterminated in the United States by the plume hunters as thoroughly as was the bison by the hide-hunters'.[110]

[107] Ibid., p. 54.
[108] 'The Osprey and the Egret', *New York Daily Tribune*, 31 January 1897.
[109] RSPB, *Feathers and Facts*, pp. 55–6.
[110] 'Killing Birds for Their Plumes', *New York Tribune*, 21 February 1897.

With the evidence of cruel methods of slaughter mounting, the millinery trade adopted a new tactic. Under pressure from critics, milliners conceded that egret feathers might perhaps be obtained with some degree of cruelty in certain places, including possibly Florida (although no one ever quite admitted this). Consumers did not need to worry, however, because: (1) artificial substitutes were now available and (2) these same feathers could now be acquired cruelty-free from other parts of the world. This led to the advent of two new fictions: the 'artificial osprey' and the 'moulted plume'.

The first of these phenomena, the so-called artificial osprey, made its debut in the mid-1890s and was designed to assuage the consciences of female consumers, who might feel bad about the idea of killing a beautiful bird in the name of fashion. Made, supposedly, from 'quills, ivorine, silk, wood' and 'the feathers of poultry, etc.', these 'fake' feathers looked identical to the real ones and appeared to answer the call of conservationists for a stylish alternative to the plumage of rare birds. With artificial feathers now available, caring women could buy aigrette-adorned hats without fear or guilt, confident that their purchases were not pushing an exotic species to the brink of extinction. Moreover, the artificial feathers were less expensive than the genuine ones; as several milliners argued, 'they could not be sold so cheaply if they were real feathers'.[111]

Bird protection organisations, however, were suspicious about these 'artificial ospreys' and quickly proved them to be fraudulent. One expert, Sir Edwin Ray Lankester, stated in a 1903 interview with the *Daily News* that '[a]n osprey has never been imitated, and whatever the shop keeper may say, it is always the parent bird, slain at the breeding season, which supplies women's hats and bonnets'.[112] Another scientist, Sir William Henry Flower of the Natural History Museum, examined several feathers sold as 'artificial' in the 1890s and reported that '[i]t did not require very close scrutiny to see that they were unquestionably genuine'. The sole exceptions were a few plumes 'made from the flight feathers of the rhea, or South American ostrich'. Here, however, 'the remedy was almost worse than the disease, for the rhea is a bird of the utmost scientific interest and importance, and its extinction in a wild state seems to be rapidly approaching'.[113] Milliners were thus either passing off real feathers as fakes or substituting the feathers of another endangered bird for those of the egret. Both practices were highly dangerous from a conservation perspective, with the promise of cruelty-free plumes

---

[111] RSPB, *Feathers and Facts*, p. 39.   [112] Ibid., p. 40.
[113] 'The Plume that Brings Death', *Washington Post*, 3 March 1907.

muddying the ethical waters and tempting buyers who would not have purchased feathers they believed to be genuine.

The second – and more potent – myth circulated with regard to egret feathers focused not on the authenticity of the plumes but on the manner in which they were obtained. In a new twist to the 'artificial osprey' story, milliners now admitted that egret feathers did indeed belong to real birds, but suggested that they were procured without violence. This was possible either because egrets moulted naturally at a certain time of year, or – in other versions of the tale – because they plucked out their own feathers in order to line their nests in the breeding season. Rather than killing egrets, therefore, plumage hunters simply picked up their shed feathers from the ground and delivered them to local dealers. Indeed, so advanced was the egret industry in some parts of the world that the birds were farmed in so-called '*garceros*' (heronries) and their discarded plumes collected on a regular basis.

The precise location of these heronries remained somewhat vague and tended to shift over time. The earliest reports of their existence placed them in China, where egret feathers were supposedly 'picked up on the walls' and sent to Europe. After the Chinese trade in egret feathers collapsed, however, due to the virtual extermination of the birds, tales started to emerge of parts of India where vast tracts of the country were 'white with shed feathers, lying in sheets like snow'.[114] One writer, Mrs Isobel Strong, published an article in the *New York Tribune* in 1899 in which she relayed Mrs Robert Louis Stevenson's claim (based on conversations with an Indian missionary) that feathers were 'picked up from the ground during the breeding season of herons' in India.[115] Another commentator, a traveller to Nigeria, described how he had seen 'moulted plumes' strewn across the jungle floor, just waiting for some enterprising individual to gather them.[116]

By the first decade of the twentieth century, the Asian and African heronries had receded and South America had become the favoured site for discarded egret plumes. In a letter to *The Times* in 1900, a Mr K. Thompson claimed that in both Nicaragua and Venezuela egrets were 'so shy and difficult to approach' that, instead of shooting them, feather hunters simply went 'round and pick[ed] up the cast plumes'.[117] A second writer, Leon Laglaize, was even more expansive. In a somewhat mysterious letter from Buenos Aires, dated 10 July 1908 and addressed to an unknown recipient, Laglaize, who signed himself 'Fellow of the

[114] RSPB, *Feathers and Facts*, pp. 41–2.
[115] 'Egret Farms a Myth', *New York Tribune*, 28 December 1899.
[116] RSPB, *Feathers and Facts*, p. 42.   [117] Ibid., p. 43.

Entomological Society of Paris', suggested that Venezuelan egrets plucked out their own feathers in order to line their nests, making them easy to gather after the nestlings had fledged. Laglaize described how, in Apure State in Venezuela, large landowners prohibited the shooting of egrets on their estates but instead hired the local Indians to paddle around the swamps in canoes, collecting moulted plumes. 'After the breeding season, when the young ones leave their nests, the abandoned nests are searched and a valuable amount of feathers is collected,' asserted Laglaize. 'The feathers have been skilfully rolled in to furnish and soften the interior of the nest, having been pulled off by the bird itself before laying the eggs.'[118] It was a similar story '[i]n some parts of Brazil and mostly in the provinces of Matto Grosso [sic] and Goyaz [Goiás]', where 'the great landowners on the banks of the Rio Alto, Paraguay, and Rio Tacuarí hire the right of picking up the feathers on their estates, under contract that no birds shall be shot'.[119]

Alongside the tales of wild egrets shedding their plumes, rumours also circulated of farms in Africa and the USA where domesticated egrets were shorn humanely of their precious tail feathers. An 1897 article in the New York Times reported the existence of some 400 egrets in a heronry outside Tunis, fed on 'minced horse and mule meat twice a day' and relieved of their dorsal plumes twice a year, in May and September.[120] An 1899 article in the Washington Post likewise detailed the plans of a certain A. Bienkowski to start an egret farm outside Yuma in Arizona, where the birds were to be kept on '160 acres of marshy land along the river bottom', their wings clipped to prevent them from flying away.[121] Clearly inspired by the success of ostrich farming in South Africa, the idea of domesticating the egret appealed to those seeking a quick profit, and it appeared to offer the possibility of 'improving' the species by selectively breeding the birds to propagate those with 'the most abundant plumes, the most accommodating appetites and the best tempers and constitutions'.[122] Although it meant loss of liberty and possibly some painful plucking, such a move also seemed like good news for the egret, which, like the ostrich, might now secure its future through domestication. Who could object to treating the exotic egret in the same manner as traditional poultry?

[118] Ibid. p. 44.   [119] 'Moulted Plumes', RSPB Pamphlet No. 60, 1908, pp. 7–8.
[120] 'Raising Egret Herons', New York Times, 21 November 1897.
[121] 'A Millinery Farm', Washington Post, 27 August 1899.
[122] 'Raising Egrets for Plumage', London Express, cited in New York Tribune, 8 January 1905.

Unfortunately, neither the 'moulted plumes' nor the egret farms turned out to have any basis in reality. In the case of the 'moulted plumes' scenario, this was deemed biologically untenable, since egrets neither plucked out their own feathers nor used them to line their nests. A few plumes might be shed naturally by the birds, or perhaps pulled out during fights between the males, but these were neither of the quality required nor available in the quantity necessary to furnish the volume of feathers used by the millinery industry. One critic, Albert Pam, a member of the council of the Zoological Society of London (ZSL) with experience travelling in Venezuela, testified before a House of Lords Committee that 'if you wished to collect feathers you would have to walk several hundred yards for each individual plume you picked up, and in the jungle of the Amazon it would be an extremely difficult occupation'.[123] Another traveller, Edward McIlhenny, was even more emphatic, informing US conservationist William Temple Hornaday 'that it is impossible to gather at the nesting places of these birds any quantity of their plumes'.

I have nesting within 50 yards of where I am now sitting dictating this letter not less than 20,000 pairs of the various species of snowy herons nesting within my preserve. During the nesting season, which covers the months of April, May and June, I am through my heronry in a small canoe almost every day, and often twice a day. I have had these herons under my close inspection for the past seventeen years, and I have not in any one season picked up or seen more than half a dozen discarded plumes.[124]

Claims of feather-shedding egrets were therefore either disingenuous or, more likely, false, betraying the ignorance or deceit of their originators.

Reports from reliable individuals on how egret feathers were actually gathered in Venezuela were much more disturbing and did not correlate with Laglaize's rosy picture. Consulted by the RSPB, one informant, the British ambassador to Venezuela, asserted that egret feathers were not picked up by Indians in canoes but collected by local peons – mostly ranch hands – who shot the birds to fulfil their assigned quotas. There had recently been a move on the part of some landowners in Apure State to protect egrets by outlawing unauthorised shooting on their lands, but this did not extend to public lands, where the birds could be killed without limit. The stories of moulted plumes, moreover, were entirely unfounded, as 'dead feathers' (those dropped naturally by living birds) fetched very little on the market due to their inferior quality.[125] The same

---

[123] RSPB, *Feathers and Facts*, p. 46.   [124] Hornaday, *Our Vanishing Wild Life*, p. 129.
[125] 'Moulted Plumes', RSPB Pamphlet No. 60, 1908, pp. 4–5.

was true in Santa Fe, Argentina, where, according to His Majesty's Consul at Rosario, no egret farms existed and laws to prevent shooting outside the close season were inadequately enforced.[126]

Another witness, an ex-plume collector, offered an even more harrowing account of how egrets were killed. Testifying for the Audubon Society before the New York State Legislature in 1911, A. H. Meyer described the heartless cruelty of Venezuelan feather hunters, who left mutilated birds to die and even used the corpses of half-dead egrets to lure other egrets to a similar fate.

> The natives of the country, who do virtually all of the hunting for feathers, are not provident in their nature, and their practices are of a most cruel and brutal nature. I have seen them frequently pull the plumes from wounded birds, leaving the crippled birds to die of starvation, unable to respond to the cries of their young in the nests above, which were calling for food. I have known these people to tie and prop up wounded egrets on the marsh where they would attract the attention of other birds flying by. These decoys they keep in this position until they die of their wounds or from the attacks of insects. I have seen the terrible red ants of that country actually eating out the eyes of these wounded, helpless birds that were tied up by the plume-hunters.[127]

This was a far cry from the idyllic picture of Indians paddling round in canoes gathering cast-off feathers and suggested a much darker reality.

As for the alleged egret farms, these appeared to be, if not complete fictions, then a considerable embellishment of the truth. Individual examples of domesticated egrets may have existed, but the fabled egret farms were never more than grand designs. When Philip Lutley Sclater of the ZSL investigated the supposed Tunis egret farm, he found no trace of it and learned that the enterprise had ended in failure.[128] Similarly, when ornithologist Herbert Brown enquired into the egret farm at Yuma, he discovered that it consisted of a single 'little white egret' kept as a pet at the Southern Pacific Hotel in the town.[129] South America, the continent most frequently associated with egret domestication and farming, did offer some evidence of tame egrets, but these animals were reared not for their feathers, as milliners suggested, but for the exotic pet trade. US naturalist George K. Cherrie, curator of birds at the Brooklyn Institute of Arts and Sciences, reported that it was 'not uncommon' in Venezuela 'to see two or three Egrets picketed in front of a rancho, a string two or three feet long being tied around one leg and attached to a stake'. These

---

[126] Ibid., p. 6. For further discussion of the local impact of the feather trade in Venezuela and Colombia, see 'Commodities and Fashion Objects' in Quintero Toro, *Birds of Empire*, pp. 13–37.

[127] Hornaday, *Our Vanishing Wild Life*, p. 130.   [128] RSPB, *Feathers and Facts*, p. 42.

[129] 'Egret Farms a Myth', *New York Tribune*, 28 December 1899.

captive birds, however, were transported by river steamer 'to Bolivar or Port-of-Spain, Trinidad, to be disposed of to tourists' and were not the source of the mythical 'moulted plumes'; 'Egret "farming" is no more an industry than is Parrot "farming".'[130] Milliners had therefore extrapolated small realities into large-scale falsehoods. There were no egret farms and no moulted plumes, only thousands of dead egrets shorn of their feathers.

The long battle between milliners and bird protectors over the source of egret feathers highlights the great value attached to this commodity in the first decade of the twentieth century and the tactics deployed by both sides to defend their interests. For milliners, desperate to ensure the survival of their business, it was important to deflect allegations of cruelty and unsustainable hunting. Defenders of the trade in egret feathers therefore argued, initially, that this trade was not to blame for the disappearance of egrets in Florida, which, if it had happened at all, had done so for other reasons. When humanitarian propaganda gave the lie to these claims, however, the milliners changed tack; instead, they tried to pass genuine feathers off as 'artificial' ones to reassure concerned customers. When these fake feathers were in turn exposed as genuine, they circulated information suggesting that the means of procuring egret feathers had changed and that the birds were now farmed rather than hunted. None of these claims were true, but the existence of such falsehoods shows that milliners were at least conscious of shifting public opinion and understood the need to present their products in a more ethical light – in itself arguably something of a victory for conservationists.

For bird defenders, the fate of the egret was equally critical, though for different reasons. If the milliners' claims were believed, women, who otherwise might have been dissuaded from buying egret feathers, would continue to do so, reassured that they were not colluding in cruelty. They would thereby perpetuate a cruel fashion, undoing all of the campaigners' good work. Conservationists duly made extensive efforts to discredit the milliners' false claims, often relying on science to expose the truth. Faced with the charade of the 'artificial osprey', the SPB immediately sought the opinion of accredited ornithologists, who performed tests on the suspect feathers to reveal their authenticity. Confronted, subsequently, with the myth of the 'moulted plumes', the organisation summoned the counter-testimonies of experienced travellers, whose long years in the field and scientific credentials made them trustworthy witnesses. In one 1909 pamphlet, entitled 'How the Osprey Feathers Are

---

[130] 'The Egret Hunters of Venezuela', *Bird Lore*, Vol. 2, 1900, p. 51.

Procured', the RSPB cited the observations of two British ambassadors, 'Mr J. Quelch, B.Sc ... formerly Curator [of the] British Guiana Museum', 'Mr G. E. Dresser, author of "The Birds of Europe"', 'Mr Frank Chapman, Curator of the American Museum of Natural History at New York', 'Mr Gilbert T. Pearson, Secretary of the National Association of Audubon Societies' and 'Mr. H. E. Mattingley, [contributor to] the *Emu*, the organ of the Australasian Ornithologists' Union', all of whom dismissed the myth of the 'moulted plume'.[131] Conservationists also used logic, photographic evidence and careful investigation to expose the inconsistencies of milliners' claims, showing how they had inflated one-off cases into untenable assertions and employed contradictory arguments. In the case of Laglaize, they went for the man, revealing that the Frenchman had no connection with any Parisian scientific society and was either a stooge or a fabrication of the millinery industry.[132] The weight of evidence and the authority of their sources ultimately enabled the SPB and its American counterpart to win through and expose the 'moulted plumes' story as a hoax. The battle was, nonetheless, an important one for conservationists, who realised that even false stories could badly derail their campaigns.

Conclusion

The struggle to end the plumage trade was a long and protracted battle. Conservationists worked hard to suppress this cruel fashion, lobbying for new legislation and campaigning tirelessly to change public opinion. Their eventual success demonstrated the effectiveness of their tactics and offered inspiration to other conservation movements, some of which consciously imitated their techniques. In that sense, the work of the SPB and Audubon Societies was pioneering, providing a model for similar campaigns.

The drive to suppress 'murderous millinery' is also significant because of the opposition it provoked; this highlighted the challenges faced by many past and present conservation movements. The millinery industry proved to be a powerful adversary, with plenty of financial and political support behind it. Unwilling to give up their business without a fight, its

---

[131] 'How the Osprey Feathers Are Procured', RSPB Pamphlet No. 61, 1909, pp. 2–4.

[132] According to William Hornaday, 'when Prof. Henry Fairfield Osborn cabled to the Museum of Natural History in Paris, inquiring about Mr. Laglaize, the cable flashed back the one sad word: "*Inconnu!*"' Hornaday, *Our Vanishing Wild Life*, p. 128. For an illuminating discussion of the establishment of scientific credibility in the Victorian era, see Stuart McCook, '"It May Be Truth, but It Is Not Evidence". Paul du Chaillu and the Legitimation of Evidence in the Field', *Osiris* 2nd Series 11 (1996), pp. 177–97.

defenders mounted stiff resistance to the bird protection movement, resorting to dubious science and outright lies to discredit the claims of conservationists. This is best demonstrated by the terse debates over the beleaguered egret, which either did not need or did not deserve protection from plumage hunters. Missouri senator James A. Reed, for instance, opposing the 1913 Tariff Bill, questioned 'why there should be any sympathy or sentiment about a long-legged bird [the egret] that lives in swamps and eats tadpoles and fish and crayfish and things of that kind'. He suggested that plumage hunting should carry on, but that 'a foundling asylum' should be established for starving chicks so that 'humanity' could continue to 'utilize the bird for the only purpose it has that evidently the Lord made it for, namely *so that we could get aigrettes for bonnets of our beautiful ladies*'.[133]

Another particularly notable aspect of the 'murderous millinery' debate was the prominent role of women. On the one hand, the emergence of new fashions reflected the growing importance of females as consumers and their capacity to shape and sustain new trends and industries. On the other, of course, it exposed women to severe censure from conservationists and humanitarians, who preached the need for them to use their new powers of consumption ethically and responsibly. Educational campaigns against the use of plumage frequently equated female feather wearers with savages and harpies and appealed to the maternal instincts of women, who, it was assumed, would not continue to wear birds if the inherent cruelty of their acquisition were made known to them. Both the SPB and the Audubon Societies, meanwhile, boasted a large female membership and were led by women who wanted not only to save the birds but to rehabilitate the reputation of their gender. A member of the Bristol branch of the SPB urged women to abjure feathers 'for the justification of our sex'. As she explained: 'It has been said that "women won't raise a finger to stop any cruelty so long as it ministers to fashion". Now, we don't deserve that; let us prove it.'[134]

Also central to the campaign to protect the birds was a problem that would dog many other conservation movements and complicate the effective enforcement of conservation legislation: how to protect species that migrated across state and national boundaries. Birds did not respect human borders, so species that were protected in one country could be legally killed once they left it. Commerce, too, took place on a global scale, so dead birds could also traverse frontiers. To put an end to the plumage trade, therefore, protection agreements had to be reached on an

---

[133] 'The Feather Proviso', *Bird Lore*, Vol. XV, 1913, p. 331.
[134] 'Society for the Protection of Birds', *Bristol Mercury*, 22 March 1894.

international level and bans needed to be placed on the export and import of endangered species. The 1913 Tariff Act and the 1921 Plumage Act eventually achieved this, but the enactment of these measures was slow and hard-fought.

Finally, the contrasting fate of two birds, the egret and the ostrich, illustrates another key element of nineteenth-century conservation campaigns: the role assigned to domestication. The egret, despite claims to the contrary, was not suited to large-scale domestication and was hunted nearly to extinction. The ostrich, on the other hand, was successfully domesticated and farmed, a process many contemporaries regarded as the bird's salvation. The underlying assumption here was that animals were put on earth for the benefit of mankind – a view clearly articulated by Senator Reed – and that the survival of a species depended on its surrendering to human control. This assumption was expressed most forcefully by manufacturers, who needed to justify the use of animals in their produce, but it was shared, to some degree, by conservationists, who often proposed domestication as the best way of preserving threatened species. Ostriches thus survived the plumage trade because they could be farmed; egrets teetered on the brink of extinction because they could not.

2   The Seal and His Jacket

[The seal] is a victim of mixed metaphor at every period of his life. His father is a 'bull', his mother is a 'cow', he himself is a 'pup', he is born in a 'rookery' and killed in a 'pod' – a truly strange Pilgrim's Progress: bovine, canine, corvine and cruciferous!   'The Seal and His Jacket', *The Leisure Hour*, February 1890, p. 254

In April 1875, a woman wrote to the RSPCA's monthly magazine, *The Animal World*, to condemn 'the dreadful cruelties' committed by the sealing industry. The writer, identified only by her initials, M.S.H., had recently read an article on the subject in the same magazine. She had been shocked to discover that the fashionable furs sold to upper-class ladies came at such a cost to their original owners, and declared that, although financially 'quite able … to purchase as many sealskin mantles and jackets as a lady could desire', she would 'not buy one, nor would I accept a gift of one, so grieved am I at the thought of the butchery performed to procure them'. To bring the full horror of the industry home to readers, M.S.H. described the brutal way in which the seals were killed, noting that many were skinned alive. She also conjured a heart-wrenching image of the young 'cubs', whose 'piteous cries' rang out across the ice as they beheld their mothers 'bathed in blood'. Since most of the sealskins obtained with such cruelty were used to produce fashionable apparel for women, M.S.H. appealed to her fellow females to boycott sealskin products, certain that only ignorance made women accessories to such cruelty. 'Luxury and vanity harden the heart; but could ladies once view a seal massacre, they would, as I do, reject the fur procured at such a price, as bloodstained a trophy as an Indian scalp gained in warfare!'[1]

M.S.H.'s letter offers an emotive insight into the passions raised by sealing in the later nineteenth century – in this case, specifically its animal welfare implications. A thriving business in this period, the sealskin industry derived its raw material from the northern fur seal (*Callorhinus*

---

[1] 'Cruelty to Seals', *The Animal World*, April 1875, p. 62.

ursinus) and employed workers in Alaska, California and London. It supplied thousands of skins every year to the North American and European markets, providing fur for jackets, gloves, pelisses and other upmarket consumer goods. As M.S.H. noted, the majority of these goods were bought and worn by women, whose lavish tastes were indirectly to blame for the slaughter of seals in distant lands.

This chapter explores the development of the seal industry in the period from c.1870 to 1914 and assesses its ecological impact. Long prized for its luxurious coat, the fur seal had already been wiped out in the southern Pacific Ocean through indiscriminate culling and, by the 1880s, was under threat in the Bering Sea. Anxious to prevent its extinction, the US Government looked for ways to limit the slaughter, introducing quotas for the number of seals that could be killed each year and banning 'pelagic sealing' (the hunting of seals in the sea). The chapter considers the wider diplomatic clashes that resulted from the decline of the fur seal population and the complications surrounding animal protection legislation, particularly when the animal in question lived in the ocean and crossed territorial boundaries. I also examine the increasingly important role played by scientific experts in informing and framing conservation programmes – and what happened when scientists disagreed.

Related to, but distinct from, these broader environmental concerns, animal lovers such as M.S.H. also began to question seal hunting from a humanitarian perspective. Sealing was an inherently cruel practice, both in Alaska and in the North Atlantic, where British, German and Scandinavian sealers slaughtered hair seals (seals with coarse hair rather than fur) for their valuable oil (used for heating, lighting and as a lubricant). Seals were typically clubbed to death to prevent excessive damage to their fur, and horror stories circulated about animals being skinned alive. Moved by these harrowing tales, animal welfare organisations in both Britain and the USA campaigned vociferously against the use of sealskin, demanding a complete end to the slaughter. Unlike scientists and conservationists, who were primarily concerned about the sustainability of the sealskin industry, humanitarians were mainly worried about the suffering of individual animals and would have found this repulsive even if those animals were not threatened with extinction. The fact that fur seals were killed to make luxury goods, moreover, made the seal cull appear even more heinous, for the suffering in question was inflicted to serve frivolous ends; as M.S.H. concluded, 'used for seal oil and killed at the proper season, all would be different'.[2] Addressing these

---

[2] Ibid.

moral qualms, the final part of the chapter examines the tactics and language employed by welfare campaigners and considers their impact on consumer behaviour.

More Than One Way to Skin a Seal ...

The fur seal belongs to the family *Otariidae*, or 'eared seals'. It inhabits the Pacific Ocean and differs from its North Atlantic cousins in having visible ear lobes, a longer neck, and limbs more distinct from the torso. Fur seals spend most of their lives at sea, feeding on fish, cephalopods and crustaceans, but come ashore annually to breed between April and June.[3] While Atlantic seals were hunted for their oil, stored in the layer of fat beneath their skin, Pacific seals were prized primarily for their dense fur undercoat, 'distributed in delicate, short, fine hairs ... all over the body' (Figure 2.1).[4] This coat was most plentiful in three- to four-year-old males, from whom the best sealskins were obtained.

Fur seals once existed all around the Pacific and on the Atlantic coast of South America. By the late nineteenth century, however, most of these seal colonies had been wiped out or severely diminished by indiscriminate hunting; the only remaining locations for sealing were a small island in the Rio de la Plata, owned by Uruguay, the Russian Commander Islands and two islands in the Bering Sea named Saint George and Saint Paul. First discovered in 1786 by the Russian explorer Gehrman Pribilof, the latter were sold by Russia to the USA in 1867, along with the rest of Alaska, and were leased by the US Government to the Alaska Commercial Company. The Alaska Commercial Company enjoyed a monopoly over the islands, recruiting Aleuts from the Alaskan mainland to hunt the seals there and paying them 40 cents for each skin collected.[5]

Sealing in the Bering Sea was conducted in two main ways, with differing ecological effects. First, the Alaska Commercial Company carried out what was known as a seal drive. This involved waiting for the immature seals to come ashore to breed, which typically happened in July, and descending on the 'hauling ground' where the young, non-breeding males congregated. The company's employees would then drive the animals 'for a mile or more ... up to the village', coerce them into a makeshift corral and bludgeon them to death, first knocking them

---

[3] Roger L. Gentry, *Behavior and Ecology of the Northern Fur Seal* (Princeton: Princeton University Press, 1998), p. 6.

[4] John Willis Clark, 'Sea-Lions', *The Contemporary Review*, December 1875, p. 28. Sealskin has a density of c.47,000 fibres/cm^2. See Gentry, *Behavior and Ecology of the Northern Fur Seal*, p. 14.

[5] 'The Seal and His Jacket', *The Leisure Hour*, February 1890, p. 256.

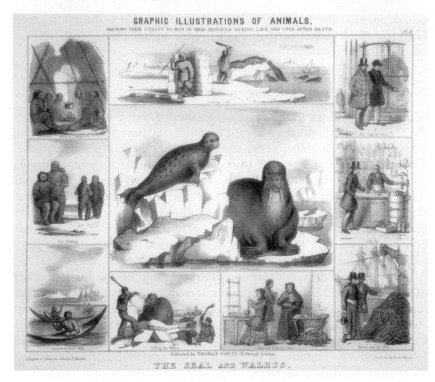

Figure 2.1 Benjamin Waterhouse Hawkins, 'Graphic Illustrations of Animals: The Seal and the Walrus', c.1850. Courtesy of Oxford Science Archive/Heritage Images

on the head to stun them, then sticking knives into their hearts 'at a point between the fore-flippers' to finish them off.[6] Once the animals were dead they were bled, to prevent them from overheating, before being skinned and salted.[7] The US Government took $2 per skin in taxation and imposed a limit of 100,000 skins per year from 1870 onwards, a quota lowered to 60,000 skins per year in 1890, when the lease was reissued to the North American Company.[8] The seal drive was conducted by native Aleuts, originally transplanted to the Pribilof Islands

[6] Ibid.; Clark, 'Sea-Lions', p. 45; 'The Seal and His Jacket', *The Leisure Hour*, February 1890, p. 256.
[7] 'The Seal and His Jacket', *The Leisure Hour*, February 1890, p. 259.
[8] Clark, 'Sea-Lions', p. 47; M. Douglas Flattery, *The Truth about the Fur-Seal Question* (Danville: Edward Fox, 1897), p. 11.

by the Russians, who reportedly had 'a way of handling the seals that they understand'.[9]

The second method of catching seals was to hunt them at sea, a practice known as 'pelagic sealing'. Employed primarily by Canadian sealers, pelagic sealing entailed shooting or harpooning seals while they were swimming in the ocean. Less discriminate than the seal drive, it was considered to be more ecologically damaging because it affected females as well as males and could consequently reduce the potential for breeding. One critic estimated that '80 or 90 per cent of the skins taken by the marauders are females, the killing of which means the death of two seals and the loss of so many members of the breeding pack'.[10]

Once the seals had been caught, either on land or at sea, the skins were salted for preservation and shipped by steamer to London via San Francisco. In the British capital, sealskins were sorted according to quality and purchased at quarterly sales by furriers, who then put them through a complex manufacturing process.[11] As an article in *The Leisure Hour* explained:

The salt is first washed off; the fat is then removed, the skins are then washed and the grease and water removed by the knife; and then they are tacked on frames to keep them smooth, and are dried in a moderate heat. They are then soaked in water and thoroughly cleansed with soap, and then, after drying, the warm skin is placed over a beam, and the hair removed with a knife. When all the hair is out, the skins are dried, then damped, and then shaved. They are then stretched, worked and dried and softened in a fulling mill, or trodden with bare feet in a barrel, the workmen dancing upon them to break them into leather, while the grease is absorbed with hard wood sawdust that from time to time he dusts in. The skins are then dyed, the dye being put on with a brush. They are then dried and dyed again and dyed again until they are of the desired colour, sometimes eight or even twelve coats of dye being required. They are then shaved down to the proper thickness, softened again in a hogshead and sometimes run in a revolving cylinder with fine sawdust to clean them.[12]

The method for separating the soft undercoat from the outer hair was invented in 1796 by London furrier Thomas Chapman, but an effective technique for dyeing the skins was found only in the 1870s.[13] After that,

---

[9] David Starr Jordan, *Observations on the Fur Seals of the Pribilof Islands, Preliminary Report* (Washington DC: Government Printing Office, 1896), p. 39.

[10] 'The Fur-Seal Fisheries of Alaska', *The Gentleman's Magazine*, March 1891, p. 256.

[11] 'Sealskin Coats, Alive and Dead', *Chambers's Journal of Popular Literature, Science, and Art*, 27 November 1886, p. 625. The majority of sealskins were auctioned in the October sale.

[12] 'The Seal and His Jacket', *The Leisure Hour*, February 1890, p. 259.

[13] Anon., *The Most Severe Case of Mr Thomas Chapman Who First Discovered the Means of Making the Fur of the Seal Available* (London: C. Cox, 1818), pp. 1–2.

all manner of sealskin 'cloaks, jackets, muffs, dainty little hats, collars, cuffs, bags [and] portemonnaies' started rolling off the production lines and seal complemented sable, beaver and chinchilla as a valued source of fur.[14]

Of all the sealskin items on the market, the sealskin jacket was the most popular, quickly becoming a mainstay in the fashionable woman's winter wardrobe. Retailing at between 20 and 30 guineas, new sealskin jackets were expensive and functioned as a marker of wealth and status.[15] Those with shallower purses, however, could obtain these prized items second-hand at a more affordable price or have existing jackets remodelled; in 1871, a vendor from London placed an advertisement for a 'splendid new sealskin jacket, trimmed with flouncings ... of chinchilla. Cost 25 guineas; price £15, as owner is going to India'.[16] Like other fashionable items, sealskin cloaks were subject to frequent changes of style, which encouraged those who could afford it to update their wardrobe every year, stoking demand. In 1886, jackets with 'long pointed fronts and close-fitting backs' were in vogue; in 1890, the fashion was for longer jackets with 'high collars that can be worn up and down, and sleeves high at the shoulders'.[17] 'The cut of the sealskin jacket, the orthodox tinge of colour in skunks are matters as important to the furrier as the study of public opinion to the politician'.[18] Sealskin jackets were thus at the forefront of female fashion in the late nineteenth century, a symbol of affluence and a driver of feminine consumption. *The Times* even reported the bizarre case of Mrs Montgomery, who was 'hugged by a Polar bear in Lime-street station' while wearing a sealskin jacket, the bear having supposedly mistaken her for a seal (Figure 2.2)![19]

Driven to Extinction?

While the sealskin industry was a thriving global business, its long-term survival was in doubt by the late 1880s. The growing demand for seal

---

[14] 'Revolting Cruelty to Seals', *The Animal World*, April 1875, p. 52.

[15] An opponent of sealskin noted that the ability to 'pay 20 guineas for a jacket of sealskin' marked an individual off as a 'rich woman'. See 'Price of a Sealskin Jacket', *Liverpool Mercury*, 14 December 1897.

[16] *The Bazaar*, 13 September 1871, p. 288. Others resorted to theft. In 1892, Bessie Hersee, 'a married woman residing at Garth, was brought up in custody charged with stealing a sealskin of the value of 30 guineas, the property of Mrs Hannah Jenkins, Proprietess of the Railway Hotel, Bangor'. See 'Alleged Theft of a Sealskin Jacket at Bangor', *North Wales Chronicle*, 16 January 1892.

[17] 'Sealskin Garments', *Birmingham Daily Post*, 29 December 1886; 'What to Wear', *Sunderland Daily Echo*, 8 November 1890.

[18] 'The Fur Trade', *The Star*, 7 January 1888.

[19] 'Hugged by a Bear', *The Times*, 8 January 1876.

Figure 2.2 'A Lady Hugged by a Polar Bear', *Illustrated Police News*, 22 January 1876. © British Library Board. All Rights Reserved/ Bridgeman Images

pelts was outstripping the supply of the animals, raising worrying questions about the possible extinction of the species. At the same time, a small but growing number of individuals were criticising sealing from an ethical perspective, arguing that it was not only unsustainable but cruel and immoral. These two concerns coalesced to bring the sealing business into question. Could the fur seal population be maintained and could the brutality of culling be mitigated? What measures could be instituted to protect a marine mammal in international waters?

The Alaskan seal crisis was not the first time that overhunting had threatened the fur seal population. On the contrary, the dangers of uncontrolled sealing were amply demonstrated by the dismal record of earlier seal colonies. Seals had once existed in other areas of the Pacific, Atlantic and Indian Oceans, including the Juan Fernández Islands, the Falklands, Tristan da Cunha, the Kerguelen Islands, South Georgia, Australia, New Zealand and the Pacific coast of South America. In all of these locations, however, the animals had been virtually exterminated by excessive culling in the late eighteenth and early nineteenth centuries, leaving the American-owned Pribilofs, the Russian Commander Islands and Cabo Corrientes in Argentina as the only remaining preserves for the

species.[20] The devastating impact of early sealing ventures was highlighted in contemporary reports, which told a bleak tale of indiscriminate
killing, glutted markets and culpable waste. One commentator
recorded how

[in the] two years 1814 and 1815 no less than 400,000 skins were obtained from
Pentipod, or Antipodes Island alone, and necessarily collected in so hasty a
manner that many of them were imperfectly cured. The ship 'Pegasus' took
some 100,000 of these in bulk, and on her arrival in London, the skins having
heated during the voyage, had to be dug out of the hold and were sold as
manure – a sad and reckless waste of life.[21]

Even in the Pribilofs the fur seal had been pushed to the brink of extinction by the turn of the nineteenth century, forcing the Russian authorities
to institute a temporary ban (or *zapooska*) on sealing in 1806–7 and a
full-blown moratorium between 1834 and 1841.[22]

The lessons of this bleak history did not go unnoticed by the US
authorities, who took measures to ensure that sealing in the Pribilofs
was a sustainable industry. On inheriting the territory from the Russians,
the US Government established a quota for the number of seals that
could be killed every year, limiting the catch to 75,000 skins on St Paul
and 25,000 on the smaller island of St George. A close season was
instituted from November to May each year to prevent the disturbance
of female seals and pups in the rookeries and the use of firearms was
prohibited anywhere on the islands to avoid frightening seals in the
breeding grounds. Only three- and four-year-old males ('holluschickies'
or 'bachelors' in contemporary parlance) – the seals with the most highly
prized fur – were slaughtered in the annual seal drive, and these were
separated from other seals before culling began. Older bulls and young
males caught up in the drive were identified by the hunters and released
unharmed.[23] Together, these precautions ensured that on-land sealing in

[20] Briton Cooper Busch, *The War against the Seals: A History of the North American Seal Fishery* (Montreal: McGill-Queen's University Press, 1985), pp. 3–37.
[21] Alexander Walker Scott, *Mammalia Recent and Extinct* (Sydney: Thomas Richards, 1873), p. 19. For a discussion of sealing in nineteenth-century Patagonia, see John Soluri, 'On the Edge: Fur Seals and Hunters along the Patagonian Littoral, 1860–1930' in Martha Few and Zeb Tortorici (eds), *Centering Animals in Latin American History* (Durham: Duke University Press, 2013), pp. 243–69.
[22] Henry W. Elliott, *Report on the Condition of the Fur-Seal Fisheries of the Pribilof Islands in 1890* (Paris: Chamerat et Renouard, 1893), p. x. On the Russian management of fur-bearing animals in Alaska – including the fur seal and the sea otter – see Ryan Tucker Jones, *Empire of Extinction* (Oxford: Oxford University Press, 2014).
[23] D. O. Mills, 'Our Fur-Seal Fisheries', *The North American Review* 151 (September 1890), p. 302.

the Bering Sea was a controlled activity, believed by most (though not all) contemporaries to provide a sustainable form of wildlife management.

The problem in the 1880s was that tight terrestrial controls were being undermined by an increase in pelagic sealing. Practised predominantly by Canadians from British Columbia, pelagic sealing was considered a much more wasteful form of culling, as it killed females, pups and unborn young indiscriminately and mortally wounded many seals whose skins were not subsequently collected. One critic, Dr H. H. MacIntyre, reported that every year the employees of the Alaska Commercial Company found 'embedded in the blubber of animals killed upon the islands large quantities of bullet, shot and buckshot', indicating that they had been injured at sea.[24] Another US citizen, Darius Ogden Mills, estimated that '[e]very skin placed upon the market by [pelagic sealers] represents the destruction of six or eight seals – an utterly unjustifiable inroad into the vitality of the herds'.[25] The depredations of pelagic sealers raised fears that seals might change their migration routes and breeding grounds, leading to their being killed by predators (such as sharks, whales and polar bears).[26] Contemporaries also expressed concern that females might be killed during the breeding season when they swam out to sea to find food for their pups. Although only 'nominal from the year 1868 to 1880', pelagic sealing intensified significantly from 1881 onwards, causing a major depletion of the seal population.[27] US agent Edwin Sims claimed that 10,000 skins were taken by pelagic sealers in that year and a substantial 61,838 in 1894.[28] Ichthyologist David Starr Jordan, writing in 1896, estimated that, since pelagic sealing began, '600,000 fur seals have been taken in the North Pacific and in the Bering Sea', which meant 'the death of not less than 400,000 breeding females, the starvation of 300,000 pups and the destruction of 400,000 pups still unborn'.[29] According to naturalist Henry Elliott, who visited the islands of St Paul and St George in the summer of 1890, there were only 959,000 seals present during the breeding season, just a third of the number he had seen two decades earlier in 1874.[30]

[24] 'The Fur-Seal Fisheries of Alaska', *The Gentleman's Magazine*, March 1891, p. 256.

[25] Mills, 'Our Fur-Seal Fisheries', p. 303.

[26] 'The Pacific Fisheries Dispute', *The Times*, 5 March 1887.

[27] Cooper Busch estimates that there were eighteen to twenty pelagic sealing vessels in operation each season in the early 1880s, fifty in 1888, nearly 100 in 1891 and 124 in 1892. See Cooper Busch, *The War against the Seals*, p. 136.

[28] Edwin W. Sims, *Report on the Alaskan Fur-Seal Fisheries, 31 August 1906* (Washington DC: Government Printing Office, 1906), p. 8.

[29] Jordan, *Observations on the Fur Seals of the Pribilof Islands*, p. 29.

[30] Elliott, *Report on the Condition of the Fur-Seal Fisheries*, p. 91. Elliott's figures, have, however, been questioned; his 1874 figure was based on flawed calculations and was

With the fur seal population seemingly in freefall, the US Government sprang into action. Pelagic sealing in all forms was strictly outlawed in US waters and the administration sent naval cruisers to the Bering Sea to seize US and Canadian ships catching seals off the Alaskan coast. In 1886, for instance, the US Revenue cutter *Corwin* seized and fined three Canadian schooners, the *Thornton*, the *Onward* and the *Caroline*.[31] The US justified its actions on the grounds that Russia had ceded its ownership of the whole of the Bering Sea when it relinquished Alaska, giving the USA the right to police the northern Pacific. The Canadians, however, disputed the legitimacy of these seizures, insisting that the USA had no jurisdiction over ships sailing in what they regarded as international waters. With both sides refusing to back down, a diplomatic crisis erupted that threatened to seriously undermine Anglo-American relations. Britain, representing its British Columbia colonists, argued that the USA had never enjoyed sovereignty over the entire Bering Sea and that American seizures of Canadian ships were unlawful.[32] The US Government accused the Canadians of zoological piracy, claiming that they were poaching US property in US waters.

As the debate evolved, more subtle internal differences also emerged. Within the USA, objections were raised to the cession of Alaska to a single company, with some contemporaries promoting the territory as a suitable location for settlement and others arguing that protecting the seals was damaging to US salmon fisheries in the northern Pacific. A letter published in *The Oregonian*, for instance, complained that Alaska was 'sealed [pun probably unintended] like a closed book and handed over to the care and custody of a private corporation to draw millions from while American citizens were carefully excluded or hunted like pirates if they ventured within its limits'.[33] Another writer pointed out that 'the seal-herd in their migration ... must necessarily destroy hundreds of millions of fish', which could otherwise be a source of profit to the 'immense fishing and canning industries [that] have sprung up along the Pacific coast'.[34] There was an argument, therefore, that preserving the fur seal was not in the interest of all US citizens and that Alaska might be used in a way that would benefit a larger number of people.

almost certainly too high, perhaps by a margin of 300 per cent. See Cooper Busch, *The War against the Seals*, p. 120.
[31] Ibid., p. 145.   [32] British Columbia joined the Dominion of Canada in 1871.
[33] 'The Pacific Fisheries Dispute', *The Times*, 5 March 1887.
[34] Flattery, *The Truth about the Fur-Seal Question*, p. 19.

In the British camp, on the other hand, there were tensions between British subjects and their Canadian counterparts, the former complaining that a few Canadian colonists were harming US–UK relations and several Canadians accusing Britain of not supporting them enough. Charles Hibbert Tupper, Canadian minister of marine fisheries, asserted that 'we [the Canadians] are in international affairs as dumb as the seals', intimating that the British Government would be more proactive if 'English vessels [from] fishing ports in the United Kingdom' were being seized in the Atlantic.[35] A contemporary article in the satirical magazine *Punch*, meanwhile, adopted a more comical approach to the seal controversy, urging both sides to stop their 'squalling and squabbling' and agree on a mutually acceptable compromise. The article featured a cartoon of a seal emerging from the ice and begging a quarrelling American and Canadian to 'Give me a "close time" and leave the "sea" an open question'. An accompanying poem, also supposedly written by the seal, asserted that 'Men can't … monopolise oceans' – an allusion to the USA's pretended rights over the Bering Sea – and implored both sides to stop 'cruelly clubbing' the seal life of the Arctic (Figure 2.3).[36] For the Canadians, sealing was a major concern, but to many Britons it was a peripheral issue and a source of diplomatic irritation.

With armed conflict on the horizon, diplomacy prevailed and arbitration talks were held in Paris in 1893. A panel of jurists, made up of representatives from France, Italy, Norway, Britain and the USA, was appointed to rule on the sealing question. The opposing sides were then given the opportunity to present their respective cases before the tribunal, the USA arguing for a complete ban on pelagic sealing and Britain defending its colonists. After several weeks of wrangling, an agreement was reached whereby a degree of pelagic sealing was allowed to continue but seals would be protected at certain times of the year and in certain locations. A sixty-mile closed zone was set up around the Pribilof Islands in which no pelagic sealing was permitted and an internationally recognised close season was established from 1 May to 1 August each year.[37] The use of firearms in pelagic sealing was forbidden at all times. Together, these measures were intended to reduce the damage caused by pelagic sealing and to ensure the continued survival of the fur seal.

[35] Charles Hibbert Tupper, 'Crocodile Tears and Fur Seals', *The National Review*, September 1896, p. 88, p. 92.
[36] 'Arbitration', *Punch*, 17 January 1891, pp.30–31.
[37] 'The Failure of the Paris Tribunal to Preserve the Seals', *The Advocate of Peace* 57:2 (1895), p. 37.

Figure 2.3 'Arbitration', *Punch*, 17 January 1891, p.31. A seal emerges from the ice and implores 'the two Jonnies' ('Brother Jonathan' (left), representing the USA, and 'John Bull' (right), representing Canada/Great Britain) to 'avast quarrelling' and 'give me a close time'.

Although it was an important step towards international wildlife conservation, the Paris Award did not reverse the decline in the seal population. Seals continued to be killed by pelagic sealers outside the sixty-mile
closed zone, and females in particular fell victim to Canadian harpooners
when they swam out to feed, often straying into unprotected waters.
Captain Shepherd of the US Reserve Marine claimed that 'I have seen
the milk come from the carcases of dead females lying on the decks of
sealing vessels which were more than one hundred miles from the
Pribilof Islands' – proof that a sixty-mile radius was not enough to
prevent suckling cows falling into the clutches of pelagic sealers.[38] As
for the Canadians, they were not happy with the agreement either, many
complaining that the terms of the Paris Award were *too* punitive. The
granting of permission to US naval vessels to conduct what one writer
described as 'inquisitorial visits and searches' on Canadian ships was
regarded as an affront to personal liberties.[39] Some commentators,
moreover, questioned whether banning firearms was truly in the best
interest of the seal, since the loud reports of rifles at least alerted swimming seals to the presence if hunters, while 'with the spear the seals are
killed silently and the herd is not frightened back into the protected
zone'.[40] With both sides dissatisfied, the Paris Award was allowed to
lapse in 1898 and was not renewed.

As the twentieth century dawned, seal numbers continued to fall,
dropping from an estimated 350,000 in 1899 to a mere 180,000 by
1906.[41] Pelagic sealing remained a serious problem, and, to make
matters worse, a new player now emerged in the picture: Japan. Having
previously confined their sealing to Robben Reef in the Okhotsk Sea, the
Japanese extended their operations to the Bering Sea in the early 1900s
and even began conducting raids on the Pribilof Islands to kill animals on
the rookeries and hauling grounds.[42] Twenty-five Japanese vessels were
spotted in the Bering Sea in 1910, each schooner carrying 'approximately
twenty-five to forty men and from five to ten boats'.[43] 'One Japanese boat

---

[38] Flattery, *The Truth about the Fur-Seal Question*, p. 8. More recent scientific research
suggests that females make foraging trips of three to nine days while nursing, which
could easily extend beyond the sixty-mile protection zone. See Gentry, *Behavior and
Ecology of the Northern Fur Seal*, p. 13.
[39] Tupper, 'Crocodile Tears and Fur Seals', p. 92.
[40] 'The Failure of the Paris Tribunal to Preserve the Seals', *The Advocate of Peace* 57:2
(1895), p. 37.
[41] Sims, *Report on the Alaskan Fur-Seal Fisheries*, p. 4.
[42] Barton Warren Evermann, 'The Northern Fur-Seal Problem as a Type of Many
Problems of Marine Zoology', *The Scientific Monthly* 9:3 (1919), p. 263.
[43] Harold Heath, *Special Investigation of the Alaska Fur-Seal Rookeries, 1910* (Washington
DC: Government Printing Office, 1911), p. 17.

seized by the Americans [in 1906] was found to contain six sealing clubs, two sealing knives, a compass, a cask full of fresh water, some ship's biscuits, a short sealing club for killing seals in the water and bamboo poles with iron hooks for hauling them aboard.'[44] As they were not bound by the Paris Award, to which they had not been a party, the Japanese violated the established sixty-mile closed zone with impunity and were accused of killing seals with excessive cruelty. According to US agent Edwin Sims, 'some of the seals were only stunned and not killed before being skinned' and several old bulls were 'pounded' in the eyes with sealing clubs to stop them protecting the females.[45]

Desperate to halt the decline in the seal population, the US Government considered alternative ways of restricting sealing. One proposed solution was to brand female pups to make their fur worthless to pelagic sealers. Attempted on several occasions, this was found to be of limited effectiveness, partly because the technique injured the animals and partly because it did not devalue the fur sufficiently to act as a deterrent. As Stanford zoologist Harold Heath remarked, 'The brands ... were in some cases fatal and are supposedly all that the young seal is able to survive, and yet not over one tenth or at most one eighth of the fur is destroyed.' The resulting reduction in value of the pelt would 'probably not amount to more than $10, and two San Francisco furriers place it as low as $5'.[46]

Another strategy pursued by the USA was to attack the profits of pelagic sealing. In December 1897, Congress passed a law banning the importation of any sealskin products made from animals caught at sea. This stipulated: 'All articles in whole or in part of sealskin must be stamped with the name of the manufacturer, with his statement under oath that the seals were not killed within proscribed water ... All skins not certified shall be seized and destroyed.'[47] Though well-meaning in intent, the law was hard to enforce and was modified in 1898 to exempt ladies crossing into the USA from Canada and female passengers arriving in the country from Europe, as long as 'they can prove that they owned their sealskins before December 30'.[48] The US authorities found themselves facing some of the problems confronting modern customs personnel, who need to be able to distinguish between legally farmed and illegally poached animal products.

[44] Sims, *Report on the Alaskan Fur-Seal Fisheries*, p. 14.   [45] Ibid., p. 22.
[46] Heath, *Special Investigation of the Alaska Fur-Seal Rookeries*, p. 14.
[47] 'The Seal Fur Trade', *The Times*, 31 December 1897.
[48] 'The United States', *The Times*, 17 January 1898.

It took a further multinational treaty to finally curb illegal hunting. Negotiated in 1911 with representatives from the USA, Britain, Russia and Japan, the Fur Seal Convention banned pelagic sealing entirely, removing the earlier caveats of the Paris Award. Russia, owner of the only other Pacific fur seal colony, supported the US stance on hunting at sea to protect its own stocks, while Canada and Japan also acceded to the agreement, partly because the dramatic decrease in seal numbers was making pelagic sealing unprofitable, and partly because they received substantial financial compensation.[49] With the treaty in place, the US Congress took additional steps to revitalise the seal population on the Pribilofs, imposing a five-year moratorium on the US seal drive for the years 1912–17 – a move advised by Elliott, although, as we shall see, considered unnecessary by other US scientists.[50] In the wake of these measures, seal numbers began to increase, rising from a mere 215,738 in 1912 to 530,237 in 1917. Conservationist William Temple Hornaday would later describe the preservation of the fur seal as 'the most practical and financially responsive wildlife conservation movement thus far con-summated in the United States'.[51]

The Science of Conservation

The sealskin controversy elucidates some of the key elements of animal conservation in the late nineteenth century and the difficulties associated with making a highly prized creature into a sustainable commodity. It also highlights several important new developments in the field of conservation, some of which continue to have relevance today.

First, one thing that stands out is the importance of science in the protection of the fur seal. To understand how best to preserve the species, and to justify proposed conservation measures, governments relied on the expertise of qualified scientists. The US Government commissioned and funded several scientific surveys of the Pribilof Islands, all staffed by recognised experts in the field of zoology.

---

[49] *Treaty Series. 1912. Convention between the United Kingdom, the United States, Japan and Russia, Respecting Measures for the Preservation and Protection of the Fur Seals of the North Pacific Ocean. Signed at Washington, July 7, 1911*, House of Commons Command Papers, Cd 6034 (London: His Majesty's Stationery Office, 1912). Both Canada and Japan received an immediate payout of $200,000, as well as 15 per cent from the Russian and American seal harvests.

[50] For a detailed analysis of the treaty's negotiation and the politics surrounding the moratorium, see Kurk Dorsey, 'Putting a Ceiling on Sealing: Conservation and Cooperation in the International Arena, 1909–1911', *Environmental History Review* 15:3 (1991), pp. 27–45.

[51] 'Saving the Fur Seal: Recovery of a Valuable Industry', *The Times*, 31 August 1920.

A survey party dispatched in 1896, for instance, featured two men from the United States Museum: Dr Leonard Stejneger, curator of reptiles, and Mr Frederic A. Lucas, curator of comparative anatomy.[52] A subsequent survey was carried out in 1910 by Harold Heath, professor of invertebrate zoology at Stanford University.[53] These men conducted careful fieldwork on the islands and offered a detailed understanding of seal behaviour and ecology. They used the latest technology to support their studies and stressed their sustained and close contact with the seal colonies to give credibility to their findings. As the naturalist Henry Elliott remarked in the introduction to his report on the condition of the fur seals in the year 1890, 'I have been exceedingly careful in gathering my data ... and to secure these data I have literally lived out upon the field itself, where those facts can be gathered honestly.'[54]

The application of scientific techniques to the study of the Pacific fur seal is clearly demonstrated by the work of the 1896 commission, under the command of David Starr Jordan. Sent to the Bering Sea by acting secretary to the Treasury Charles S. Hamlin with a detailed agenda, the commission engaged in a wide-ranging research programme designed to better understand the breeding habits of the fur seal and the consequent impact of both the seal drive and pelagic sealing on the survival of the species. Among other investigations, the team of scientists conducted a census of the seals in the rookeries to ascertain the number of pups and breeding females; studied the nursing practices of cows, noting that orphaned pups were not fed by other nursing females; examined the stomach contents of dead seals to learn more about their diet and to determine whether pups of a certain age were capable of feeding themselves if deprived of their mothers' milk; sought to establish whether the seal herds of the Pribilof Islands and the Russian-owned Commander Islands intermingled (they did not); and assessed whether sealing on land or at sea had any impact on the distribution of seals within the rookeries and hauling grounds. Careful observation, mapping, hydrography, photography and dissection were some of the methods employed to answer these questions, with detailed evidence supplied to support each finding. To show that many seals wounded by pelagic sealers were not subsequently caught and skinned, for example, and that many were female, Jordan cited the cases of 'a wet [i.e. nursing] cow' found at the bay of Polovina on 23 July 1896 'with bloody shot holes in her shoulder' and a 'very fat' cow with 'a large unborn pup' 'washed on shore ... near

---

[52] Jordan, *Observations on the Fur Seals of the Pribilof Islands*, p. 5.
[53] Heath, *Special Investigation of the Alaska Fur-Seal Rookeries*, front cover.
[54] Elliott, *Report on the Condition of the Fur-Seal Fisheries*, p. xv.

Sea Lion Neck' with 'a large number of buckshot holes in her back'.[55] To prove that pups required milk until they left the breeding grounds in November, the scientists killed a selection of animals from 29 August onwards and examined the contents of their stomachs; in the vast majority of cases, these were found to be either empty (in the case of orphans) or 'full of milk' well into October.[56] Such information underlined the methodical approach of the commission and lent credence to its conclusions.

Science did not necessarily provide definitive answers, however; despite its supposedly impartial methodology and carefully evidenced approach, different naturalists often came to different conclusions. These were not mere technicalities, moreover, but could have a major bearing on proposed conservation measures.

Take the contentious case of the seal drive. Elliott, who visited the Pribilofs for extended periods in 1872 and 1890, concluded that driving bachelor seals from the hauling grounds to the killing grounds caused them serious injuries that might have an impact on their subsequent fertility. As he put it, 'I am now satisfied that they sustain in a vast majority of cases internal injuries of greater or lesser degree that remain to work physical disability or death thereafter to nearly every seal thus released, and certain destruction of its virility and courage necessary for a station on the rookery.'[57] This led him to recommend a moratorium of at least seven years on the land cull to allow the seal population time to recover.[58] Visiting the islands six years later, by contrast, Jordan and his team found no evidence that male seals were physically impaired by the seal drive and denied that it impacted at all on their ability to reproduce. The 1896 commission theorised that Elliott's 'mistake' stemmed from 'the supposition that owing to the exposed position of the testes in the male animal they were liable to injury when he was in motion'. It claimed that this was not the case, for 'it is found as a matter of fact that the testes are under the control of the animal and are withdrawn into the body cavity when he is in motion, thus being entirely protected from injury'.[59] This prompted the conclusion that the seal drive had no lasting impact on the survival and fertility of male seals and that pelagic sealing alone was responsible for the recent dramatic decline in the seal herd. As Jordan phrased it, '[T]he treatment of the bachelors [in the seal drive], whatever it might be, would affect the breeding rookeries no more than the treatment of horses on the London omnibus lines affects the royal

[55] Jordan, *Observations on the Fur Seals of the Pribilof Islands*, p. 44.   [56] Ibid., p. 33.
[57] Elliott, *Report on the Condition of the Fur-Seal Fisheries*, p. viii.   [58] Ibid., p. 201.
[59] Jordan, *Observations on the Fur Seals of the Pribilof Islands*, p. 38.

stables.'[60] Kurkpatrick Dorsey identifies in this scientific dispute the beginnings of two distinct approaches to conservation: conservation to ensure the continued viability of the sealing industry (economic conservation) and conservation for its own sake (i.e. saving a species to ensure the continued existence of the animal).[61]

At an international level, scientific disagreements were even more stark. US scientists, despite their differences over the seal drive, almost universally agreed that pelagic sealing was the primary cause of seal depletion. The Canadians, however, took a different view and sought scientific evidence to support their claims. In 1893, for instance, as the Paris Tribunal convened to rule on the future of sealing in the Bering Sea, the *Canadian Record of Science* re-published an essay by New Zealand lawyer and part-time naturalist Frederick Revans Chapman entitled 'Notes on the Depletion of the Fur-Seal in the Southern Seas'. The article described the annihilation of the fur seal in the southern hemisphere and recounted numerous occasions on which sailors had landed on antipodean islands to commit indiscriminate slaughter. It did not make any explicit reference to the situation in the Bering Sea, but it was co-opted by the *Canadian Record of Science* as evidence that Canadian sealing methods were not to blame for the fur seal's decline. As the journal's editor, G. M. Dawson, remarked in a pointed footnote:

It must be remembered in reading Mr Chapman's paper, that the pursuit of fur-seals in the Southern Hemisphere has been entirely confined to the killing of these animals on shore, at their breeding stations. Pelagic sealing has never been practised in the South, where vessels have been employed merely as the means of reaching the otherwise inaccessible resorts of the seals. Thus Mr Chapman's observations, insofar as they bear on the question of the preservation of the fur-seal of the North Pacific, go to show the extreme importance of protecting the littoral breeding resorts of the animals from all disturbances.[62]

So, according to this interpretation, it was the seal drive, and not pelagic sealing, that was doing most damage to the seals of Alaska.

While science generated both answers and controversy, there was also an important legal and diplomatic dimension to the whole sealing debate as lawyers and politicians struggled to enact national and international measures that would save the seals from extinction. Understanding the seal's breeding habits and migration routes was one thing, but devising a legal framework for their protection posed further challenges. How could an animal that straddled international boundaries be preserved without

---

[60] Ibid., pp. 42–3.   [61] Dorsey, 'Putting a Ceiling on Sealing', pp. 27–45.
[62] Frederick Revans Chapman, 'Notes on the Depletion of the Fur-Seal in the Southern Seas', *Canadian Record of Science*, 1893, footnote, p. 447.

the agreement of multiple nations? Could any one nation assert legal ownership over a species that graced its shores for only part of the year? Who was to enforce any international agreement that was instituted? These were all contentious issues, and ones that still have resonance for wildlife conservation today.

The first major question that confronted nineteenth-century law-makers was the legal status of the seal itself. This was problematic because seals were amphibious mammals that spent part of their lives on land and part at sea, in international waters – a source of some classificatory confusion. As one contemporary observed: '[The seal] is a victim of mixed metaphor at every period of his life. His father is a "bull", his mother is a "cow", he himself is a "pup", he is born in a "rookery" and killed in a "pod" – a truly strange Pilgrim's Progress: bovine, canine, corvine and cruciferous!'[63] The slippery status of the seal was a source of concern for US legislators, who wanted to lay claim to the animal but found their ownership rights challenged at both national and inter-national level. In 1906, for instance, when US agent Edwin Sims pro-posed attaching metal tags to the young seals on the Pribilof Islands, 'on which might be stamped the words "Property of the United States"', he faced objections that no animal could be considered the property of a nation unless some efforts were made to domesticate and farm it.[64] Maurice Douglas Flattery, a lawyer from Kentucky, contended that 'a seal cannot be said to be a domestic animal by any stretch of the imagination', for the Aleuts 'never give them food of any kind, although the bulls, on whose vitality and virility the increase of the herd depends, remain on the Islands several months without food or drink, so that when they leave the breeding grounds after the season is over they are emaci-ated and almost devitalised'.[65]

In part to counter this narrative of non-domesticity, some American scientists proposed a more interventionist approach to fur seal manage-ment. Writing in 1896, following his tour of the Pribilofs, Jordan recom-mended improving the rookeries on the islands by 'repairing and obliterating' the 'death traps' in which fur seal pups were often trampled by large bulls (usually smooth, sandy areas in which there were no rocks or gullies in which to hide). He also suggested that '[t]he rookery grounds themselves could be extended on St Paul and St George by blasting off the cliffs and strewing the flats with bowlders [sic]', thereby

---

[63] 'The Seal and His Jacket', *The Leisure House*, February 1890, p. 254.
[64] Sims, *Report on the Alaskan Fur-Seal Fisheries*, p. 29.
[65] Douglas Flattery, *The Truth about the Fur-Seal Question*, pp. 34–5.

increasing the area where seals were able to haul up and breed.[66] Returning to the issue in 1910, Jordan elaborated on these ideas, arguing that a 'naturalist' should be dispatched to the Pribilofs to 'take charge of the killing and treat the rookeries just as a forester does a grove of valuable trees', weeding out the 'superfluous males' in the same way as 'the undergrowth and young trees which crowd the others too much are eliminated from the forest'.[67] Writing in the same year, the commissioner of fisheries, George Bowers, advocated raising motherless pups in a special 'orphan asylum' on the Pribilofs and even translocating live fur seals to the US mainland, where they might be naturalised in the fresh-water lakes of New York, New Hampshire and Minnesota – a scheme inspired by Lieutenant Judson Thurber's successful hand-rearing of two orphan fur seal pups, Bismarck and Mamie, and their subsequent trans-fer to Washington DC.[68] Fur seals might thus be farmed like domestic cattle, culled like surplus steers or selectively logged like a well-regulated timber plantation. As Jordan reflected, 'There is probably no wild animal whose conditions of life could not be artificially improved by human influence if it were thought worth the while.'[69]

If the fur seal's wildness posed one legal quandary to legislators, a second major issue was the animal's mobility. Unlike trees, seals did not remain in one place throughout their lives but roamed widely across the Pacific Ocean.[70] This meant that international cooperation was essential for their protection, for whatever US citizens did or did not do to manage the seal population while it was on the Pribilof Islands, their efforts would be futile if the animals were killed outside US waters. Such complications were not, of course, confined to seals; they affected many other species also threatened with extinction. Migratory birds, protected in the USA (or at least in some states) from the late nineteenth century, were often killed when they over-wintered in South America, necessitating similar transnational discussions.[71] Elephants slaughtered for their ivory in

---

[66] Jordan, *Observations on the Fur Seals of the Pribilof Islands*, p. 36.
[67] 'Kill Superfluous Bulls Only Way to Preserve Herd, Say Dr Jordan and Dr Lucas', *New York Tribune*, 24 July 1910.
[68] 'May Raise Fur Seals Here', *New York Sun*, 2 July 1910, cited in William T. Hornaday, *Scrapbook Collection on the History of Wild Life Protection and Extermination*, Vol. 5, Wildlife Conservation Society Archives Collection, 1007-04-05-000-a.
[69] Jordan, *Observations on the Fur Seals of the Pribilof Islands*, p. 36.
[70] For a discussion of the seal's migratory nature, and the crucial role played by migrating marine fauna in the shaping of the Pacific world, see Ryan Tucker Jones, 'Running into Whales: The History of the North Pacific from Below the Waves', *American Historical Review* 118:2 (2013), pp. 349–77.
[71] William Hornaday advocated 'the enactment of a law providing federal protection for all migratory birds', complaining that 'it is robbery, as well as murder, for any southern state to slaughter the robins of the northern states, where no robins may be killed'. William

Africa also showed little respect for colonial boundaries, limiting the effectiveness of sanctuaries (see Chapter 3). The sealing crisis thus reflected a wider move towards preservation in the late nineteenth and early twentieth centuries, and a growing recognition that wild animals did not confine themselves to artificial human borders. As Henry Elliott observed, there were limits to the protection that could be afforded a species 'entirely beyond our control, over which we have no knowledge of its coming or of its going ... which we cannot confine to certain fields, or even protect from its natural enemies while at sea, where it is nearly half of the time of its existence'.[72]

A third and related issue concerned the seal's value and how this ought to be assessed. As Dorsey has shown, the success of the 1911 treaty negotiations was contingent on paying compensation to Canada and Japan for the sealskins they would lose by abandoning pelagic sealing. Determining the correct level of compensation, however, was difficult for there was no clear agreement on the value of a seal. Should a male seal be accorded the same value as a female (given her reproductive role)? Would the value of a sealskin change if the population of fur seals increased or decreased? How could a price be put on a 'moving, eating, reproducing resource' that 'did not fit into any standard economic framework'?[73] Over and above the question of compensation, some contemporaries wondered whether it made economic sense to protect the fur seal at all, given the negative impact of preservation measures on other equally valuable animals. Writing in 1897, Flattery argued that seals were 'most voracious animals' that consumed huge quantities of fish during their migrations along the Pacific coast. Protecting them might therefore be a mistake from an economic perspective, since 'what the lessees and the United States Treasury would gain by the increase of the seals would be at the expense of the fishing and canning industries, which would probably lose in proportion as the seals increased numerically'.[74] Debating the proposed moratorium on the land drive in 1912, meanwhile, Jordan and his colleague George Clark claimed that a complete cessation of the killing would decimate the introduced Arctic blue fox population on the Pribilofs, which depended on 'the carcasses of the seals' for their winter food. This was bad for the foxes, which, according

Temple Hornaday, *Our Vanishing Wild Life: Its Extermination and Preservation* (New York: Clark and Fitts, 1913), pp. 266–7.

[72] 'Preservation of the Fur Seal', *Congressional Record*, Sixty-Second Congress, Second Session, p. 4, cited in Hornaday, *Scrapbook Collection on the History of Wild Life Protection and Extermination*.

[73] Dorsey, 'Putting a Ceiling on Sealing, p. 40.

[74] Flattery, *The Truth about the Fur-Seal Question*, p. 20.

to Jordan, 'were beginning to starve and eat one another', but it was also bad for the US Government, which had extracted some 40,000 fox pelts from the Pribilofs over the previous forty years.[75] The different measures proposed to protect the fur seal thus had consequences for the wider North Pacific ecosystem (a term not used by any contemporary commentators) and needed to be considered within that context.

Finally, those who wanted to protect the seal – or other forms of wildlife – needed to think hard about the most appropriate means of enforcing conservation legislation. It was all very well to declare a species protected, but how were agreed quotas or boundaries to be policed? Two main alternatives presented themselves. On the one hand, the seas of the Bering Sea could be patrolled and vessels suspected of breaching the terms of international agreements (going too close to the shore, hunting during the proscribed season, using firearms, etc.) could be seized and impounded and their owners imprisoned or fined. On the other hand, the enforcement could focus on the skins themselves, with illegally caught skins being barred from the market. The Jordan commission advocated 'an international arrangement whereby all skins of female fur seals should be seized and destroyed by the customs authorities of civilised nations, whether taken on land or sea'.[76]

Though promising, both methods had limitations. Patrolling the Bering Sea necessitated international consent if the USA were to avoid causing a diplomatic incident, as they had done in the early 1890s. It also required serious investment to fund the patrols, just as contemporary game sanctuaries in Africa required sufficient and competent wardens to successfully catch and punish poachers. Identifying illegally obtained skins, meanwhile, was not always straightforward, and again required international collaboration if the banned products were not simply to find another way to the consumer. Both enforcement measures were further complicated by clauses in the 1893 and 1911 treaties exempting native peoples from the ban on pelagic sealing on condition that they refrained from using steam-powered vessels or modern firearms.[77] This exemption was intended to permit the continuation of hunting for subsistence purposes, but it created a potential loophole for commercial

---

[75] David Starr Jordan and George A. Clark, 'Truth about the Fur Seals of the Pribilof Islands', *Economic Circular No. 4*, 20 December 1912, pp. 6–7, cited in Hornaday, *Scrapbook Collection on the History of Wild Life Protection and Extermination*.

[76] Jordan, *Observations on the Fur Seals of the Pribilof Islands*, p. 69.

[77] *Treaty Series. 1912. Convention between the United Kingdom, the United States, Japan and Russia, Respecting Measures for the Preservation and Protection of the Fur Seals of the North Pacific Ocean. Signed at Washington, July 7, 1911*, House of Commons Command Papers, Cd 6034 (London: His Majesty's Stationery Office, 1912), p. 39.

sealers and misrepresented the true relationship between fur seals and First Nation peoples, who, as Robert Irwin demonstrates, frequently worked on pelagic sealing vessels. The Canadian authorities, moreover, seem to have done little to enforce the technological restrictions imposed on indigenous peoples, turning a blind eye when the latter pushed spears through bullet holes to make it appear that a seal had been hunted legitimately, or had their canoes towed out into the open sea by steamboats so that they could access the seal migration routes.[78] Policing the commerce in sealskin therefore posed major legal, logistical and financial challenges and was hampered by differing interpretations of the science and the law in relation to fur seals.

'Quivering Nostrils' and 'Painful Cries'

So far, we have examined the seal industry from an environmental perspective. There was, however, another angle to the sealskin debate, and this came from the perspective of animal welfare. Scientists and conservationists were concerned primarily about the survival of the fur seal as a species and were happy to let the killing of the animals continue as long as it was managed in a sustainable way. Indeed, scientists often killed animals in the name of research. Humanitarians, however, took a more uncompromising approach, arguing that all killing was cruel, whether sustainable or otherwise. This was a minority viewpoint at the turn of the twentieth century, but one espoused by an increasingly vocal group of campaigners. It put the emphasis on the suffering of the individual animal rather than on the long-term survival of the population.

Humanitarian opposition to sealing manifested itself in a series of emotive articles and pamphlets, most of them targeted at the consumers of sealskin products. These lingered over the brutal methods used to kill seals and emphasised the pain experienced by the animals. The RSPCA published several articles on seal culling in the late 1870s, claiming that 'it is a common sight to see seals swimming in the sea after having been stripped of their skins'.[79] Thirty years later, in 1910, campaigner Joseph Collinson penned a highly graphic and emotive description of the seal cull, citing chilling testimony from individuals who had witnessed the slaughter. One witness, Mr W. G. Burn Murdoch, described the 'pitiful' sight of a seal, 'its nostrils wide and quivering, its dark ox-like eyes trembling in agony as the knife tears down its white skin'. Another,

[78] Robert Irwin, 'Canada, Aboriginal Sealing, and the North Pacific Fur Seal Convention', *Environmental History* 20 (2015), pp. 57–82.
[79] 'Skinning Animals Alive for Profit', *The Animal World*, April 1878, p. 54.

Norwegian explorer Captain Borchgrevink, claimed that seals were often skinned while still alive – deliberately so, for 'in the utmost agony the wretched beast draws its muscles away from the sharp steel, which tears away its skin, and thus assists in parting with its own coat'.[80] Such descriptions were designed to shock readers and to bring home to them the full horror of the annual cull. As the humanitarian Henry Salt remarked, the fur industry 'makes patchwork ... not only of the hides of its victims, but of the conscience and intellect of its supporters'.[81]

Perhaps predictably, anti-sealing campaigners devoted particular attention to the fate of the seal pups, condemned to die a long and painful death after the killing of their mothers. Collinson, eager to demonise the sealing industry, cited statistics from a US Government report, which claimed that 16,000 young seals were 'found dead from starvation' on the Pribilof Islands in 1896 alone.[82] Naturalist Frank Buckland wrote a deeply moving account of British and Scandinavian sealing in Newfoundland (in this instance of the hair seal), in which he likened the wails of the famished seal pups to the cries of human infants:

It is horrible to see the young ones trying to suck the carcases of their mothers, their eyes starting out of the sockets, looking the very picture of famine. They crawl over them until quite red with blood, poking their noses, no doubt wondering why they are not getting their usual food, uttering painful cries the while. The noise they make is dreadful. If one could imagine himself surrounded by four or five thousand human babies all crying at the pitch of their voices, he would have some idea of it.[83]

This kind of language, heavy with pathos, was calculated to appeal to the maternal instincts of female readers. As the primary consumers of sealskin products, women were seen as the indirect cause of seal slaughter, as well as the potential saviours of the animals.

---

[80] Joseph Collinson, *How Sealskins Are Obtained* (London: Animal's Friend Society, 1910), p. 3.

[81] Henry Salt, *Animal Rights Considered in Relation to Social Progress* (New York: Macmillan, 1894), p. 67.

[82] Collinson, *How Sealskins Are Obtained*, p. 6.

[83] 'Proposed Close Time for Seals', *The Animal World*, April 1873, p. 49. Contemporaries frequently confused the two forms of sealing. In 1889, for instance, Canon Bell of Cheltenham delivered a rousing sermon condemning the use of sealskin jackets and alleging that each one 'represents some half-dozen dams who have more or less been skinned alive, while their little ones have been left to die in all the slow agony of starvation'. The *Warehouseman and Drapers' Trade Journal*, however, refuted these claims, insisting that 'His description applies, so far as it is correct at all, to the chase of the hair seal, which is currently sought for its blubber, to be rendered down for oil, and of which the skin is useless as fur'. See 'Sealskin Jackets', *York Herald*, 28 December 1889.

While most of the above descriptions focused on pelagic sealing, the Alaskan seal drive did not entirely escape criticism, for it still entailed a degree of butchery. Collinson, who opposed killing in all forms, recounted emotively how the seals on the Pribilof Islands 'are driven and forced onwards, panting and helpless, mostly over rough, stony ground'.[84] The Canadian Charles Hibbert Tupper also insisted – no doubt for partisan reasons – that the 'greatest injury' to seals was done not at sea but on land, where they were 'subjected ... to all possible cruelty and torture'.[85] US commentators, of course, were inclined to defend the seal drive, insisting that it was the most humane way of slaughtering seals and caused 'less needless pain' than one might see in the average 'shambles'.[86] Elliott's critique of the drive, however, while primarily concerned with issues of sustainability, intimated that the process inflicted a degree of suffering on the seals, which were often 'heated to the point of suffocation, gasping, panting', and were sometimes driven multiple times in a single season, as happened to one five-year-old bull 'with mottles on its dark fore-flippers' that was 'driven up in this way' three times in 1890.[87] Language like this humanised the seals and raised doubts about their welfare. What humane person could fail to be moved by accounts of 'the flying of the eyes from the struck seal, the crush of the skull, the flow of blood, the sobs of the dying and the brutality of the heartless and careless men'?[88]

Faced with such brutality, animal welfare campaigners advocated a boycott of sealskin products, insisting that the cruelty involved in procuring this coveted fabric outweighed the desire for fashionable jackets and purses. Rather than looking for ways to sustain the supply of sealskin, as conservationists were doing, they instead attempted to reduce the demand for this luxury commodity, putting the suffering of the individual animal above the survival of the species. Since most (though by no means all) consumers of sealskin were ladies, humanitarians directed their appeals primarily at women, whom they saw as ultimately responsible for the bloodshed. One article in *The Animal World* bluntly informed female readers that '[t]he simple fact stands thus – that a sealskin jacket is made of sealskin; that to get the skin we must knock the seal on the nose, and that if the fur is to have a proper gloss on it the seal must be skinned as nearly as it may be alive'.[89] Another contribution,

---

[84] Collinson, *How Sealskins Are Obtained*, p. 4.
[85] Tupper, 'Crocodile Tears and Fur Seals', p. 88.
[86] Jordan, *Observations on the Fur Seals of the Pribilof Islands*, p. 37.
[87] Elliott, *Report on the Condition of the Fur-Seal Fisheries*, p. viii, 150.
[88] Collinson, *How Sealskins Are Obtained*, p. 5.
[89] 'Revolting Cruelty to Seals', *The Animal World*, April 1875, p. 52.

in the form of a poem, exposed the hypocrisy of female consumers –
specifically mothers – who doted on their pets but were complicit in the
suffering of other equally sentient creatures:

> The duchess rides in skins of seals,
> While on the air is borne the peals,
> Of pain from seals they skin alive,
> That she may sport them in her drive;
> Yet strange it is, the noble dame,
> Would weep, and vow it was a shame,
> If any dared her pug to hurt,
> Whether 'twere done for use or sport,
> And noble lords trimm'd out in seal
> Must not believe that seals can feel,
> Or is't they care not how much pain,
> Is caused to seals being flay'd ere slain?
> But, English mothers, how can you,
> Be human, and encourage, too,
> The wearing seal got by such means?
> Can you not hear their dying screams?[90]

By closing the distance between the living animal and the fashionable
product, campaigners hoped to make consumers understand that their
fashion choices had a very real humanitarian cost, one that could be
avoided only through their complete abstinence from sealskin goods.
This message hit home for at least one subscriber to *The Animals'
Friend*, who, on reading 'about the cruelty used in connection with the
killing of the poor seals ... made up my mind never to wear a sealskin coat
again, and immediately disposed of the two I had in wear'.[91]

Opponents of sealing also tapped into deeper qualms about the state of
Western civilisation, suggesting that the killing of animals for fashion
undermined human progress and was comparable with other, better-
known acts of barbarity. John Willis Clark drew parallels between sealing
and vivisection, then a deeply contentious issue, alleging: 'We, who pride
ourselves so much on our humanity, suffer barbarities to be committed
on Jan Mayen Island that would disgrace a savage, and compared with
which the so-called "cruelties of vivisection", about which we hear so
much, become kindly treatment.'[92] Collinson, meanwhile, condemned
the hypocrisy of British women, who 'glibly speak with pharisaic satisfac-
tion of the "lack of refinement" in the Spanish woman, who is the over-
eager spectator of the bull-fight', but would happily purchase 'a fashionable sealskin coat'. He also compared the suffering of seals with that of

---

[90] 'A Protest in Rhyme against Seal Cruelty', *The Animal World*, March 1890, p. 39.
[91] 'The Cost of a Skin', *The Animals' Friend*, 1907, p. 128.   [92] Clark, 'Sea-Lions', p. 48.

Figure 2.4 'National Zoological Park, Northern Fur Seal Pup', c.1910.
Smithsonian Institution Archives, Image SIA_000095_B49_F05_003

animals in Chicago slaughterhouses, where 'there seems to be no doubt
that pigs are scraped alive'.[93] In some cases, these comparative cruelties
were designed to show how sealing was even more shocking than other
abuses; in others, they were used to denounce a whole raft of different
abuses at once. Furthermore, there was a distinct sense that complicity in
the killing of seals undermined Britain's perceived status as a leader in
animal welfare – and, indeed, in moral reform more widely. It was
'London, the heart of humane England and civilisation' that was 'the
great emporium for fur sealskins, the preparation of such peltry being a
speciality of English – and pre-eminently of London – workmen'.[94] The
blood of the butchered seal thus left an indelible stain on British and
American consciences.

Public sympathy for the fur seal may have been made more tangible to
readers by the growing numbers of seals in British and US aquariums in
this period, which served literally to put a face to the sealskin jacket
(Figure 2.4). In 1875, four Patagonian fur seals – Minnie, Fan, Billy
and Kate – were exhibited at London Zoo by their keeper, François Le

---

[93] Collinson, *How Sealskins Are Obtained*, p. 7.   [94] Ibid., p. 8.

Compte, who had taught the animals to catch fish in their mouths and kiss him on the lips.[95] In 1903, a mother and baby sea lion from Patagonia were exhibited across Britain in Bostock and Wombwell's Menagerie, housed in a portable 'aquatic tank', while in June 1911 a young seal named Minnie charmed visitors to New York Aquarium by pushing a match 'around in the water with the point of her nose'.[96] While these animals were not exhibited with an explicit conservation message, more perceptive viewers would doubtless have made the mental link between the playful pinnipeds in the zoo or aquarium and the luxury clothing derived from their hides. When Bostock and Wombwell's seals visited Northampton, for instance, the *Northampton Mercury* remarked: 'Not many years ago the sea lion was found in great quantities on islands off the coast of South America, but having been hunted without mercy, for their furs, they are now nearly extinct.'[97] Describing the young fur seal Minnie in the New York Aquarium, meanwhile, the *New York Times* observed: 'On her shining back she wears a home grown sealskin sack that any dealer would pay $40 for, and which would be worth $150 after it had been dressed' – a pretty stark conflation of the animal and the sealskin coat.[98] Seeing a captive pinniped in a zoo or menagerie could thus evoke sympathy for the animal's wild counterparts, prompting some to reflect on their sartorial choices.

Finally, if the suffering of innocent seals was not enough to persuade consumers to renounce their sealskin jackets, humanitarians highlighted the human suffering associated with the fur industry. The Aleut seal killers may have been part of the problem when it came to seal butchery, but campaigners against sealing were clear that 'responsibility for the sealer's deeds rests ultimately on the wealthy wearers of the skin', and that they were also responsible for the 'degradation' of the slaughterers. As Frank T. Bullen remarked:

Lady, with the hundred-guinea sealskin coat, know for certain that the men who looked death between the eyes, and brutalised themselves lower than the shark to wrench that coat from its rightful owner, got less than a hundred pence for so doing. The bulk of the money went to the city magnates and full-fed speculators who never gave its origin a second thought.[99]

[95] 'A Winter Visit to the Zoological Gardens', *Daily News*, 6 January 1875; 'Sea Lions at Brighton', *The Times*, 16 October 1875; Clark, 'Sea-Lions', p. 23.

[96] 'Bostock and Wombwell's Menagerie', *Sheffield Daily Telegraph*, 24 November 1903; 'You Are Invited to Attend the Evening "At Home" with the Fish Family', *New York Tribune*, 25 June 1911.

[97] 'Bostock and Wombwell's Zoo', *Northampton Mercury*, 4 December 1903.

[98] 'You Are Invited to Attend the Evening "At Home" with the Fish Family', *New York Tribune*, 25 June 1911.

[99] 'The Price of Sealskin', *The Humanitarian*, October 1911, p. 173.

As for the (mainly female) fur-pullers employed to prepare the skins for manufacture, they also suffered both physically and financially, often contracting 'consumption owing to the stoppage of the respiratory organs by the bits of fluff which fly off the skins they are handling and invade the nostrils and air passages, so entering the lungs'. They also had to endure 'the stench which rises from the animal skins', making the work deeply unpleasant.[100] To ignore such suffering, both human and animal, was to show an extreme level of callousness and to put personal vanity above the collective good. For women, moreover, the ostentatious wearing of fur could even put their rationality into question at a time when female suffrage was forcing its way onto the political agenda. As humanitarian Gertrude Mallet commented:

[W]hat a paralysing effect it must have on the minds of men, who are being asked to extend the political franchise to the larger half of humanity on the ground that they are reasonable beings, to see a woman going about carrying a whole animal, head and claws and tail, for her muff, or with a pitiable bird cocked aloft on the crown of a beehive or other savage-shaped hat.[101]

Sealskin was thus a moral as well as an environmental ill, harming people and animals alike.

Conclusion

Sealskin was big business by the late nineteenth century. A series of technological innovations had converted what was once a crude, unworkable material into a desirable luxury product, making sealskin accessories a must-have item in a fashionable woman's wardrobe. Obtained in the Bering Sea, sealskins were processed in London, the centre of the world's fur trade, and distributed across the globe. As one contemporary remarked, '[E]ven American skins caught in American vessels or on American shores, intended for clothing American bodies, are sent over [to Britain] to be prepared.'[102] The sealskin trade was thus a truly global industry, with employees in Alaska, California and Great Britain.

In order to meet the growing demand for sealskin, it was necessary to carefully manage the fur seal population. Excessive culling had already had devastating effects in other parts of the Pacific and even in Alaska, where the Russians had been forced to institute a complete ban on killing, or *zapooska*, between 1834 and 1841.[103] By the 1890s, the seal

---

[100] 'Humane Dress', *The Humanitarian*, January 1911, p. 99.   [101] Ibid., p. 100.
[102] 'The London Fur Trade', *Chambers's Journal of Popular Literature, Science, and Art*, 6 October 1894, p. 625.
[103] Elliott, *Report on the Condition of the Fur-Seal Fisheries*, p. x.

population was again in crisis, falling from an estimated 4,000,000 in 1867 to just 180,000 in 1906.[104] Pelagic sealing was seen as the primary cause of this decline, but it was difficult to police. A series of transnational treaties sought to save the seal from extinction, the last of which, the 1911 accord, is still seen as a landmark in marine wildlife conservation.

The case of the fur seal illuminates some of the complexities of wildlife protection in the late nineteenth century. As a wild creature, the fur seal challenged notions of domestication and could not be claimed as the property of an individual owner or nation state. As a migratory animal, it could not easily be confined to a sanctuary or game reserve and protected from poachers. Only international agreement could ensure the survival of the species, but whether this was to take the form of a close season, a restricted area for hunting, the confiscation of female seal pelts or the outright banning of pelagic sealing remained open to debate. There was also the question of the role of the consumer in wildlife protection. Was it the Aleut hunter who was primarily responsible for the abuse or disappearance of fur seals? Or the female consumer, whose love of fashion was driving vulnerable animals to extinction?

The fur seal controversy also illuminates three distinct approaches to wildlife conservation that continue to have resonance today. The first of these, the utilitarian approach, sought to preserve the fur seal primarily because it was useful to man. Proponents of this view advocated sustainable hunting, not because they valued the seal in itself but because without it the sealskin industry would disappear, with potentially damaging impacts for its employees – particularly the Aleuts. They wanted to protect certain forms of wildlife whose value to man was clear, but they were largely indifferent to broader ecological issues. Rather than leaving nature to itself, moreover, utilitarians advocated human intervention where necessary to increase the yield of the seal herd and manage it more effectively. David Starr Jordan, a proponent of this viewpoint, suggested that measures should be taken to change the topography of the rookeries to prevent seal pups from being accidentally crushed by marauding bulls. He also argued that the land drive was good for the seal population, since it kept down the number of males and reduced the potential for fighting, with consequent casualties among females and pups. As he expressed it, 'The herd should be treated as a breeding herd of cattle or horses would be. It should be under the immediate control each year of a competent naturalist, who should devote his energies to the study of the needs of the

---

[104] Sims, *Report on the Alaskan Fur-Seal Fisheries*, p. 3.

herd, its preservation, increase and possible improvement.'[105] Such a view – and particularly such language – explicitly equated fur seals with domestic animals and suggested that they should be managed in a similar way.

The second approach towards wildlife, the conservationist approach, sought to preserve animals for their own sake, or for the sake of posterity. Advocates of this approach regretted the disappearance of once common species such as the bison and the passenger pigeon and lobbied for the protection of other endangered creatures. They stressed man's steward-ship of the natural world and underlined the importance of passing on natural treasures to future generations. Commenting on his first sight of the Pribilof rookeries, for instance, Professor Harold Heath remarked that 'such a show of mammalian life is to be met with nowhere else on the face of the earth, and from several points of view it would indeed be a calamity if the seal meets the fate of the manatee, the sea otter or the buffalo.'[106] While utilitarians tended to promote human intervention to 'improve' upon the natural world, moreover, conservationists were scep-tical of the value of human meddling and felt that this undermined the sanctity of nature. Critiquing George Clark's proposal to slaughter twenty-nine out of every thirty 'superfluous' young male seals, Elliott protested that this contravened natural selection and undermined the notion of the survival of the fittest:

What warrant has Mr Clark to assert that the creator of [the fur seal] has not brought it into a stable existence, but that, unless man steps in with his superior sense, it will perish by its own volition? Think of the expression on the faces of Thomas Huxley and Charles Darwin were they to reappear in the flesh and read this argument of the 'academic secretary of Stanford University'![107]

Interfering in the breeding processes of other species was thus a futile exercise and an example of misguided human self-importance. To quote Elliott's ally, William Temple Hornaday: 'Man never yet has succeeded in making the slightest impression on any wild seal, sea-lion or walrus, save with club, knife or gun.'[108]

Finally, we see in the sealskin debate the emergence of a third approach towards wildlife management, one that focused on animal

---

[105] Jordan, *Observations on the Fur Seals of the Pribilof Islands*, p. 36.

[106] Heath, *Special Investigation of the Alaska Fur-Seal Rookeries*, p. 17.

[107] 'Sea Butchers', *The Globe*, 10 December 1910, cited in Hornaday, *Scrapbook Collection on the History of Wild Life Protection and Extermination*.

[108] William T. Hornaday, *The Last Fight for the Persecuted Fur Seal* (New York: Office on Game Protection and Preserves, 1912), p. 10, cited in Hornaday, *Scrapbook Collection on the History of Wild Life Protection and Extermination*.

welfare. This approach, exemplified by Collinson, emphasised the suffering of the individual animal over the survival of the species and concentrated on cruelty. For welfare campaigners, it was not important that an animal was rare or endangered, but simply that it was suffering. Thus, while Jordan advocated the employment of a 'naturalist' to study and improve the herd, Collinson called for the recruitment of a government inspector to 'give the seals protection from wantonness in their slaughter'.[109] This mirrored contemporary campaigns by the Humanitarian League for public slaughterhouses (to prevent concealed cruelty to cattle), for an end to vivisection (also conducted behind closed doors), and for the better treatment of performing animals, often abused offstage during training.[110] All three of these viewpoints, and particularly the latter two, continue to inform twenty-first-century approaches to wildlife conservation.

[109] 'A Protest against the Seal Butchery', *Humanity: The Journal of the Humanitarian League*, July 1896, p. 131.
[110] On the aims and structure of the Humanitarian League, see Dan Weinbren, 'Against All Cruelty: The Humanitarian League, 1891–1919, *History Workshop Journal* 38 (1994), pp. 86–105.

3   Is the Elephant Following the Dodo?

Next to those shameful accounts of reckless elephant slaughter which too often disgrace the columns of sporting papers, the saddest sight which can be seen by a lover of Nature is one of our fashionable ivory shops in London or elsewhere, the windows of which are full of worthless trifles, the trophies torn by foolish fashion from those magnificent and irreplaceable herds of African elephants slain by selfish and stupid men in their green and warm fastnesses of the Veldt and the Karoo.   'A Plea for the African Elephant', *Manchester Courier*,
17 October 1896

In 1879, the South Shields naturalist William Yellowly published an article in the *Newcastle Courant* in which he advocated the domestication of the African elephant (*Loxodonta africana*). Yellowly favoured the project because he believed that elephants offered the best solution to the problem of transport in the interior of Africa, where 'no traveller dare take his horses or wagon oxen' for fear of the deadly tsetse fly. He also perceived domestication as the potential salvation of the elephant itself, at a time when the animal was being hunted to the verge of extinction for its ivory, 'from which are made billiard balls, knife handles and other things'. Countering claims that the African elephant was temperamentally unsuited to domestication, Yellowly pointed to the docility and tractability of African elephants currently living in Britain, among them 'two fine specimens' in Mrs Edmonds' travelling menagerie, 'now exhibiting in the Northern counties of England', the largest of which, Emperor, 'has been in her menagerie about twelve years, and is amongst the first imported into the country from Abyssinia' (Figure 3.1). As evidence of the species' perilous predicament, the naturalist cited Dr David Livingstone's sobering observation that 'within a year and a half of his discovery of Lake Ngami, no fewer than 1,100 elephants were slaughtered round about the lake and the adjacent country'.

William Yellowly's article appeared at a time when there was growing concern for the future of the African elephant. Long slaughtered 'for

Figure 3.1 Detail from 'Knocking Down a Menagerie', *The Graphic*, 9 August 1884. The elephant depicted here is clearly an African one, as evidenced by the large ears and pronounced tusks.

sport, as it is called', the elephant was also the victim of a growing demand for ivory, which was being consumed in Europe and North America in ever larger quantities. In order to meet this demand, thousands of the animals were killed in Africa each year, causing their population to plummet; as Yellowly noted, 'at the beginning of the present century, they were numerous in Cape Colony; now they are rarely found south of the Orange River, a wholesale war of extermination having been waged against them simply for the sake of their tusks'. Yellowly's article illustrates this growing concern for the survival of a majestic species and offers a novel but widely proposed solution to the problem: domestication of the elephant for use as a beast of burden. His comments on Mrs Edmonds' elephants, whom he had inspected in person, highlight the importance of the domestic zoo and menagerie as a testing ground for imperial acclimatisation projects. His reference to British India, where 'there is now a law … prohibiting the wanton destruction of these animals', underlines the trans-imperial dimensions of animal preservation and domestication and the role of knowledge transfer – practical, scientific and legal – in exploiting and conserving the natural world.[1]

This chapter examines one of the most high-profile and widely used animal products of the Victorian era – ivory. Employed to make all manner of consumer goods, ivory was heavily sought after in the nineteenth century and was worked on an industrial scale. In the early nineteenth century, much of the ivory consumed in Europe came from historical stockpiles, gathered over centuries by African societies and purchased – or more often seized – by Arab traders for sale on the international market. By the 1870s and 1880s, however, these stockpiles had been exhausted and elephants began to be slaughtered in large numbers for their tusks – with devastating consequences for the species.[2] The chapter explores the complex networks that brought ivory from the African savannah and forest to the cutlers of Sheffield and piano-makers of London and considers the severe environmental impact of the ever increasing demand for tusks. It goes on to examine the measures taken to protect the African elephant, which ranged from hunting licences and game reservations to export bans on underweight ivory.

One suggested solution to the unregulated culling of African elephants was Yellowly's idea of domesticating or taming the animals. This scheme has received little attention from historians but was supported by many eminent naturalists and imperialists, among them Philip Lutley Sclater of the ZSL, big game hunter Sir Samuel Baker and imperial hero General Charles Gordon, governor of Khartoum. It also received strong support from British subjects in Asia, who had seen elephants at work in the colonies and believed that their African brethren might be put to similar use; Singapore resident Mrs A. H. Brackenbury observed trained Asian elephants stacking teak in the timber yards of Moulmein in Burma and wondered whether their African equivalents might not be bred 'on farms, as ostriches are bred' and employed in 'navvy work, for which they are probably as well suited (education being supplied) as their Asiatic cousins'.[3] Although ultimately unsuccessful, domestication generated considerable interest in the 1880s and 1890s and was attempted on a

---

[1] 'Notes for Naturalists', *The Newcastle Courant*, 11 April 1879. Yellowly did not give the name of Mrs Edmonds' elephant, but a later article on the sale of the menagerie referred to the animal as 'the great African elephant, Emperor'. The accompanying illustration clearly depicts an African elephant, rather than the more common Asian species. See *The Graphic*, 9 August 1884.

[2] E. D. Moore, *Ivory: Scourge of Africa* (New York and London: Harper and Brothers Publishers, 1931), p. 166.

[3] 'Elephants Moving Timber at Moulmein, Burmah', *The Graphic*, 4 August 1883.

number of occasions. Here, I assess the pros and cons of domestication, as perceived by contemporaries, and situate this approach within the wider landscape of resource management, empire and wildlife conservation.

The Ivory Trade

Elephants were the source of one of the most valuable materials in nineteenth-century Europe: ivory. Surprisingly versatile, ivory was the plastic of its day. It was 'used extensively as an inlaying or a veneer for various articles of fancy furniture', employed in making the 'scales of various kinds of mathematical instruments', carved to manufacture 'artificial teeth, the tablets for miniature paintings and combs' and even used by confectioners as a stiffener for jellies.[4] Most importantly, ivory was the material of choice for cutlery handles, billiard balls and piano keys, products that experienced a sharp increase in popularity over the course of the nineteenth century as essential accoutrements for the bourgeois home.[5]

To meet the growing demand for this coveted material, British manufacturers turned to Africa and Asia. Within Africa, the best quality of ivory was believed to come from animals living within ten degrees of the equator, with the most prized 'soft' ivory emanating from Central East Africa.[6] Tusks brought from East Africa were generally sourced from indigenous elephant hunters (known as *Makua*) and transported to trading hubs such as Zanzibar in caravans organised by Swahili and Omani merchants, such as the infamous slave trader Tippu Tip.[7] Tusks brought from Cameroon were transported across the desert on

---

[4] 'The Origin, Nature and Uses of Ivory', *Penny Magazine of the Society for the Diffusion of Useful Knowledge*, 15 June 1839, p. 231; 'A London Ivory Sale', *The Leisure Hour*, January 1892, p. 627.

[5] In 1859, for instance, the Berlin-based piano-maker Bechstein manufactured only 176 instruments; just twelve years later, in 1871, it was producing 4,855. See Jonas Kranzer, 'Tickling and Clicking the Ivories: The Metamorphosis of a Global Commodity in the Nineteenth Century' in Bernd-Stefan Grewe and Karin Hofmeester (eds), *Luxury in Global Perspective: Objects and Practices 1600–2000* (Cambridge: Cambridge University Press, 2016), p. 252. See also David A. Shayt, 'The Material Culture of Ivory Outside Africa' in Doran H. Ross (ed.), *Elephant: The Animal and Its Ivory in African Culture* (Los Angeles: Fowler Museum of Cultural History, 1992), pp. 367–81; 'Hunting, Wildlife and Imperialism in Southern Africa' in William Beinart and Lottie Hughes, *Environment and Empire* (Oxford: Oxford University Press, 2007), pp. 67–8.

[6] Moore, *Ivory: Scourge of Africa*, pp. 219–20.

[7] Abdul Sheriff, *Slaves, Spices and Ivory in Zanzibar* (Athens: Ohio University Press, 1987), pp. 77–115.

the backs of camels by Hausa traders and often arrived cracked due to improper care.[8] As an article in the periodical *Once a Week* explained:

The camels are unladen at night and the tusks placed on the ground. The next morning while still moist with the heavy dew, they are placed unwiped on the camels' backs again, and are again exposed to the hot sun of the desert. This alternate heat and moisture causes numerous cracks, which, however, strange to say, are more numerous inside than they are on the outer covering or bark.[9]

A much smaller quantity of ivory was imported from India and Ceylon (Sri Lanka) (where only male elephants possess tusks), in this case removed from the animals while alive. According to US writer Charles Frederick Holder:

All the tuskers which are taken by [elephant catcher George] Sanderson and others in Asia have their tusks shortened or cut, the tips being valued ivory. The end is bound with a brass ring to prevent the tusk from splitting: Jumbo's tusks were cut off in this way. As the tusk is continually growing, the pulp being converted into ivory, the trimming operation can be repeated at certain intervals, generally every eight or ten years.[10]

American Consul Mr Webster disaggregated the origins of ivory imports, noting that, of the 12,435 hundredweight (cwt) of ivory imported into Britain in 1880: 2,972 cwt came from the British East Indies (mostly East African ivory re-exported from Bombay); 2,310 cwt from the west coast of Africa; 2,003 cwt from Egypt; 1,114 cwt from British possessions in South Africa; 693 cwt from Aden; 612 cwt from France; 431 cwt from Holland; 411 cwt from Malta; 361 cwt from Portuguese possessions in West Africa; 162 cwt from British possessions in West Africa; and 1,267 cwt from all other countries.[11] These figures were not static, however, for political instability, colonisation and the destruction of more accessible populations of elephants all affected the dynamics of the ivory trade. The war in the Sudan in the 1880s, for example, disrupted the supply from Egypt, forcing English ivory merchants to seek new outlets at the port of Massawa on the Red Sea. The depletion of West African elephant herds by the late nineteenth century likewise impacted the acquisition of ivory, shifting the focus of the trade

---

[8] Edward Alpers, 'The Ivory Trade in Africa: An Historical Overview' in Ross (ed.), *Elephant*, pp. 367–81.

[9] 'African Elephants', *Once a Week*, 28 July 1866, p. 112.

[10] Charles Frederick Holder, *The Ivory King: A Popular History of the Elephant and Its Allies* (New York: Charles Scribner's Sons, 1902), pp. 220–1. The regular removal of the tips of the tusks of Asian timber elephants was done to make them less dangerous to their human handlers. See Thomas R. Trautmann, *Elephants and Kings: An Environmental History* (Chicago: University of Chicago Press, 2015), pp. 323–4.

[11] 'The Ivory Trade', *Chambers's Journal*, 1 May 1886, p. 287.

to the Congo Basin and East Africa.[12] By 1891, the island of Zanzibar off the coast of German East Africa (modern-day Tanzania) was supplying 75 per cent of the world's ivory, much of it re-exported to Europe from Bombay (now Mumbai).[13]

Once ivory arrived in Europe, it was sold at auction to the highest bidders. London was the main hub of the trade (one journalist estimated that it controlled five-sixths of the trade in 1883), although smaller quantities of ivory were also auctioned in Liverpool, Antwerp and Rotterdam.[14] A visitor to a London ivory sale in 1892 described the scene before him, expressing his amazement at the volume of ivory on display:

The floor is crowded with ivory of all sorts and sizes, in tusks and sections, and odds and ends, some of it in huge teeth weighing 70 lb [pounds; approximately 30 kilograms] each, some mere trifles of 20 lb [9 kilograms] apiece, some mere pigmy 'scrivelloes', and crooked, cracked, hollow, decayed and broken. On every lot is a big clumsy number and every assemblage of lots has a notice board giving the broker's name and the first and last numbers of the lots he has to sell. The wilderness of teeth seems all in movement round the gigantic pair of travelling scales in the centre; the curving tusks are like so many worms, all strangely scratched and scribed, and are of all colours, from white through to brown to almost black; and an expert can tell at a glance where each came from, and can sort the lots from the pink Calcutta to the black West Coast which comes wrapped up in the raw hides bearing the mysterious name of 'schroons'.[15]

Buyers from across Europe and the USA attended these sales, selecting the grade of ivory most suited to their needs. French merchants purchased ivory for businesses in the Norman town of Dieppe, famed for its ivory carving expertise.[16] The New York firm F. Grote and Co. imported ivory for all manner of products, from billiard balls to spatulas,[17] while Hamburg ivory dealer Heinrich Adolf Meyer sold ivory in Germany for the manufacture of 'inlets in brushes and combs, the frontispieces of prayer books, knife handles and fans, piano keys, stick handles, lids for boxes, billiard balls and artistic carvings'.[18] Within Britain, most ivory ended up in the steel-producing city of Sheffield, where it was transformed into cutlery handles by firms such as Joseph Rodgers and Sons. As early as 1856, Sheffield manufacturer Mr Dalton estimated that 'annual consumption [of ivory] in the town of Sheffield alone is about

---

[12] 'The Scarcity of Ivory', *Music Trade Review*, November 1883, p. 87.
[13] Alpers, 'The Ivory Trade in Africa', pp. 367–81.
[14] 'The Scarcity of Ivory', *Music Trade Review*, November 1883, p. 87.
[15] 'A London Ivory Sale', *The Leisure Hour*, January 1892, p. 624.   [16] Ibid., p. 626.
[17] Holder, *The Ivory King*, pp. 221–2.
[18] Bernhard Gissibl, *The Nature of German Imperialism: Conservation and the Politics of Wildlife in Colonial East Africa* (New York: Berghahn, 2016), pp. 41–2.

108 tons, equal in value to 7,000*l* [shillings], and requiring the labour of 500 persons to work it up for trade'.[19] By 1882, another Sheffield cutler was getting through 522 tusks in a fortnight – necessitating the slaughter of at least 261 elephants.[20]

Like sealskin and ostrich feathers, the growing popularity of ivory resulted in part from technological improvements that allowed ivory to be sawn, etched, sanded, dyed and polished more efficiently. New techniques emerged for bleaching piano keys and carving toys and umbrella handles, and new lathes were engineered for turning perfectly shaped billiard balls. A visitor to Staight's piano factory in Feltham, west London, described the complex process of transforming raw tusks into polished piano keys, noting how the tusks were first cut up with steam-powered saws and then 'dipped in a solution' to bleach them – an operation performed by 'a machine with rack and pinion motion which went through two or three operations at one time'.[21] Scientific advances in other areas further enabled buyers to better assess the quality of the ivory on sale and to more accurately distinguish fake from true ivory. In the 1880s, 'the electric light' was 'beginning to be used to test the soundness of the tusks'.[22] In 1895, meanwhile, a Belgian chemist invented 'a ready means of distinguishing between animal and vegetable ivory' by dipping the material in 'concentrated sulphuric acid'; if the material was unaffected, it was animal ivory, but if it developed 'a rose tint' after several minutes' immersion it was vegetable ivory.[23] Purchasers of ivory could thus be more confident of the quality and authenticity of their investment and could select the type of tusk best suited for their purpose.

The burgeoning ivory trade was therefore a product of imperial expansion and technological innovation. Increased penetration of Africa gave European merchants access to new sources of ivory, while new machinery – the product of the Industrial Revolution – made it possible to process the material more quickly and in much larger quantities. From a luxury product, used in Asia and Africa for primarily decorative uses, ivory was transformed into an everyday necessity for the average Briton, who ate his dinner with ivory-handled knives and forks, potted ivory balls at the billiard table and drummed out a tune on an ivory-keyed piano.[24]

---

[19] 'Ivory', *Reynold's Miscellany*, 27 September 1856, p. 136.
[20] 'Editorial', *The Times of India*, 13 June 1882.
[21] 'Staight's Ivory Works at Feltham', *Music Trade Review*, 1 December 1889, p. 130.
[22] 'The Ivory Trade', *Chambers's Journal*, 1 May 1886, p. 287.
[23] 'Ivory', *Chambers's Journal*, 16 November 1895, p. 727.
[24] For a discussion of how other Asian luxuries underwent a similar transformation in the eighteenth century, see Maxine Berg, *Luxury and Pleasure in Eighteenth-Century Britain* (Oxford: Oxford University Press, 2005).

The Elephant Is Doomed

The ivory trade had one fatal flaw, however. New technologies and expanding commercial links may have increased the demand for ivory, but the supply, unfortunately, was finite. As the nineteenth century wore on, contemporaries started to realise that elephants were in trouble, raising concerns about the long-term viability of the trade. Could ivory be produced at a rate sufficient to meet the growing demand? Who, or what, was to blame for the decrease in elephant numbers? Was the African elephant doomed to extinction? In answering these unsettling questions, commentators focused on three key factors that made the elephant exceptionally vulnerable: the huge quantities of ivory now being consumed in Europe and North America; the fact that at least one elephant had to die for every pair of tusks brought onto the market; and the slow reproduction rate of large pachyderms, which meant that they were being slaughtered much more rapidly than they were able to procreate.

First and foremost, the elephant was a victim of the exponential rise in ivory exports in the years after 1800. At the end of the eighteenth century, 'the annual average importation into England was only 192,600 lb'. By 1827, it had risen to '364,784 lb, or 6,080 tusks, which would require the death of at least 3,040 male elephants'; by 1863, it had reached 1,000,000 lb, which, 'estimating each tusk at 60 lb', equated to 'the lives of 8,333 male elephants'.[25] Since it was not possible to obtain the tusks without killing the elephant (bar some exceptions in India, where the tusks of tame elephants were occasionally trimmed), growing demand for ivory inevitably meant increased rates of slaughter, pushing elephant populations closer to the precipice. As *The Friendly Companion* noted in 1887, 'there is no more possibility of obtaining ivory without slaying the elephant than there is of collecting whalebone without harpooning the whale, or sealskins without clubbing the seals'.[26] The introduction of firearms – among both European big game hunters and indigenous Africans – accelerated the killing, accounting for the lives of ever more elephants.[27]

The growing pressure on the elephant population was reflected in the disappearance of the animals from parts of Africa where they had once

[25] 'Ivory', *The London Reader*, 12 September 1863, p. 573.
[26] 'The Supply of Ivory', *The Friendly Companion*, 1 December 1887, p. 321.
[27] 'Indian and African Elephants', *The Graphic*, 28 September 1878.

roamed in large numbers and in the increasing number of small tusks on the market, which suggested that surviving elephants were being slaughtered at a younger age than previously. Writing in 1887, one magazine reported that '30 years ago, all the region between [Lake] Tanganyika and the coast was rich in ivory', yet 'only six years ago [explorer] Mr Joseph Thomson announced that over that vast territory scarcely a tusk was to be got; the land had been utterly despoiled'.[28] A year earlier, *Chambers's Journal* had remarked that '[t]he large proportion of very small tusks which are now brought to market annually is a sure indication of the increasing number of elephants that die young'. Earlier in the century, hunters had focused their efforts on the largest elephants, with the most impressive tusks, but now, as supplies dwindled, any elephant was fair game. To illustrate the point, *Chambers's Journal* compared the current underweight offerings available at auction with the tusks that had appeared in a Sheffield showroom some years before, one of which was 'nine feet long, twenty-one inches in girth' and 160 lb in weight. 'The value of the tusk was one hundred and thirty pounds, and it is said that an animal large enough and strong enough to carry such a pair would attract far more attention than Jumbo did.'[29]

As noted above, the elephant's plight was exacerbated by the species' slow rate of reproduction, which prevented it from rapidly replacing individuals lost to the hunter's bullets. Female elephants do not typically become fertile until around fourteen years of age (often not mating until their late teens), and gestation takes twenty-two months. Over a lifetime of sixty to seventy years, they will produce a maximum of six calves.[30] At this rate, elephant births could not possibly keep up with the shocking number of elephant deaths and the long-term future of the animal looked bleak. As one contemporary reflected, with reference to the parallel environmental issue of deforestation:

It is the timber problem over again. It takes as many minutes to cut down a tree as it took years to grow it, and it takes considerably more years for an elephant to grow his tusks than it takes minutes to kill him ... Nowadays the ordinary tusks average about three to the hundredweight, so that 15,000 elephants have to be killed to furnish the British market, and some say 75,000 are killed a year. And as the elephant does not begin to breed until he is thirty years old, and averages but one youngster every ten years after that until he is ninety, the rate of increase is much too slow to overtake the slaughter.[31]

---

[28] 'The Supply of Ivory', *The Friendly Companion*, 1 December 1887, p. 323.
[29] 'The Ivory Trade', *Chambers's Journal*, 1 May 1886, p. 287.
[30] Dan Wylie, *Elephant* (London: Reaktion Books, 2008), p. 51.
[31] 'A London Ivory Sale', *The Leisure Hour*, July 1898, p. 626. On the problems posed by deforestation in British colonies see Richard Grove, *Green Imperialism: Colonial*

The fact that female elephants themselves were not exempted from the slaughter made things even worse, for killing a female meant killing her calf, who could not survive long without its mother. Edward Buxton, a leading conservationist, cited the 'disgraceful' case of 'an Englishman in the new British Protectorate of Somaliland' who 'slaughtered thirty elephants, every one of them females or very young males'. He also condemned 'a certain Count' who 'recently boasted of having killed four elephants in four minutes, all cows and calves'.[32] German ivory hunter August Knochenhauer, meanwhile, suggested that female elephants were more likely to be killed than males, since African hunters 'would wait until matriarchal herds congregated and male elephants joined them for mating' before launching their attack. 'The tuskers would, from "decades of experience", always escape the attack of the hunters, who then resorted to the indiscriminate slaughter of the maternal herds.'[33].

Ultimately responsible for the elephant's precipitous decline, of course, was the insatiable demand for ivory products, which showed no sign of abating. 'Fashion will not be denied, and so long as society requires ivory paper-knives, pen-knives, billiard-balls and Christmas cards, so long will the slaughter of the largest quadruped for the smallest result continue.'[34] Critics of the ivory trade recognised the connection between consumption in Europe and North America and elephant deaths in Africa and raised the unsettling prospect of extinction. *The Graphic* warned that 'if matters go on much as they are now, this noble quadruped, the majestic living reminder of the old days, when frogs were as big as bullocks and elks towered like giraffes, will become as extinct as his hairy-coated brother, the mammoth'; *The Review of Reviews* claimed that, '[u]nless measures are taken at once, the African elephant will be as extinct as the dodo in thirty years' time'.[35] Whether consumption habits could be changed in time to reverse this inexorable decline was another matter, and depended on whether people valued the survival of an iconic species over the goods derived from its tusks. As the *Portsmouth Evening News* meditated: 'The elephant ... is doomed. But he will be regretted. Indeed, if the recent national grief for the loss of one elephant, albeit the biggest of its kind, furnishes any index of human sentiment with regard to this "huge earth-shaking beast", we may look for great sorrow and

---

Expansion, Tropical Island Edens and the Origins of Environmentalism, 1600–1860 (Cambridge: Cambridge University Press, 1995).

[32] 'Imperial Game Reserves', *The Spectator*, 23 May 1903.

[33] Gissibl, *The Nature of German Imperialism*, p. 76.

[34] 'Exit Elephant', *Portsmouth Evening News*, 2 December 1882.

[35] 'Indian and African Elephants', *The Graphic*, 28 September 1878; 'Is the Elephant Following the Dodo?', *The Review of Reviews*, September 1899, p. 287.

lamentation in the future, when the last Jumbo shall have fallen.'
Humanity, in short, had a tough choice to make. 'Which will society
elect to keep – its elephants or its ivory knife-handles?'[36]

Saving the Elephant

Alternatives to Ivory

While many may have opted for the knife handles, a substantial number
of Victorians did regard the elephant as worthy of preservation. The
question, however, was how the animal could be saved when the uses
of ivory were so extensive and the demand for the material so high. As
with sealskin and birds' feathers, the answer lay partly in policing the
hunting of elephants and partly in changing consumer behaviour. And as
with these other commodities, it required the collaboration of scientists,
artisans, colonial governors and animal welfare organisations.

One solution to the problem of elephant decline was, of course, to
reduce the consumption of ivory. This meant alerting consumers to the
environmental impact of their actions and trying to persuade them to
renounce ivory products – easier said than done, given the material's
wide range of functions. To make the consuming public aware of the
consequences of their actions, campaigners for elephant preservation
targeted purchasers (especially females) and emphasised the cruelty
inherent in killing elephants. In 1856, a critical article on ivory in
Reynold's Miscellany remarked: 'But few ladies, as they twirl their fans
or run their fingers over the keys of a piano, are aware of the manner in
which this article is procured, the quantities of it which are annually sold
and the number of noble animals which are yearly slain for the purpose of
supplying the constantly increasing demand.'[37] A later article in the
RSPCA's The Animal World magazine was even more emphatic, insisting
that 'nothing can be deprecated so much as extinction for the paltry
object of supplying ivory for pianoforte keys, fans, brush-handles etc.
Noble indeed!'[38] By showing consumers that elephants were intelligent
animals, capable of suffering and human-like emotions, those lobbying
for the animal's preservation hoped to induce guilt and make clear the
true cost of ivory. One ex-big game hunter, Edouard Foa, stressed the
gentleness and 'maternal love' of the elephant, describing how 'when

[36] 'Exit Elephant', Portsmouth Evening News, 2 December 1882.
[37] 'Ivory', Reynold's Miscellany, 27 September 1856, p. 136.
[38] 'Domestication of Elephants', The Animal World, December 1883, p. 178.

taking any long, dangerous journey, the mother pushes her little calf in front of her, holding him up with her trunk'.[39]

The plight of African elephants in the wild was further brought home by the growing number of elephants on show in zoos and menageries back in Europe and the interactions people enjoyed with them. Commodities of a different kind, these creatures were often named, petted and anthropomorphised, eliciting public affection and triggering concern when they appeared to be mistreated. When Wombwell's elephant Chubby developed a large tumour on his 'off side thigh', for instance, his 'friends' in Hull persuaded a local veterinary surgeon to operate, restoring the sickly pachyderm's 'appetite and good looks'.[40] When London Zoo's famous African elephant Jumbo was sold to P. T. Barnum in 1882, meanwhile, elephant-lovers protested against his departure, bombarding zoo officials with angry letters and regaling Jumbo himself with buns and other delicacies.

Locks of hair were sent to him, apparently not in derision; a long gold neck chain, a loving cup, expensive bouquets of flowers, snuff and cigars were delivered under affectionate epistles, under an equivalent impression that Jumbo, although an elephant, was really equal to humans in his tastes; and immense numbers of pumpkins and other gourds – even oysters, wedding cake and champagne – were consigned to him. More strange still, a box about two feet square full of 'corrective pills' was sent to him to 'prevent disorder during his voyage'.[41]

While, as Susan Nance argues, many of those who cared about Jumbo were able to dissociate the plight of this individualised, celebrity animal from that of his wild cousins, some commentators did seek to link the two, trying to raise awareness of the African elephant's near extinction.[42] *The Spectator* commented: 'It is for elephants' general good that they should be greatly sought after and fetch high prices and draw great crowds, and so justify careful feeding, good treatment and generous keep … Otherwise their only destiny would be knife handles.'[43] *The Animal World*, meanwhile, took advantage of the furore over Jumbo to provide its readers with 'a little knowledge respecting the structure, habits, disposition, sagacity and utility of elephants', including the disturbing fact that 'we annually import upwards of 10,000 cwts of elephants' tusks into England for the manufacture of ivory articles, to supply

---

[39] 'Is the Elephant Following the Dodo?', *The Review of Reviews*, September 1899, p. 287.
[40] 'Operation on an Elephant', *Leeds Mercury*, 26 December 1857.
[41] 'Jumbo', *The Animal World*, March 1884, p. 34.
[42] Susan Nance, *Animal Modernity: Jumbo the Elephant and the Human Dilemma* (New York: Palgrave Macmillan, 2015), p. 32.
[43] 'Jumbo', *The Spectator*, 25 February 1882.

which (only English wants), 25,000 elephants are killed every year!'[44] Like modern-day zoo animals, menagerie inmates in nineteenth-century Britain could, to a degree, act as stand-ins for their entire species, personalising their plight and even informing conservation measures in the wild.

Some British consumers, then, may have been persuaded to boycott ivory on humanitarian grounds, moved by the spectacle of suffering animals. For most, however, the pervasive need for ivory made doing without the product difficult, and the only real prospect of reducing consumer demand for elephants' tusks lay in finding a viable alternative for the material. In the twentieth century, plastic would fulfil many of the uses of ivory, significantly reducing its commercial importance. In the second half of the nineteenth century, before plastic arrived on the scene, contemporaries experimented with a range of other materials, both natural and synthetic, hoping that a suitable substitute could be found in time to save the elephant.

Natural sources of ivory offered one potential solution. The elephant was not the only creature to produce the material, and hippopotamus, narwhal and walrus teeth also featured in nineteenth-century ivory sales. Walrus ivory was deemed by one commentator to be 'poor stuff, the outer part of the tooth being alone of any good, the middle being more like coarse bone'. Hippopotamus ivory, however, elicited greater interest and was used extensively in the first half of the nineteenth century to make false teeth.[45] In 1836, Messrs Taylor and Co., 'Surgeon Dentists and Artists', charged clients 6 shillings per tooth for 'enamelled Hippopotamus Teeth' and '£3 for a full set, top or bottom', promising that the dentures would 'retain their colour unchanged to the latest period of use'.[46]

Hippopotamus ivory obviously had the same downside as elephant ivory, in that its acquisition meant killing the animal from which it came (although, as we shall see, conservationists were rather less concerned about preserving the hippo than they were about preserving the elephant). This was not the case with two other natural sources of ivory: so-called fossil ivory and vegetable ivory. The former was found extensively in Siberia, on the banks of the Lena River, and emanated from the tusks of 'a species of elephant and mammoth now extinct'.[47] Although

---

[44] 'Probiscidians', *The Animal World*, April 1882, p. 50.
[45] 'A London Ivory Sale', *The Leisure Hour*, January 1892, p. 627.
[46] 'Messrs. Taylor and Co., Surgeon Dentists and Aurists', *Hampshire Advertiser*, 4 June 1836.
[47] 'Ivory', *The London Reader*, 21 March 1874, p. 484.

generally considered 'too brittle for fine work', about 14 per cent of mammoth ivory was deemed 'good' by ivory trader Mr Westendorp, and a further 17 per cent 'could be used in some way'.[48] Today, mammoth ivory presents a quandary for conservationists, who recognise its potential value as a replacement for elephant ivory but worry that a legal trade in prehistoric tusks may provide a cover for the illegal trade in elephant tusks.[49]

The second potential ivory substitute, vegetable ivory, was the fruit of a species of South American palm called *Phytelephas macrocarpa* (literally 'plant elephant'). Discovered by Europeans in the eighteenth century, these fruits were 'filled with [a] crystalline liquid, like water, that is often used by travellers throughout these forests when drinking water is lacking'. According to Spanish botanist Hipólito Ruíz, who visited Peru in the 1770s: 'After a few days this water turns milky and acid, and later it coagulates into a sweet, tasty beverage that gradually becomes solid and hard. It is finally converted into a kind of ivory and is called *marfil vegetal* (vegetable ivory).'[50] From the 1840s, vegetable ivory was exported in increasing quantities from the province of Chocó in Colombia and used in Britain and the USA 'in making buttons, umbrella handles, knobs for doors, work-boxes, inlay work, reels for spindles, toys and trinkets'.[51] Due to their small size, however, tagua fruits could be used only for small items; they were never employed for making piano keys, billiard balls or cutlery handles.

Finally, the second half of the nineteenth century witnessed the creation of a number of synthetic ivory substitutes, designed explicitly with the aim of countering elephant decline. In 1863, Birmingham inventor Alexander Parkes exhibited a substance called Parkesine at the Sheffield Literary and Philosophical Society, made from 'a compound of oil, chloride of sulphur and collodion', which he claimed could be used to make 'knife handles and scales, buttons, book backs, dishes, models in

---

[48] Holder, *The Ivory King*, p. 226.

[49] See, for instance, Jani Actman, 'Woolly Mammoth Ivory Is Legal, and That's a Problem for Elephants', *National Geographic*, 23 August 2016; Neil Connor, 'Booming Trade in Mammoth Ivory Fuels Fears over Elephants', *The Telegraph*, 2 May 2017. It is possible to distinguish mammoth ivory from elephant ivory by examining patterns known as Schreger lines, but this is difficult to do in smaller items.

[50] Hipólito Ruíz, *The Journals of Hipólito Ruíz, Spanish Botanist in Peru and Chile 1777–1788*, translated by Richard Evans Schultes and María José Nemry von Thenen de Jaramillo-Arango (Portland: Timber Press, 1998), p. 289.

[51] Claudia Leal, *Landscapes of Freedom: Building a Postemancipation Society in the Rainforests of Western Colombia* (Tucson: University of Arizona Press, 2018), pp. 78–88; 'Vegetable Ivory', *The Bazaar*, 9 September 1910, p. 693.

relief, combs and various other articles'.[52] Seven years later, the American Hyatt brothers patented celluloid, an improved version of Parkesine made from 'dry pryoxyline and gum camphor, together with such colouring or other material as may be desired'. Though initially only available in thick sheets, by the early 1880s celluloid could be obtained in 'thin veneers' and was described as both 'durable' and 'capable of [the] most beautiful variety of shades, from mottled to imitate the finest tortoise shell, to malachite'. The *Music Trade Review* hoped that celluloid might be the salvation for dwindling pachyderm populations, enabling 'the elephant ... to enjoy a respite like that which has come to the whale on the discovery of petroleum'.[53] In reality, however, celluloid's physical limitations – not least its flammability – prevented it from immediately supplanting elephant ivory, and it was only with the invention of Bakelite in 1907 that synthetic plastics started to replace ivory on a large scale.[54]

Preservation Measures

While reducing the consumption of ivory offered one source of salvation for the elephant, contemporaries recognised that this alone would never be enough to stop the killing. Something also had to be done to protect elephants in the wild, where they were being slaughtered at an alarming rate. As humanitarians worked to reduce the demand for ivory in Europe, therefore, colonial administrators shifted their attention to conservation in the field, drawing up a series of game protection measures. In 1900, representatives from Britain, Germany, Spain, Belgium, France, Italy and Portugal met in London and signed the Convention for the Preservation of Wild Animals, Birds and Fish in Africa, an agreement intended to ensure 'the preservation throughout their possessions in Africa of the various forms of animal life existing in a wild state which are either useful to man or are harmless'.[55] Three years later, a group of

---

[52] 'Conversazione of the Literary and Philosophical Society', *Sheffield and Rotherham Independent*, 30 January 1863.

[53] 'Celluloid versus Ivory', *Music Trade Review*, 6 October 1882, p. 37.

[54] On the dangers posed by celluloid, see 'Explosive Fakes: Plastic Combs and Artificial Silk' in Alison Matthews David, *Fashion Victims: The Dangers of Dress Past and Present* (London: Bloomsbury, 2015), pp. 176–89. On the invention of Bakelite, see Daniel Crespy, Marianne Bozonnet and Martin Meier, '100 Years of Bakelite, the Material of 1,000 Uses', *History of Science* 47 (2008), pp. 3322–8.

[55] 'Convention for the Preservation of Wild Animals, Birds and Fish in Africa, signed at London, May 19, 1900', *Journal of the Society for the Preservation of the Wild Fauna of the Empire*, Vol. I, 1904, pp. 29–37. The Convention was never ratified by all of the attendees, but its existence established the principle of wildlife conservation on an international plane and shaped subsequent wildlife management policies in many European colonies.

ex-big game hunters and colonial officials established the Society for the Preservation for the Wild Fauna of the Empire (SPWFE), lobbying the British Government to set and enforce regulations to protect species threatened with extinction and publishing a journal dedicated to the subject of game conservation. From these discussions, three measures emerged as the most promising strategies for saving threatened species – including the African elephant: the issuing of hunting licences; the creation of sanctuaries for game; and the introduction of export bans on certain animal products, specifically underweight ivory.[56]

The first of these measures, hunting licences, represented an attempt to regulate the behaviour of sportsmen in the field. Introduced in the final decades of the nineteenth century, the licensing system divided African fauna into different classes, or 'schedules', and required would-be hunters to purchase a special licence if they wished to kill any of the species listed as protected. In 1896, for instance, the hunting regulations introduced into German East Africa stipulated that 'members of sporting expeditions' must pay 500 rupees to shoot elephants and rhinoceroses, while 'those who are not natives have also to pay 100 rupees for the first elephant killed and 250 for each additional one'.[57] To encourage the successful reproduction of threatened species, several colonial authorities banned the killing of females and established a close season during the breeding period. To curb the mass destruction of game, some British colonies also outlawed 'unsporting' behaviour, such as the shooting of animals from the safety of steamers and rail cars.[58] Together, these measures were intended to stem the slaughter of African wildlife and to keep killing within carefully prescribed boundaries. In reality, however, the system was hard to police and the increasing availability of modern weapons among Africans meant that regulating the activities of European sportsmen was not always sufficient. Writing in 1899, the governor of Nyasaland, Alfred Sharpe, contended that

the African native throughout the continent since the introduction of firearms, urged on by the high value of ivory in European markets, has slaughtered elephants wherever he could find them, regardless of size or of sex; and so long as ivory of all descriptions is a valuable trade article, elephants will continue to be

---

[56] For a broader discussion of early twentieth-century preservation efforts, see 'From Preservation to Conservation: Legislation and the International Dimension' in John MacKenzie, *The Empire of Nature: Hunting, Conservation and British Imperialism* (Manchester: Manchester University Press, 1988), pp. 200–24.

[57] 'Big Game Preservation in East Africa', *The Times*, 24 October 1896.

[58] 'An Ordinance for the Preservation of Wild Animals and Birds', *Journal of the Society for the Preservation of the Wild Fauna of the Empire*, Vol. I, 1904, p. 22.

indiscriminately killed, until, in many portions of Africa, they will be totally exterminated.[59]

Six years later, an article in *The Times* reported that the native chiefs of Uganda 'see no reason why they should restrict themselves to [the two elephants they were permitted to kill with a sportsman's licence] if more can be killed without detection, since elephant ivory has a very definite money value to the Uganda chief'.[60] Licences alone, therefore, could not eliminate poaching, particularly when the demand for ivory was so high.

Sanctuaries offered another potential solution to the decimation of Africa's fauna. Inspired by the creation of national parks in the USA, sanctuaries first came into existence in the 1890s and constituted designated reserves in which protected animals could not be culled (or where their killing was at least closely monitored). In 1896, Member of Parliament James Bryce proposed the establishment of 'a reserve or sanctuary' for African elephants in British Somaliland, in the belief that, without such protection, 'the wild elephants now to be found in the Somaliland Protectorate' will 'very soon be driven out or extirpated'.[61] In the same year, German authorities in Dar es Salaam set up 'special game preserves' in East Africa where 'no shooting whatever will be allowed ... without special permission from the Government'.[62] Often situated close to rivers or railways, from which they could be effectively surveyed, sanctuaries provided wild beasts with a place where they could graze unmolested, creating the conditions in which tottering populations could recover. As with licences, however, reserves had to be adequately demarcated and policed if they were to be meaningful, and this entailed a level of expenditure that many colonial administrations were unwilling to bear. The SPWFE, a major champion of reserves, constantly urged the British Government to invest more money in the policing of sanctuaries, insisting that, 'if it is worth establishing a reserve at all, it is worth spending a little money to see that it is properly watched', but their pleas often fell on deaf ears.[63] Writing in 1906, F. J. Jackson reported that the Sugota reserve in Uganda was 'at present outside the limits of effective administration, and until recently continued to be the happy hunting

[59] Alfred Sharpe, 'The Preservation of African Elephants', *The Edinburgh Magazine*, January 1899, p. 89.

[60] 'London, Tuesday, December 26 1905', *The Times*, 26 December 1905.

[61] 'Parliament: House of Lords', *The Times*, 1 July 1896.

[62] 'Big Game Preservation in East Africa', *The Times*, 24 October 1896.

[63] 'The Year', *Journal of the Society for the Preservation of the Wild Fauna of the Empire*, Vol. II, 1905, p. 10.

ground of Somali, Baluch and other traders, who killed large numbers of elephants without let or hindrance'.[64]

Given the difficulties in enforcing licences and sanctuaries, conservationists also proposed a final measure to protect elephants in the wild: a ban on the export of underweight ivory. Designed to prevent the killing of females and young animals, the export ban sought to make the slaughter of immature animals unprofitable and thereby discourage it. As Alfred Sharpe explained:

If all the Powers holding territory in Africa would agree to strictly prohibit the export of tusks under a certain weight – say 14 lb [6.4 kilograms] – (or portions of such tusks), and would faithfully carry out such agreement, all small ivory would become valueless to the owner … Not many cow tusks exceed 12 lb [5.4 kilograms] in weight, and one result of this prohibition would be that in [the] course of time, as soon as the news had spread throughout tropical Africa that small tusks were no longer of any value, neither cow elephants nor undersized beasts would be shot for their ivory.[65]

What exactly the minimum weight for tusks should be remained open to debate. The Convention for the Preservation of Wild Animals, Birds and Fish in Africa set it at 5 kilograms or less, an amount accepted in the 1900 Game Regulations for Rhodesia, which stated: 'If any person is found in possession of any elephant tusks weighing less than 11 lbs [5 kilogram] … he shall be guilty of an offence … and the tusk or ivory shall be forfeited.'[66] The first Belgian game ordinance, however, lowered the minimum weight to a measly 2 kilograms, unlikely to protect females and young elephants, while in 1905 Colonel Delme-Radcliffe recommended raising the minimum weight to 'something about 40 lbs [18 kilograms], which would show that nothing but bulls had been killed, and which would protect the females entirely, and also protect the smaller bulls'.[67] Whatever the agreed weight, the crucial point was that all of the imperial powers must subscribe to an export ban, for 'if the control in any one colony were lax, the illicit ivory trade would find its way there, the

[64] 'Extracts from Blue Book Issued November 1906', *Journal of the Society for the Preservation of the Wild Fauna of the Empire*, Vol. III, 1907, p. 36.
[65] Sharpe, 'The Preservation of African Elephants', p. 91.
[66] 'Convention for the Preservation of Wild Animals, Birds and Fish in Africa, signed at London, May 19, 1900', *Journal of the Society for the Preservation of the Wild Fauna of the Empire*, Vol. I, 1904, p. 33; 'Game Regulations for North-Eastern Rhodesia, 1903', *Journal of the Society for the Preservation of the Wild Fauna of the Empire*, Vol. II, 1905, p. 75.
[67] Gissibl, *The Nature of German Imperialism*, p. 253; 'Minutes of Proceedings at a Deputation from the Society for the Preservation of the Fauna of the Empire to the Right Hon. Aldred Lyttleton (His Majesty's Secretary for the Colonies)', *Journal of the Society for the Preservation of the Wild Fauna of the Empire*, Vol. II, 1905, p. 17.

revenue of the other colonies would suffer and no benefit would accrue to the African elephant'.[68] As was the case in the sealskin controversy, therefore, international collaboration was crucial to the success of conservation measures, since both elephants and their tusks crossed territorial boundaries and weak enforcement in one region would undermine anti-poaching measures throughout the continent.

Looking at these turn-of-the-century game preservation measures as a whole, a number of points stand out. First, the successful preservation of elephants, whether through licences or sanctuaries, depended on adequate policing, which in turn depended on adequate funding. Both were often lacking, limiting the effectiveness of these schemes. As S. H. Whitbread complained, 'It is not enough to colour spaces on a map and to add in a footnote "spaces coloured pink are Game Reserves". Reserves must be watched and policed to ensure that their limits are maintained inviolate and their regulations observed; this means men, and men means money.'[69] Licences and sanctuaries, moreover, often had decidedly racist undertones, with high licence fees excluding native hunters (theoretically at least) and reserves often displacing Africans from their homes. This imperialist style of conservation would have a long and contentious legacy in postcolonial Africa.[70]

Second, even when reserves were patrolled diligently (which they generally were not), the animals in them were not fully protected, since they often strayed beyond the boundaries into territories where they could legally be shot. To site sanctuaries correctly, and to ensure that they were the right size, conservationists required knowledge of the foraging habits and ranges of different species, which could be acquired only through careful observation. Once again, however, this was frequently lacking, and the boundaries of reserves often bore much greater relation to human political frontiers than to elephants' migration patterns. As Bernhard Gissibl has shown, moreover, the mere existence of a reserve could justify the continued exploitation of animals outside its borders, rather than bringing about a wholesale change in attitude

---

[68] 'Big Game Preservation in East Africa', *The Times*, 24 October 1896. In the twentieth century, when elephants were once again under threat, large quantities of ivory were smuggled out of Africa through Burundi, the one state that had declined to sign the 1973 Convention on International Trade in Endangered Species of Wild Fauna and Flora. See Cynthia Moss, *Elephant Memories* (Chicago: University of Chicago Press, 2000), p. 299.

[69] S. H. Whitbread, 'The Year', *Journal of the Society for the Preservation of the Wild Fauna of the Empire*, Vol. III, 1907, p. 12.

[70] On the postcolonial legacy of conservation, see Roderick P. Neuman, *Imposing Wilderness: Struggles over Livelihood and Nature Preservation in Africa* (Berkeley: University of California Press, 1992).

towards wildlife. In Eduard von Liebert's game ordinance of 1896, for instance, 'the reserves were hardly more than conservationist fig leaves with paradoxical effects: simply by existing as areas coloured green on the map, they could be taken as a justification to exploit wildlife even more recklessly in the rest of the colony'.[71]

A third point worth noting is that European conservation efforts were couched in the language of imperial stewardship and were heavily influenced by preceding developments in the USA – notably the creation of Yellowstone National Park in Wyoming. The grave plight of the American bison, brought to the brink of extinction by overhunting for hides, struck a chord with British conservationists and led some to advocate the preservation of African fauna – not purely for commercial reasons, but as part of an imperial inheritance owed to future generations.[72] As Lord Curzon told the Colonial Secretary in 1906:

We are the owners of the greatest Empire in the universe; we are continually using language which implies that we are the trustees for posterity of the Empire, but we are also the trustees for posterity of the natural contents of that Empire, and among them I do undoubtedly place these rare and interesting types of animal life to which I have referred.[73]

This bears a striking resemblance to Harold Heath's comments about the northern fur seal (see Chapter 2). It also, however, betrays what Gissibl has identified as a deeper European desire to preserve Africa and its megafauna as a pristine Eden, untouched by civilisation (regardless of the needs of native peoples).[74]

Fourth, however, despite such assertions of nature's inherent value, conservationist leanings did not, in fact, extend to all species. On the contrary, contemporaries generally drew a clear distinction between animals that were 'useful to man or ... harmless', which should be protected, and those that were not, which should be actively persecuted. The 1903 game regulations for North Rhodesia, for instance, explicitly advocated the killing of 'vermin', such as lions, leopards, hyenas, hunting

---

[71] Gissibl, *The Nature of German Imperialism*, p. 93.

[72] S. Whitbread observed, for instance, that 'it took but a few years to exterminate the vast masses of the American bison which used to darken the prairie and were counted one of the wonders of the New World. And so it will be with Africa unless the authorities will be wise in time.' See 'The Year', *Journal of the Society for the Preservation of the Wild Fauna of the Empire*, Vol. III, 1907, p. 12.

[73] 'Minutes of Proceedings at a Deputation from the Society for the Preservation of the Wild Fauna of the Empire to the Right Hon. The Earl of Elgin, His Majesty's Secretary of State for the Colonies, 15 June 1906', *Journal of the Society for the Preservation of the Wild Fauna of the Empire*, Vol. III, 1907, p. 24.

[74] Gissibl, *The Nature of German Imperialism*, pp. 216–17.

dogs, otters, baboons 'and other harmful monkeys', large birds of prey 'except vultures, the secretary bird and owls', crocodiles, poisonous snakes and pythons.[75] Several commentators also called for the destruction of the hippopotamus, which destroyed crops and was prone to capsize boats in rivers. Lord Cromer reported that: 'In the narrow rivers of the Bahr-el-Ghazal they swarm and are a positive pest, damaging the crops near rivers, and constantly making unprovoked attacks on small boats, dugouts, etc. Quite recently a Berthon boat, carrying the mail for the north, was attacked and sunk, the mail and two rifles lost and the two men in the boat narrowly escaping.'[76] Some animals were thus more equal than others when it came to conservation measures, and there was a clear hierarchy of priority determined by utility.[77] This highly anthropocentric approach to conservation reflected the wider conviction that animals must be useful to humans in order to survive.

Domestication

The elephant, of course, *was* useful to humans on account of its ivory, but only became so after death. To shift it into the different, and (from the animal's point of view) more desirable category of being useful while alive, another unusual scheme for saving the African elephant emerged in Britain in the late 1870s. Championed by some of the most prominent agents of British imperialism (and by French, German and Belgian imperialists as well), this scheme focused on the domestication (or, more accurately, taming) of the African elephant, which, it was argued, could not only be preserved but converted into an active ally in the exploration and colonisation of the so-called Dark Continent.[78] Proponents of the scheme took inspiration from Asia, where elephants were widely used as beasts of burden, and suggested that Indian elephants might be

---

[75] 'Game Ordinances of Rhodesia', *Journal of the Society for the Preservation of the Wild Fauna of the Empire*, Vol. II, 1905, p. 74.

[76] 'Extract from Lord Cromer's Report for Egypt and the Sudan for the Year 1902', *Journal of the Society for the Preservation of the Wild Fauna of the Empire*, Vol. I, 1904, pp. 66–7.

[77] For a discussion of how this doctrine of utility operated in a contemporary Indian context, see Viajaya Ramadas Mandala, 'The Raj and the Paradoxes of Wildlife Conservation: British Attitudes and Experiences', *The Historical Journal* 58:1 (2015), pp. 75–110. Mandala shows how efforts were made from the 1870s to protect elephants, which, though once perceived as a threat to agriculture, now played a crucial role in transportation, logging and railway construction and were protected by the Elephant Protection Act of 1879. By contrast, tigers and other 'vermin' were ruthlessly exterminated, with bounties placed on their heads.

[78] On the challenges facing African explorers during this period, see 'Logistics' in Dane Kennedy, *The Last Blank Spaces: Exploring Africa and Australia* (Cambridge: Harvard University Press, 2013), pp. 129–58.

transplanted to Africa as a preliminary step and used to tame their African counterparts.[79] If living African elephants could be made useful to man, the theory went, the demand for the ivory of dead animals would diminish, benefiting humans and pachyderms alike: elephants would exchange one valued commodity – their ivory – for another – their labour.

The first stirrings of interest in this novel acclimatisation/domestication scheme arose in the autumn of 1878 when General Charles Gordon conducted an experiment with elephants in the Sudan. Some years previously, Gordon had received a gift of five elephants from India which he had installed at his residence in Cairo. On hearing that the animals were underemployed and 'nearly eating their heads off in idleness', Gordon suggested sending them on to Khartoum, where he was now acting as *Khedive*, along with 'one smaller African elephant from the Gizereh Gardens'. He then commissioned a Dalmatian officer named Mr Marco to test their abilities as beasts of burden by walking them to the settlement of Lodo, several hundred miles to the south.

Following Gordon's orders, Marco travelled first along the eastern banks of the White Nile until he reached Hellet-Kaka, and from there on to Fashoda, where the Indian mahouts absconded and were replaced by Africans attendants. From Fashoda, the party progressed south through Shillook country, until they reached the Sobat River, and from here they marched to Bahr – a gruelling thirty-one-day trek on meagre rations. The elephants reached Lodo in a further ten days and were subsequently forwarded to Dufli, where they were 'employed in carrying all kinds of heavy goods'. The expedition was deemed to have shown that 'the Indian elephant can live in Africa', that 'it need not be fed in the luxurious manner that is thought indispensable in India' and 'that Indian attendants are not required' – the latter having been successfully replaced by African soldiers.[80]

Details of Gordon's experiment were published in *The Times* on 14 September 1878, where they elicited a steady stream of enthusiastic correspondence. One interested reader, naturalist Frank Buckland, contended that elephants could be used 'for the purpose of transport during the Zulu war, especially in places where horses cannot go on account of the tsetse fly'.[81] Several months later, a Royal Geographical Society

---

[79] Asian elephants are technically tame, not domesticated, since they are born and reared in the wild and captured when adult for use by humans. See Richard Bulliet, *Hunters, Herders and Hamburgers* (New York: Columbia University Press, 2005), p. 94.

[80] 'Central Africa', *The Times*, 14 September 1878.

[81] 'African Elephants', *Once a Week*, 1 April 1879, p. 31.

report speculated that 'the introduction of a few Indian elephants, with their drivers, might materially facilitate the transport of material' for the construction of a telegraph line in Central Africa, aiding 'commercial intercourse ... in regions infected by the "Tsetse fly" and in those that are covered with tall grass not easily penetrated by men on foot'.[82] Another contemporary, H. B. Cotterill, perceived the scheme as a virtual panacea that would reduce 'the difficulty of transporting heavy loads – such for instance, as the sections of a boat'; end 'the interminable delay occasioned by the caprice of chiefs or the desertion of carriers'; and prevent the elephant itself from being 'mercilessly shot down for the sake of a few pounds of ivory'.[83] Although some commentators were more sceptical, warning that the elephant's 'skin is easily chafed by harness and his feet are liable to sores which render him non-effective for months', the overall feeling was positive.[84] The domestication of the African elephant had the potential both to aid further exploration and to save the animal from extinction.

Two men were particularly inspired by the prospect of domestication and expatiated at length on its possible applications. The first of these was Philip Lutley Sclater, secretary of the Zoological Society. Personally acquainted with elephants through his work at London Zoo, Sclater was convinced that the animals were 'the proper beast[s] of burden for Africa' and well suited to its difficult climate and terrain. Various European animals had been tried in African exploration, but all had perished from the bite of the deadly tsetse fly. Elephants offered what promised to be a more economical and reliable alternative, improving transport and communication in the African interior. As Sclater reflected:

The colony of English missionaries just planted on the banks of [Lake Tanganyika] will certainly require regular supplies from and intercourse with their base of operations at the sea-board. Instead of despatching caravans of some 200 or 300 negroes [sic], over-weighted by the burden of their own food and requiring two or three months to cross a distance of 500 or 600 miles, it would surely be more economical and more expeditious to import a few elephants from India for use on this route. There can be little doubt that a small party mounted on elephants would accomplish the distance in very much less time and would carry a far greater amount of supplies than the usual huge caravan of bearers on foot, and at the same time be better able to defend itself from the attacks of the natives, who levy blackmail on the route.[85]

---

[82] 'The Proposed African Overland Telegraph', *Proceedings of the Royal Geographical Society and Monthly Record of Geography* 1:4 (1879), p. 268.

[83] 'The African Elephant', *The Times*, 7 January 1879.

[84] 'London, Friday September 27', *The Standard*, 27 September 1878.

[85] 'The African Elephant', *The Times*, 2 January 1879.

While Sclater initially advocated using Asian elephants, as Gordon had done, his ultimate goal was the domestication of the African elephant. To kick-start the domestication process, the zoologist suggested sending 'a detachment of elephant catchers from one of the Kheddah departments of India' to Zanzibar, where African elephants were 'still abundant', and using Asian elephants to capture and domesticate their African cousins. He theorised that 'a few thousand pounds spent in this way would be more likely to solve the vexed problem of opening the interior of Africa to trade and civilisation than the despatch of further expeditions over-weighted by what they have to carry on foot', while at the same time 'rescuing from utter extinction one of the finest and noblest of animals by rendering its existence beneficial to mankind'.[86]

Sir Samuel Baker, another domestication enthusiast, fully supported Sclater's proposal, which he believed could accelerate exploration, settlement and conquest in the sub-Saharan region. A renowned big game hunter, Baker had spent many years in Africa and lamented the lack of 'dependable transport' in the African interior, where 'horses, camels and donkeys ... all die from the attacks of flies'. He eulogised the good qualities of the elephant, which could 'wade through deep morasses' and 'defend itself against insect plagues by coating its skin with dust', and he expressed the hope that 'the tame Indian elephants, and one young African, that have been sent up from Cairo by General Gordon to Dufli' would form 'the nucleus for an extensive improvement in communication'. Articulating the racial prejudices of his era, Baker blamed the current failure to domesticate the African elephant not on the animal, which was eminently tractable, but on the 'bloodthirsty nature' of the native African, who would sooner kill a wild beast than tame it. If the Africans would not make use of the elephants themselves, however, Europeans could use the animals against the natives, who would surely flee in terror at the sight of 'a few elephants armoured with raw hide or thin steel plates, with castles and riflemen upon their backs'. Such a technique had already been used with success by Baron Napier of Magdala during his Abyssinian campaign in 1867, and Baker hoped it might be deployed again.[87]

While Baker envisaged using the elephant in war, others anticipated more pacific outcomes, perceiving the domestication of the African elephant as the salvation of the species itself and as a counter to the human slave trade, which was intimately connected to the trade in ivory. If African elephants were to be given a useful function, commentators

---

[86] 'Elephants in African Exploration', *The Times*, 20 September 1878.
[87] 'Use of Elephants in Africa', *The Times*, 24 September 1878.

reasoned, they would be more desirable alive than dead and the demand for their ivory would decrease. As a result, the estimated 30,000 Africans conscripted each year by Zanzibari traders to transport this valuable material would no longer be enslaved, and humanity as a whole would benefit.[88] As Buckland put it, 'the employment of native elephants could not fail to give a great blow to the African slave trade, because the ivory trade is now a cloak for the slave trade, and because African elephants might eventually supersede slaves as beasts of burden'.[89] Such an outcome appealed to many in the imperial establishment, who viewed the abolition of slavery as one of the main justifications for the British intervention in Africa – not least General Gordon, who branded the ivory trade as 'but the slave trade under another name'.[90] The caveat to this, of course, was that the survival of the African elephant depended on the animals becoming useful to mankind in another way, and, as Sclater put it, 'mak[ing] them work for their livelihood in their own country'.[91] Elephants would thus have to earn their right to exist.

The Belgian Expedition of 1879

The elephant acclimatisation programme reached its climax in 1879 when an expedition was launched to assess its viability. Financed by Leopold II of Belgium (better known for his appalling atrocities in the Congo Free State),[92] this expedition was led by an Irishman named Captain F. Falkner Carter and was set to travel from the coast of East Africa into the interior. The party included four Indian elephants, whose adaptability to Africa's climate and terrain was to be monitored carefully. If the elephants survived the journey, it was intended to use them to capture and tame some of their African counterparts, thus establishing a reliable supply of beasts of burden for travel and exploration.

The Belgian expedition got under way in May 1879 when Colonel Mignon selected four animals for the experiment from the Poona elephant-catching establishment in Maharashtra. The elephants were transported to the port of Bombay, where they were winched, with some difficulty, aboard the ship *Chinsura*. From there they were shipped across the Indian Ocean to Dar es Salaam, calling in at Aden and Zanzibar on

---

[88] Moore, *Ivory: Scourge of Africa*, p. 59.

[89] 'African Elephants', *Once a Week*, 1 April 1879, p. 31.

[90] 'The Ivory Trade', *The Anti-Slavery Reporter* 21:6 (1879), p. 142.

[91] 'The African Elephant', *The Times*, 2 January 1879.

[92] For a chilling analysis of Leopold's genocide in the Congo, see Adam Hochschild, *King Leopold's Ghost: A Story of Greed, Terror and Heroism in Colonial Africa* (Basingstoke: Pan Macmillan, 2006).

the way. The animals chosen for the expedition were Sundhar Gaj ('Beautiful Elephant'), Nadir Baksh ('Wonder Inspiring'), Phul Masla ('Flower Garland') and Susan Kali ('Budding Lily').[93]

In June 1879, the *Chinsura* anchored in Masani Bay, just off the East African coast. Here, the party encountered its first major problem, as no equipment had been loaded to carry the animals from the ship to the shore. With the *Chinsura* unable to go any closer to land for fear of grounding, it was decided to make the elephants swim the 2.5-mile crossing. Initially disaster threatened, as the female elephant Susan Kali thrashed about helplessly, making 'the most frantic efforts to get on board the ship again'. After nearly three hours in the water, however, she finally reached the beach, 'where her mahout was ready to receive her with *jhools* and warm clothing and bread mixed with *mussala* and a bottle of rum'. The rest of the elephants executed the crossing more calmly the following day, towed towards the shore with the assistance of a small boat.[94]

After their shaky landing, the elephants commenced their arduous journey from Dar es Salaam to Mpwapwa, deep in the interior of German East Africa (now Tanzania). Three of the pachyderms were 'laden with packs', while the fourth, 'used by HRH the Prince of Wales when in India', carried Captain Carter and his colleague Mr Mackenzie.[95] The elephants coped well with the gruelling trek, fording multiple rivers and surviving the onslaught of the tsetse fly, which 'swarmed on the elephants till the blood trickled down their flanks in a constant stream'.[96] The trip was not, however, without incident, for the animals repeatedly sank into bogs and had to be rescued. Expedition member L. K. Rankin chronicled how 'one male elephant – Sundar Gaj – had a very bad temper, and was given to bolting into the jungle'. He had to be coaxed back into line by Susan Kali, who 'would fondly rub her head against his and then give him a lead up to the objectionable bog or ford'.[97]

Despite these challenges, the party made it to the initial destination of Mpwapwa on 3 August seemingly in good shape, and Rankin and Carter dispatched an upbeat report to Leopold trumpeting the experiment's success. Almost immediately, however, a series of disasters struck the

[93] 'Royal Belgian Expedition to Central Africa', *The Times of India*, 20 August 1880.
[94] 'The King of the Belgians' Expedition: Landing the Elephants in Africa', *The Times of India*, 22 July 1879.
[95] L. K. Rankin, 'The Elephant Experiment in Africa; A Brief Account of the Belgian Elephant Expedition on the March from Dar-es-Salaam to Mpwapwa', *Proceedings of the Royal Geographical Society and Monthly Record of Geography* 4:5 (1882), p. 274.
[96] Ibid., p. 277   [97] Ibid., p. 275.

expedition. The first was the unexpected death of Sundar Gaj. He had just returned from Lake Kimagai with another elephant, Nadir Baksh, 'laden with miombo [grass]' and 'seemingly well', but at dawn the following day his Indian mahout discovered him dead. Rankin, who performed an autopsy on Sundar Gaj, concluded that he was 'a victim of too-herculean labours and of an insufficiency of food', a diagnosis apparently confirmed by the death, shortly afterwards, of a second animal from the same cause.[98] In a further twist of fate, a third elephant died later on in the journey, leaving only one, 'a large old female', alive by the time the party reached Karema on the banks of Lake Tanganyika. The expedition itself then came to an abrupt and brutal end when Captain Carter and his deputy Mr Cadenhead were murdered at Mpimbore by 'a band of 3000 Ruga-Rugas (slave and ivory traders)'.[99] Early promise thus culminated in tragedy and failure.

While the Belgian expedition ended ignominiously, this did not entirely dampen hopes that Indian elephants could be acclimatised in Africa. On the contrary, Rankin, in a subsequent article on the subject, asserted that the experiment had been a success, ascribing the deaths of three of the elephants to error and mismanagement and insisting that, if such mistakes could be avoided in future, acclimatisation was still possible. Explaining the death of Sundar Gaj, for instance, Rankin surmised that he and the other elephants had been underfed, overloaded and not given enough time to acclimatise to African conditions. As he put it: 'They had been stall-fed in India, on white bread, etc., the fat of the land in short, and then after only a short gradual reduction at Dar-es-Salaam, had had to forage for themselves, very little corn and rice being brought for them.' Moreover, 'whereas, according to [the Indian elephant catcher George] Sanderson, 700 lbs ... is the limit-weight an elephant should carry for a prolonged time, these bore at first 1,200, then 1,500, and at one time 1,700 lbs; while they daily climbed the most tremendous hills, which no mule, I think, could have climbed under a proportionate burden'.[100] If such errors were corrected, Rankin believed that elephant acclimatisation and domestication were still a possibility, the Belgian experiment having proven:

1) Their immunity to the tsetse fly, after 23-days' exposure to the insect; 2) their maintenance during one month mostly upon the uncultivated food of the country, and therefore at little cost; 3) their ability to march over all styles of

[98] Ibid., p. 287.
[99] 'Geographical Notes', *Proceedings of the Royal Geographical Society and Monthly Record of Geography*, New Monthly Series 2:10 (1880), pp. 626–7.
[100] Rankin, 'The Elephant Experiment in Africa', p. 287.

ground, soft, stony, sandy, boggy; to conquer all eccentricities of topography –
hill, dale, river and jungle – while labouring under double their due weight
of baggage.

Elephant carriage was, furthermore, a comparatively cheap form of
transport, the 'cost of 4 elephants, including keep and fines for damage
to crops' coming to the 'trifling' sum of 25.5 rupees.[101]

Expertise and Agency

The attempted acclimatisation and domestication of Asian and African
elephants required the input of a range of individuals and reflected the
importance of trans-imperial networks of expertise. As such, it provides
an interesting illustration of inter-imperial and even inter-species relationships and reveals the agency accorded to both African peoples and
African elephants in the domestication process.

First, when it came to the question of who should procure wild
elephants for taming, a variety of candidates were proposed for the
job – as well as a range of different methods. Writing in June 1879, in a
private letter, Livingstone's comrade, the missionary Dr John Kirk,
suggested that the best place in which to catch elephants was at water
holes, where they were forced to congregate during the dry season. Kirk
recounted how 'Livingstone and I once caught a calf in this way'. He
believed that the same technique would work on a 'full-grown animal',
and expressed confidence that the animals, once caught, could easily be
tamed. 'I once had an African elephant, now in India, which was quite
tame'; 'if elephants cannot be taught and tamed I shall be much sur-
prised'. Should European expertise prove insufficient, Kirk recommended employing native methods of killing elephants that might
equally well be used to take them alive. The 'elephant hunters of
Givongo', he noted, hunted elephants 'on foot with small dogs, ham-
stringing the elephants' and lighting fires to surround and subdue the
animals. These 'brave fellows' could be enlisted in the cause of domestication and 'their knowledge, hitherto used only in killing the animals,
might be of service in catching them'.[102]

Kirk's views found favour with the engineer in charge of building a
road between Dar es Salaam and Lake Nyasa, who seconded his proposal
to exploit African talent. Like Kirk, the engineer believed that catching
the animals should pose few problems, for baby elephants often loitered
round the bodies of their dead mothers when the latter were shot and

---

[101] Ibid., pp. 285–6.   [102] 'Elephants for Africa', *Lloyd's Weekly Newspaper*, 8 June 1879.

could easily be taken if anyone so desired. This would represent an improvement on the current practice, which was 'to kill the poor little beasts for the sake of their meat'. As to who should actually catch the young elephants, the engineer advised conscripting Arab hunters, for they had the most frequent contact with the animals and knew where elephant herds were to be found. He himself had drawn much of his own information from an Arab who 'goes up to Makengi most years, shooting elephants and buying ivory'; he was certain that, 'if the natives once find it is possible to capture elephants alive, and they can get more money for them than they can get for the ivory, there will be little further difficulty in procuring African elephants for training'. To set the ball rolling, the engineer personally offered 'a reward of 100 dollars to anyone who will bring me a young elephant alive and in good health'.[103]

While Kirk and the engineer sought the expertise of Arab hunters, others turned to India for assistance and advice. One writer, a journalist for *The Times of India*, recommended sending Indian trainers to Africa and using their experience to round up and tame the indigenous elephants. The journalist conceded that the mahouts on Carter's expedition had proven a disappointment, but felt that if the men were 'very carefully selected' and 'accompanied by a European possessing ample knowledge of the wants and capabilities of the elephant and the different methods of capturing and taming the wild animals', such a strategy might prove fruitful in future. It was in the interest of the Indian authorities, moreover, to provide such manpower, for commerce between the subcontinent and the African mainland was growing. 'The experiment is of so much importance in the development of the African trade which is now carried on with Bombay, and chiefly with Bombay capital, that the Government of India might be fairly asked to assist by the gift of half a dozen elephants and the loan of a few good experts.'[104]

Sir Samuel Baker also felt that India was the most appropriate source of elephant expertise, although he put more faith in European trainers than in indigenous mahouts. Having recently read G. P. Sanderson's book, *Thirteen Years with the Wild Beasts of India* (1879), Baker believed that Indian experience should be capitalised upon. To that end, Sir Samuel advised General Gordon to read Sanderson's book and suggested that Sanderson might make a direct contribution to the domestication scheme by coming to East Africa with his band of hunters and establishing a training establishment there.

[103] Ibid.   [104] 'Editorial', *The Times of India*, 13 June 1882.

Large herds of elephants are now captured by Mr Sanderson for the supply for the Indian Government, and he could, doubtless, obtain permission to afford assistance in men and animals to the Khedive for an elephant-catching establishment in Central Africa. These animals could be landed at Massoawah, on the Red Sea, and could be marched at the proper season direct to Khartoum.[105]

As Baker's comments make clear, what was required in Africa was not just the expertise of the Indian mahouts, but also that of Asian elephants. Employing techniques that were already widely used in India, the elephants shipped to Africa would be used to lure their African counterparts into a trap – the traditional *kheddah* method – and then to accustom them to human service. Although they were not necessarily willing collaborators in the initiative, the animals were evidently crucial to its success and were accorded a key role in proceedings. *The Times of India* speculated that, 'with half a dozen Indian elephants to start with [African missionaries] might soon train up a regular service of African elephants between the coast and the lake country'.[106] Baker went even further, claiming that 'Indian elephants will form a new "missionary society", which will quickly make converts and educate the now savage herds of Africa'.[107] Such language explicitly elevated the elephants to key actors in the 'civilisation' of the African elephant, according them a degree of elephantine agency as they proselytised among their less sophisticated counterparts. As Sanderson himself remarked, 'Elephants alone can deal with elephants.'[108]

The notion that elephants themselves possessed agency is further underlined by the way in which those who actually worked with the elephants talked about their charges, often assigning them names and specific characters. An article on Carter's elephants in *The Times of India*, for instance, recorded the Indian names of the four elephants, along with their English translations, and offered the exhilarating account of how Susan Kali swam to shore from the *Chinsura*.[109] L. K. Rankin, one of the survivors of the ill-fated expedition, expressed similar respect and affection for the elephants he accompanied, even hinting at their distinctive personalities: Sundar Gaj, known affectionately as 'Old Musty', was a renowned troublemaker, while Susan Kali was a much calmer character,

[105] 'Use of Elephants in Africa', *The Times*, 24 September 1878.
[106] 'Editorial', *The Times of India*, 13 June 1882.
[107] 'Use of Elephants in Africa', *The Times*, 24 September 1878.
[108] 'Domestication of Elephants', *The Animal World*, December 1883, p. 178.
[109] 'The King of the Belgians' Expedition: Landing the Elephants in Africa', *The Times of India*, 22 July 1879; Rankin, 'The Elephant Experiment in Africa', p. 282.

stroking Sundar Gaj with her trunk to counter his fits of obstinacy.[110] Rankin consistently commended the elephants' intelligence and endurance and was keen to emphasise that the pachyderms were not to blame for the expedition's ultimate failure.

[A]mid their many real trials of strength, difficulties and occasional danger, as they clambered up and down, over boulders and tree-trunks, across treacherous bogs and shifty, stony, torrent beds, and up hills which made them pause, look round for help, and trumpet with remonstrance – amid all this these noble beasts at all times exhibited unfailing judgement, patience and willingness. Their pluck under their too great labours compelled an admiring pity for them.[111]

This was more than could be said for the Indian mahouts, who were reportedly 'always sick, always whining and complaining'.[112]

If the behaviour of Indian elephants in Africa influenced perceptions of the viability of domestication, so did the behaviour of African elephants in Europe. Unknown in the continent since the Roman era, African elephants began to be imported in increasing numbers from 1865 onwards and were exhibited in zoological gardens and travelling menageries in Britain, France and Germany; Abraham Bartlett, superintendent of London Zoo, estimated that there were thirty African elephants in Europe by 1882.[113] Supporters of domestication scrutinised the temperament of these living imports and assessed their capacity for training. In a letter to *The Times*, Walter Severn marvelled at the 'tameness' of the 'specimens of the African elephant and the rhinoceros' he saw performing at the Alexandra Palace in 1878 and suggested that they might 'be turned to account in the way described by Sir S. Baker'.[114] Another commentator, a journalist for *The Graphic*, admired the 'astonishing tricks' of the African elephant Zara, who performed at the Duke's Theatre, Holborn,[115] while Liverpool animal dealer William Cross emphasised the docility of many African elephants he sold in the 1870s and 1880s, among them a 'young male African elephant' named Cetywayo (after the Zulu chieftain) and a female that he characterised as 'quite a pet'.[116] William Yellowly, as we have seen, praised the

---

[110] Rankin, 'The Elephant Experiment in Africa', p. 275.   [111] Ibid., p. 282.

[112] Ibid. p. 288.   [113] 'The Removal of "Jumbo"', *The Standard*, 8 March 1882.

[114] 'African Elephants', *The Times*, 25 September 1878.

[115] 'Theatres', *The Graphic*, 11 November 1876.

[116] 'A Young Zulu Elephant in Liverpool', *Dundee Courier and Argus*, 9 April 1879; 'Advertisements and Notices', *The Era*, 23 March 1879. Cross advertised another trained African elephant that 'sits up at table, rings the bell for [the] waiter, eats from a plate, pays [the] waiter with money it takes from a drawer, waltzes backwards and forwards ... plays musical instruments ... takes a handkerchief from its head and hands it to the keeper, and other tricks'. 'Advertisements and Notices', *The Era*, 15 February 1880.

tractability of Mrs Edmonds' African elephant, Emperor, who, in 1884, walked all the way from Liverpool to Margate, covering a distance of 283 miles in just thirteen days and consuming a daily diet of 'ten loaves ... half a bushel of oats, half a sack of bran, twelve pailfuls of oatmeal and water ... and a gallon of beer'.[117] These positive encounters with African elephants persuaded contemporaries that domestication was possible, assuaging doubts about the species' intelligence and character.

While various expatriate African elephants helped rehabilitate the image of the species, the poster boy for domestication was Jumbo, beloved resident of London Zoo. Acquired by the Zoo from the Parisian Jardin des Plantes in 1865, in exchange for an Indian rhinoceros, Jumbo grew to become a firm favourite with Londoners and received regular coverage in the contemporary press. Although best known for the furore surrounding his sale to American showman P. T. Barnum in 1882, Jumbo also made a less celebrated contribution to the imperial domestication project, offering tangible proof that his species was susceptible to training and human control. An 1879 article in *The Times*, for instance, noted that 'much interest attaches to this beast, because he is an example of the docility of the African elephant and therefore represents a hopeful means of transit, travel, trade and war'. No tradition currently existed for training elephants in Africa itself, even though elephant calves were often brought to the Red Sea port of Suakin for export to Europe, but Jumbo 'in two lessons was taught to obey his masters, and for many years he has been accustomed on public holidays to carry children on his back up and down the gardens, to behave with the greatest good humour and self-restraint among the crowd, and to accept their buns without greediness, although without limit'.[118] Writing after Jumbo's much lamented departure, the *Bristol Mercury* made a similar point, noting that Jumbo was 'referred to by writers upon some of the Anglo-Indian journals as affording proofs – may we not say *great* proofs – that the elephants of Africa are not incapable of being domesticated like the Asiatic elephant and trained to domestic service'. The newspaper claimed that the capacity of African elephants for domestication had been 'proved by the case of Jumbo' and by another elephant still in the Jardin des Plantes. It concluded that, 'if it should be found in the end that Jumbo's reputation for docility and intelligence has contributed to [the "civilising" process in Africa], the *Daily Telegraph*'s worship of the noble beast will not have been in vain'.[119] Here, then, was a clear example of what an African

---

[117] 'An Elephant's Journey', *North-Eastern Daily Gazette*, 16 April 1884.
[118] 'The Easter Holidays', *The Times*, 12 April 1879.
[119] 'News of the Day', *Bristol Mercury*, 3 August 1882.

elephant could do if raised in captivity. As in the case of ostrich acclimatisation, zoos and menageries played their part in imperial animal management schemes, providing inspiration and knowledge for overseas operations.

Domestication Reprised

Although it failed in the short term, the dream of domesticating the African elephant persisted for several decades, resurfacing at regular intervals during the 1880s, 1890s and early 1900s. As elephant numbers plummeted still further, another generation of conservationists, humanitarians and entrepreneurs reprised the idea of taming African elephants, giving some interesting new twists to the earlier domestication projects. These included relocating African elephants to another continent and even farming them for their tusks.

One more outlandish scheme for saving the African elephant came from the pen of Arnold Harris Mathew, a Roman Catholic priest who changed his name by deed poll to Count Arnoldo Girolamo Povoleri. A self-professed lover of nature, Povoleri was appalled by the 'extermination of many interesting animals, which are rapidly disappearing as victims sacrificed on the altars of folly and fashion', and called on his countrymen to come to their rescue. The African elephant was top of Povoleri's salvation list, given both the animal's nobility and its slow rate of reproduction, and he proposed a novel way of saving it: namely, gathering elephants into managed reserves and rounding them up on an annual basis to harvest their ivory. As he explained in a letter to *The Standard*:

Since an elephant's tusks, which are solid, can be cut off with a sharp saw without causing pain to the animal, and an elephant once captured by the methods employed in the Government Kheddahs in India can be easily secured for this process, it seems to me that the ivory traders would gain time, save labour and avoid the criminal folly of exterminating their source of revenue, if they could be induced to resort to this more humane method of obtaining ivory, instead of the unnecessary and brutal butchery of vast herds of valuable, inoffensive and tractable animals, which takes place year by year.[120]

Farming elephants required taming them first, but Povoleri was confident that this could be achieved. 'Several specimens have been tamed, and trained to some extent, at the Zoological Gardens in Regent's Park, amongst them the late lamented Jumbo', and '[a]t this moment there are

---

[120] 'The African Elephant', *The Standard*, 6 December 1892.

two young males of this species in the Society's collection which are trained to carry children, and are perfectly docile and obedient'.[121]

A more radical variant of Povoleri's proposal entailed not merely farming elephants in situ but moving tamed herds to a new and better protected location – in effect, naturalising African elephants overseas. One supporter of this idea, Mr Holt, suggested shipping the African elephant to South America, 'where the scenery and surroundings are sufficiently like what he is used to to prevent home sickness'.[122] Povoleri himself, in a revised version of his plan, proposed 'purchasing as many living specimens of the animal as can be obtained in Africa, and conveying them either to Cape Colony or to Algeria, to be placed upon preserves where their safety can be insured'. Moving elephants across oceans was not, of course, a cheap operation, but Povoleri hoped to raise the requisite funds from Sheffield cutlers, whose knife handles had wreaked such havoc among elephant populations. By way of encouragement, he pointed to the recent success in domesticating the ostrich, which was now generating vast profits as a farmed animal. 'The African elephant and the ostrich would stand side by side as monuments of British enterprise, rescued from extinction, and converted into valuable sources of revenue to their preservers.'[123]

In reality, ostriches were somewhat easier to farm in this way than elephants, given that 'the ostrich grows an annual crop of feathers and the elephant a single pair of tusks', and this particular project appears to have died a quiet death.[124] The desire to domesticate the elephant for work purposes, however, remained strong throughout the 1880s and 1890s, and variants of the original taming proposal reappeared repeatedly in contemporary newspapers, with both new and old voices adding to the debate. In 1893, Philip Lutley Sclater, such a vocal supporter of domestication in the late 1870s, wrote to *The Times* to reprise the idea, citing on this occasion the docility of Jumbo's successor, Jingo, who 'is daily engaged during the summer months in carrying children and other visitors about [London Zoo]'.[125] In 1895, another Briton, the well-known feminist and anti-vivisectionist Frances Power Cobbe, likewise

---

[121] 'The African Elephant', *The Standard*, 12 December 1892.
[122] *Freeman's Journal*, 23 September 1892.
[123] 'The African Elephant', *The Standard*, 12 December 1892.
[124] *Freeman's Journal*, 23 September 1892. E. D. Moore was equally scathing of Povoleri's scheme, pointing out that the elephant's tusks, once removed, would not re-grow, that the African elephant lacked 'the sheeplike disposition of his Asiatic brother', and that cutting into the tusk with a saw was liable to cause its owner 'excruciating toothache'. See Moore, *Ivory: Scourge of Africa*, pp. 210–13.
[125] 'The Preservation of the African Elephant', *The Times*, 26 October 1893.

took up the subject, publishing a letter she had received from a correspondent in Natal wanting to save 'the noblest of God's dumb creatures – the splendid, sagacious, lovable elephant'. Cobbe observed: 'It has been counted (since the days of the Carthaginians, when the last African elephants were tamed) a proof of the inferiority of the negro [sic] races that they have never mastered the noble animal, but only killed it ruthlessly for meat or for its ivory.' Invoking a blend of environmental stewardship, humanitarianism and practical imperial gain, she suggested that it would be 'a fitting inauguration of our great South African dominion if the lordly *Elephas africanus* were to be rescued from the barbarous extermination which now threatens his race, and attached – a willing servant – to the car of British prosperity'.[126]

Across the English Channel, another commentator to weigh in on the elephant domestication scheme was the German Carl Hagenbeck, who wrote an article in the *Hamburger Nachrichten* advocating preservation as a prelude to 're-domestication'.[127] Something of an expert on African elephants, having imported 170 of the animals to Germany, Hagenbeck dismissed the popular view that the animals were untameable and insisted, on the contrary, that African elephants were 'stronger, and at least as tractable, and as capable as beasts of burden, or to be ridden, as Indian elephants'.[128] As proof of his claim, he succeeded, in 1878, in training five recently arrived young African elephants to 'carry loads and allow themselves to be ridden'.[129] Concerned about the increasing mortality of elephants in Africa, and anxious to improve transportation in the region, the German government took up Hagenbeck's idea, founding a Committee for the Purpose of Taming and Preserving the African Elephant in 1895 and paying for a German officer to visit an Indian *kheddah* to master the methods of catching and training wild elephants. The Committee subsequently hired a staff of Indian trainers to establish a government elephant stud in German East Africa, funding the project with grants from the German Colonial Company and the Colonial Department of the Foreign Office.[130]

---

[126] 'The African Elephant', *The Times*, 7 November 1895. Belgian writer Paul Van Bellinghen took a similar stance, arguing that 'the African elephant, like his Indian counterpart, must become the auxiliary of the coloniser' rather than the victim of the bloodthirsty natives – a view, of course, imbued with the racist assumptions of the period. See 'L'Éléphant au Congo', *Gazette de Charleroi*, 3 December 1901.

[127] See Chapter 1 for details of Hagenbeck's ostrich farming scheme.

[128] 'The African Elephant', *Evening Telegraph and Star and Sheffield Daily Times*, 1 April 1895.

[129] Carl Hagenbeck, *Beasts and Men*, translated by Hugh S. R. Elliot and A. G. Thacker (London: Longmans, 1912), pp. 160–2.

[130] 'The Re-Domestication of the African Elephant', *The Spectator*, 6 April 1895.

107 Congo Belge   API   Les éléphants au bain.
    Belgisch-Congo   Badende olifanten.

Figure 3.2 Tame elephants bathe in the river at Api. Postcard, Belgian
Congo, 1929

In the event, none of these efforts succeeded and the African elephant
remained undomesticated. By the 1920s, however, a Belgian programme
to convert elephants into working animals finally bore fruit, demonstrating that the process was possible. Begun in 1904 by a Belgian officer,
Edmund Laplume, the scheme involved catching young elephants in the
jungle and transferring them to a special farm at Api in the heart of the
Belgian Congo. Initial domestication efforts proved costly, with twenty-
two of the 132 animals caught between 1899 and 1908 having 'to be
released, owing to their tender age' and a further eighty-six dying 'from
various causes'.[131] By 1927, however, the establishment could boast fifty
elephants, six of which were 'still under training', twenty-five 'trained,
but too young to work', and nineteen working animals (Figure 3.2).[132]
These beasts were used for a variety of tasks, including 'carrying bricks
for building', carrying 'materials in a waggon' and drawing ploughs.[133]
Visiting Api in 1927, Captain Keith Caldwell of the Kenya Game
Department witnessed 'the cotton crop' in the area being 'shifted by

[131] Philip Lutley Sclater, 'The Domestication of the African Elephant', *Journal of the Society
for the Preservation of the Wild Fauna of Empire*, Vol. IV, 1908, p. 50.
[132] 'Taming African Elephants: School Work in the Congo', *The Times*, 9 April 1927.
[133] T. Alexander, *Across the Great Craterland to the Congo* (New York: Alfred A. Knopf,
1924), p. 235.

Figure 3.3 'Elephants Pulling a Chariot'. Postcard, Belgian Congo, 1920

elephants to Titule' on 'large rubber-tired four-wheel carts' (Figure 3.3).[134] Touring the establishment one year later, fellow Briton Tracy Philipps reported that '[f]our elephants are employed on the mission, two by Mr Steenhault de Waerbeke, a planter at Dembea, and others on a cotton farm at Bambessa'.[135]

To train the elephants, the Belgians used human mahouts, initially from Burma and later drawn from the local Azande tribe. They also enlisted tamed pachyderms, known as '*moniteurs*', who were harnessed to the newly caught calves and guided them through the actions they needed to perform – another example of elephantine agency. Caldwell emphasised the humane treatment accorded the animals, who were never 'struck' but were coaxed into obedience with morsels of 'manioc, sweet potato or pineapple'.[136] Philipps corroborated this, remarking: 'All the animals were noticeably "well-groomed"'; 'What few sores they had were clean and freshly smeared with antiseptic ointment.' One elephant, Bama, had grown 'one metre 22 centimetres in 25 years' in captivity, showing that African elephants could thrive in captivity on a natural diet

[134] 'Taming African Elephants: School Work in the Congo', *The Times*, 9 April 1927.
[135] 'Farming with Elephants: An Experiment in Africa', *The Times*, 8 March 1928.
[136] 'Taming African Elephants: School Work in the Congo', *The Times*, 9 April 1927.

of 'leaves, herbage and vegetation'.[137] Leopold II's dream of domesticating the African elephant thus became reality in the early twentieth century, attracting the attention of officials from other colonies. For all its apparent success, however, the training programme remained largely a curiosity and an imperial propaganda tool, contributing very little to elephant conservation.[138]

Conclusion

The nineteenth century witnessed a dramatic rise in the demand for ivory. From a luxury product, ivory became a widely used necessity, incorporated into a broad range of manufactured items. London emerged as the centre of the world ivory trade, hosting four major ivory sales per year. Further quantities of ivory reached Europe through Liverpool, Antwerp and other continental ports. These imports passed through the workshops of multiple artisans and entered Victorian homes as piano keys, billiard balls, knife handles and even stiffeners for jelly.

Elephants paid a high price for this rocketing demand. Slow to reproduce, they were unable to quickly replace losses to the ivory trade and suffered a rapid decline in numbers. Colonial penetration and access to muzzle-loading rifles eradicated elephants progressively from areas where they had once been numerous, while the unregulated shooting of females added further to the decimation, pushing the species to the brink of extinction. As Povoleri observed in 1892:

When it is remembered that reproduction in the African elephant does not commence until the animal is over twenty years old, and that the period of gestation is three years, and, further, that the female elephant produces but one calf at a time, and that only at rare intervals, it will at once be seen that we are well within measurable distance of the date of the annihilation of the 'Ivory King' of the Dark Continent.[139]

To forestall the elephant's imminent extermination, a growing body of conservationists leapt into action, suggesting ways in which the animal might be saved. Campaigners alerted consumers to the ecological and humanitarian cost of ivory, emphasising not only its impact on the

---

[137] 'Farming with Elephants: An Experiment in Africa', *The Times*, 8 March 1928.

[138] Indeed, the collateral damage from catching the young elephants, coupled with their continuing high mortality, probably did more harm than good to local elephant populations. Continued depredations, moreover, ultimately led elephant herds to vacate the region around Api, forcing the domestication station's closure in 1932. See Violette Pouillard, *Histoire des Zoos par les Animaux: Imperialisme, Contrôle, Conservation* (Ceyzérieu: Champ Vallon, 2019), p. 282.

[139] 'The African Elephant', *The Standard*, 12 December 1892.

elephant but its close connections with the slave trade. Scientists searched for a sustainable alternative to elephant tusks, from mammoth ivory to celluloid. Hunting licences and bans on the export of underweight ivory were introduced to reduce the slaughter, while the creation of game sanctuaries provided safe havens for elephants and other endangered wildlife. None of these options were easy, and their ultimate success relied on shifting consumer attitudes, proper enforcement of game legislation and effective international cooperation. For some, however, the loss of such an iconic species was unthinkable, and urgent action essential. As Sclater stated: 'It would be a disgrace to our age to allow such a fine and noble animal as the African elephant to perish off the face of the earth, and I place it on the list of the most important African animals that should be protected.'[140]

One of the more novel means proposed to preserve the African elephant was to domesticate the species and make it useful to man. Although largely forgotten, this idea garnered a lot of attention in the 1870s and 1880s and surfaced again in subsequent decades, receiving support from colonial administrators, big game hunters and naturalists. Advocates of the scheme believed that it would simultaneously save the elephant and open up Africa to European exploration. Underlying these assumptions was the notion that only domesticated animals could survive in the modern world, and that a species valued for its body parts alone was doomed; as one contemporary put it, 'The African elephant has never successfully been tamed, and is therefore a failure as a source of energy. As a source of ivory, on the other hand, he has been but too great a success.'[141] The message, therefore, was stark: if elephants could not serve humans while they were alive, they would be exterminated.

[140] 'On the Best Mode of Preserving the Existence of the Larger Mammals of Africa for Future Ages', *Journal of the Society for the Preservation of the Wild Fauna of the Empire*, Vol. II, 1905, p. 47. Sclater's other targets for conservation were the rhinoceros, the giraffe, the eland and the zebra.
[141] Moore, *Ivory: Scourge of Africa*, p. 166.

4   Silk of the Andes

Third follows the Alpaca, or Paco … They are the fellows I should like
to see over here [Victoria]. They are something like a sheep, but a far
more graceful and elegant animal, with the large expressive eyes of the
deer. The fleece is beautifully soft and very long, almost like floss silk,
and it shines in the sun when of white or yellow colour, like silver
or gold.   'The Alpaca', *The Argus*, 28 March 1857

On 7 April 1848, a public dinner was held at the Exchange Rooms in
Bradford to celebrate 'her Majesty [Queen Victoria] having extended the
Royal patronage to the new fabric manufactured in that town from
Alpaca wool, being the first instance of the kind ever accorded to
Bradford by any British Sovereign'. The dinner was attended by 150 of
the great and good of Bradford and took place in a large hall, 'elegantly
decorated for the occasion'. Several 'specimens of Alpaca wool, of various hues' were 'arranged in festoons in different parts of the room', and
the seats of the two vice-chairmen were emblazoned with a copy of the
city's arms and 'a painted representation of the Alpaca'. Diners could
also inspect 'a bridle belonging to General O'Brien, the workmanship of
native Indians [from Peru], mounted with silver, the reins of which were
formed of the skin of the Alpaca'.

    After consuming a hearty meal, prepared by Mr Harrison of the Sun
Inn, those present listened to a series of enthusiastic toasts, each extolling
the value of the trade in alpaca wool to Bradford. The first speaker,
chairman G. Horsfall, congratulated the assembled company on 'the
general prospects of their trade' and informed them that her Majesty
had ordered 'a considerable number of piece goods … in Bradford for
one of the royal palaces'. He was followed by R. Pollock, who believed
that the Queen had done the city a favour by setting 'the nobility the
example of wearing their fabrics' – a trend others would likely emulate.
The last speaker to take the floor, William Walton, went beyond alpaca
fibre to embrace alpacas themselves. Responding to Pollock, Walton
observed that '[t]he manufacture of textures was carried out now to very
great perfection, and it behoved them to look beforehand for a sufficient

supply of wool; for if they went on progressing as of late, even the supplies of Peru would be insufficient to meet the demand'. 'If the English manufacture went on increasing so very rapidly, more wool would be required than silk, so that it was of importance, in order to extend the race of the animals, that they should be introduced into this country, and permitted to range upon the waste lands and upon the mountains of Scotland and Ireland.'[1] Not only should alpaca wool be imported into Britain, but living alpacas as well.

Thirteen years later, and half a world away, 200 members of the New South Wales elite congregated on a ranch on the outskirts of Sydney to honour another alpaca-related achievement – the successful introduction of 256 living alpacas to Australia. Arriving at the Sophienburg estate of Mr J. H. Atkinson, where four of the animals were being held, the party of visitors first met with the alpacas' keeper, Charles Ledger, who had choreographed their transportation to New South Wales. They conducted 'a careful inspection of the animals, which patiently submitted to the examination of their fleeces', quizzed Ledger about their 'habits and characteristics' and expressed their admiration for the 'health and condition' of the camelids – particularly a 'pure alpaca, almost perfectly white ... born at Liverpool the day the flock arrived there two years and a half ago'. The party then adjourned to a 'spacious tent' next to Atkinson's house, where they consumed a 'handsome luncheon' consisting of 'several joints of alpaca meat'.

As in Bradford, the repast culminated in a series of toasts, this time celebrating the prosperity alpacas would likely bring to New South Wales. One speaker, local politician Charles Cowper, asserted that '[t]hey had met upon one of the most interesting occasions which had ever been witnessed in New South Wales'. He pointed out that '[t]he alpacas, the meat of one of which they had just feasted on, had been purchased by the Government, and had been reared in the Colony', making them Australia's first native alpacas. Mr Atkinson spoke next, praising Ledger for his hard work and expressing his hope 'that he might live to see the alpacas become a very important element in the prosperity of New South Wales'. He was followed by R. Jones, who prophesied that alpaca wool would one day become 'one of the staple exports of the colony'. Ledger concluded the mutual back-patting, insisting that he 'looked forward to an occasion like this as a stimulus to exertion'. Today, they had seen 'fleeces such as Peru could never produce'; when he killed the alpaca for their luncheon, 'the fat obtained from feeding on

---

[1] 'Public Dinner at Bradford', *Leeds Mercury,* 12 April 1845.

the natural grasses of this country exceeded any he had ever seen in South America'. This proved conclusively that there was 'no country better suited than this for alpacas'. The toasts over, the party dispersed, impressed by the 'graceful beauty' of the alpacas, the 'length and fineness of their fleeces' and the flavoursome quality of their meat.

This chapter focuses on alpaca wool – a novel South American fibre that first entered British markets in the mid-1830s. Already widely used in Peru, alpaca wool first took off in Britain in 1836, after the woollen manufacturer Titus Salt discovered a bag of the fibre while walking through the docks in Liverpool. It was imported in increasing quantities from Peru in the 1840s and 1850s and used for the manufacture of shawls, cloth, ladies' dresses and umbrellas. Charting the alpaca's journey from the Andes to the outback, the chapter considers why contemporaries set so much store by the animal and how they went about appropriating it. I assess how increased access, new technologies and new markets made alpaca wool viable and profitable as a luxury fabric, and I examine the transcontinental relationships that brought alpaca fibre, and later living alpacas, to Britain and its colonies. I also highlight the important local and regional dynamics of alpaca naturalisation and the ways in which alpacas infiltrated wider discussions about identity, free trade and biopiracy.

Central to debates surrounding the alpaca was the concept of improvement. A key tenet of British agriculture throughout the eighteenth and nineteenth centuries, improvement could mean improving the land, by converting unused wilderness into valuable pasture, or improving plants and animals, through careful breeding to bring out the most desired qualities.[2] In the case of the alpaca it meant both. Walton hoped that the camelids would improve Britain and Australia by putting barren or infertile land to profitable use, while Ledger believed that acclimatisation would lead to the improvement of the alpaca itself, which would benefit from better nutrition and the breeding strategies already applied to British sheep and cattle.[3] Acclimatisers also sought to gain control over an exotic resource by extracting it from its native land and rearing it

---

[2] On the concept of improvement and its application to eighteenth- and nineteenth-century Scotland, see Fredrik Albritton Jonson, *Enlightenment's Frontier: The Scottish Highlands and the Origins of Environmentalism* (New Haven: Yale University Press, 2013).

[3] On the improvement of British livestock, see 'Barons of Beef' in Harriet Ritvo, *The Animal Estate: The English and Other Creatures in the Victorian Age* (Cambridge: Harvard University Press, 1987), pp. 45–81; Rebecca J. Woods, *The Herds Shot around the World: Native Breeds and the British Empire, 1800–1900* (Chapel Hill: University of North Carolina Press, 2017).

within the British Empire – a fate alpacas shared with other prized commodities such as tea, rubber and quinine.[4]

Focusing in turn on attempts to naturalise alpacas in Britain and in Australia, the chapter considers why British subjects in the United Kingdom, Australia and South America came to see acclimatisation as desirable and how they went about achieving it. I emphasise the scientific, commercial and social relationships that facilitated acclimatisation and show how British subjects in places as diverse as Bradford, Liverpool, Sydney and Arequipa assisted in, and hoped to benefit from, the naturalisation programme. I conclude with a look at the alpaca's wild relative, the vicuña, which was hunted to the point of extinction to secure its coveted fleece.

Highland Camelids

The alpaca (*Vicugna pacos*) originates from the Peruvian Andes. It is one of four South American members of the camel family, alongside the llama (*Lama glama*), the guanaco (*Lama guanicoe*) and the vicuña (*Vicugna vicugna*), and is now believed to be a domesticated variant of the latter (the llama is a domesticated guanaco).[5] First domesticated around 6,000 to 7,000 years ago, the alpaca was extensively farmed by the Incas, who used its fleece to weave colourful textiles.[6] In the nineteenth century, alpaca wool became one of Peru's main exports, together with sugar, cotton and guano,[7] rising in value from £122,000 per year in 1845–9 to a peak of £489,000 per year in 1870–4.[8]

---

[4] Important studies of 'economic botany' include Lucille Brockway, *Science and Colonial Expansion: The Role of the British Royal Botanic Gardens* (New Haven: Yale University Press, 2002); Richard Grove, *Green Imperialism: Colonial Expansion, Tropical Island Edens and the Origins of Environmentalism, 1600–1860* (Cambridge: Cambridge University Press, 1995); Richard Drayton, *Nature's Government: Science, Imperial Britain and the 'Improvement' of the World* (New Haven: Yale University Press, 2000); Emma Spary, *Utopia's Garden: French Natural History from Old Regime to Revolution* (Chicago: University of Chicago Press, 2000); Londa Schiebinger, *Plants and Empire: Colonial Bio-Prospecting in the Atlantic World* (Cambridge: Harvard University Press, 2004).

[5] Miranda Kadwell, Jane Wheeler et al., 'Genetic Analysis Reveals the Wild Ancestors of the Llama and the Alpaca', *Proceedings of the Royal Society, London* 268 (2001), pp. 2575–85.

[6] Luis Mengoni Goñalons, 'Camelids in Ancient Andean Societies: A Review of the Zooarcheological Evidence', *Quaternary International* 185 (2008), pp. 59–68.

[7] The guano boom was particularly profitable to Peru and lasted from c.1841 to 1870, the same period in which alpaca naturalisation was attempted. See 'The Guano Age' in Gregory T. Cushman, *Guano and the Opening of the Pacific World: A Global Ecological History* (Cambridge: Cambridge University Press, 2013), pp. 23–74.

[8] Peter Flindell-Klarén, *Peru: Society and Nationhood in the Andes* (New York: Oxford University Press, 2000), p. 166.

The first living alpacas appeared in Britain in the 1810s, during the Spanish American Wars of Independence, and were exhibited as exotic curiosities. Edward Cross exhibited a 'white and brown female alpaca' at his menagerie in London from 1810 to 1816.[9] Another alpaca, a white female, featured in Ducrow's circus, where it was 'taught to gambol, kneel and lay down at the word of command'; a third alpaca appeared in the gardens of the Zoological Society in the 1820s.[10] Noted for their silken fleeces and propensity to spit, alpacas were on show in regional zoological gardens by the 1830s and visited provincial towns regularly in travelling menageries.[11] Ducrow's alpaca even performed on stage in 1813, treading the boards in London at Covent Garden's Theatre Royal in 'The Popular Pantomime of the Red Dwarf' (Figure 4.1).[12]

While living alpacas initially functioned as sources of entertainment, alpaca wool was making important inroads into British commerce. Highly prized for its quality and softness, alpaca was mixed with silk and cotton to make a variety of garments, mostly fashionable, high-end products. Dresses, shawls and umbrellas were all made from alpaca, as were coat linings, cravats and the occasional Scottish tartan.[13] Alpaca clothing was also valued in 'tropical climates' such as India, Jamaica and Africa, where British expatriates were 'thankful to possess a black coat which, while it has the appearance of broad cloth, is not a fourth of its weight'.[14] Sourced from Indian communities in the southern highlands of Peru, alpaca wool was purchased by British merchants in the southern Peruvian city of Arequipa and shipped to Liverpool and (to a much lesser extent) London.[15] From there, most of the wool passed to Bradford, already a major player in the worsted trade and now Britain's primary site

[9] William Walton, *The Alpaca: Its Naturalisation in the British Isles Considered as a National Benefit, and as an Object of Immediate Utility to the Farmer and Manufacturer* (New York: Office of the New York Farmer and Mechanic, 1845), p. 15.
[10] William Walton, *A Memoir Addressed to Proprietors of Mountain and Other Waste Lands and Agriculturalists of the United Kingdom, on the Naturalisation of the Alpaca* (London: Smith, Elder and Co., 1841), p. 22; John Miller, *The Memoirs of General Miller* (New York: AMS Press, 1973), Vol. I, p. 234.
[11] In 1839, an 'Al pacha' was exhibited at Liverpool Zoological Gardens, along with two llamas. *List of Animals in the Liverpool Zoological Gardens* (Liverpool: Ross and Nightingale, 1839), p. 26.
[12] R. Norman, 'Grimaldi and the Alpaca in the Popular Pantomime of the RED DWARF', 11 January 1813, H. Beard Print Collection, Victoria and Albert Museum.
[13] William Danson, *Alpaca, the Original Peruvian Sheep before the Spaniards Invaded South America, for Naturalisation in Other Countries* (Liverpool: M. Rourke, 1852), p. 7.
[14] George Ledger, *The Alpaca: Its Introduction into Australia and the Probabilities of its Acclimatisation There. A Paper Read before the Society of Arts, London. Republished by the Acclimatisation Society of Victoria* (Melbourne: Mason and Firth, 1861), p. 5.
[15] Rory Miller, 'The Wool Trade in Southern Peru, 1850–1915', *Ibero-Amerikanisches Archiv* 8:3 (1982), pp. 297–311.

GRIMALDI & the ALPACA, in the Popular Pantomime of the RED DWARF, now Performing with Unbounded Applause, at the Theatre Royal Covent Garden.

Figure 4.1 An alpaca performs in 'the Popular Pantomime of the RED DWARF' with the famous clown Joseph Grimaldi. R. Norman, 11 January 1813. H. Beard Print Collection, Victoria and Albert Museum

for spinning and weaving alpaca fabric. A total of 34,652,701 lbs of wool were imported into Britain in the years 1834–59, rising from just 5,700 lbs in 1834 to a peak of 2,974,493 lbs in 1856 (Figure 4.2).[16]

Although alpaca wool was originally imported from Peru, plans arose in the 1840s to introduce the alpaca into Britain and thereby gain direct control over the trade. Advocates of the scheme anticipated multiple benefits. First, the naturalisation of the alpaca would ensure a sufficient and continuous supply of its wool, at a time when demand was outstripping supply; as Figure 4.2 shows, imports fluctuated considerably from year to year, and high internal demand within Peru limited the amount that could be exported.[17] Second, alpaca acclimatisation would permit British farmers to make better use of their land, since alpacas would be

---

[16] Ledger, *The Alpaca: Its Introduction into Australia*, p. 6.

[17] 'The Alpacas', *Sydney Morning Herald*, 21 August 1860. Wool merchant Charles Ledger estimated that 40,000,000 lb of alpaca wool was used annually for domestic consumption.

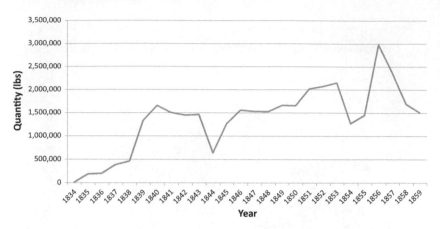

Figure 4.2 Alpaca wool imports from Peru to Britain, 1834–59. Data from George Bennett, *The Third Annual Report of the Acclimatisation Society of New South Wales* (Sydney: Joseph Cook, 1864), p. 6

able to survive in inhospitable mountain terrain unsuitable for cattle or sheep (of which 30 million acres were believed to exist in the United Kingdom).[18] Third, alpacas could be easily looked after, their thick coats keeping them warm in the harshest winter and making it unnecessary 'to smear [them] with tar and butter as the farmers are obliged to do with the flocks in Scotland'.[19] Fourth, the introduction of the alpaca would provide employment for labourers and artisans, maintaining individuals who would otherwise be 'dependent on their parishes for support'. And fifth, the alpaca might fulfil additional economic functions, its meat forming 'an excellent ingredient for a pie' and its 'strong and pliant' skin offering a suitable material for bookbinding.[20] The naturalisation of the alpaca was presented as an important boost to the British textiles industry, a boon to agriculture and an antidote to various social problems. Enthusiasts identified the Scottish Highlands, Shetland, Wales, the Cheviot Hills, Dartmoor and the mountains of Kerry and Wicklow as the most promising regions for alpaca introduction, the terrain and climate of these locations most closely resembling those of the Andes.[21]

During the early 1840s, these ideas began to gain momentum, attracting the attention of some important local and national institutions. In 1840, the Tenth Annual Meeting of the British Association for the Advancement of Science featured a lecture by William Danson on the

[18] Walton, *The Alpaca*, p. 12.   [19] Walton, *A Memoir*, p. 19.
[20] Walton, *The Alpaca*, pp. 11, 7–8.   [21] Ibid., p. 13; Danson, *Alpaca*, p. 19.

alpaca and its wool, during which samples of the wool and 'living specimens' were exhibited.[22] In 1841, William Walton published an important chapter on the subject in the *Polytechnic Journal*, which he later
revised and extended into a short book, while in 1844, the Highland
and Agricultural Society awarded 'a premium of Five Sovereigns and an
honorary silver medal' to 'Alexander Gartshore Stirling of Craigbarnet'
for 'the best pair [of alpacas] born in the kingdom'.[23] By the mid-1840s,
a number of improving landowners were rearing alpacas on their estates,
some from mere curiosity but others with the intention of establishing a
new line of business. Thomas Stevenson of Oban received several ship-
ments of alpacas from his son in Peru. Joseph Hegan, Charles Tayleure
and the Earl of Derby farmed small flocks of alpacas in Cheshire, while
Robert Bell of Listowel, Kerry, introduced alpacas to Ireland. The
Queen's consort, Prince Albert, was another high-profile alpaca fancier,
keeping two of the animals on the royal estate at Windsor.[24]

The arrival of increasing numbers of alpacas boosted confidence in the
prospects of naturalisation and raised hopes that a valuable new species
might be added to Britain's native farmstock. It also led to claims that
British-reared alpacas produced more and better wool than their Peruvian
counterparts. Menagerist Edward Cross, for instance, claimed that 'he
noticed a visible improvement in the fleece of his alpaca, which he had
shorn more than once, although the animal was kept under restraint and
subjected to an unsuitable regimen, besides breathing the impure air of a
populous town'.[25] Robert Bell contended, similarly, that the wool of his
Irish alpacas was 'very much finer than any alpaca wool I have yet seen
imported into England', prompting alpaca enthusiast Walton to suggest
that careful British husbandry would facilitate an improvement of the
species, the 'dirty and scurfy state' of imported wool being due to 'the
deciduous habits of the Indian'.[26] Views like this supported a broader
narrative of agricultural improvement, which dismissed non-European
farming techniques as primitive and anticipated major advances now that
exotic species were in British hands. Such ideas were very much in keeping
with contemporary livestock practices, which sought to raise the quality of
animals through selective breeding and diet.[27]

---

[22] 'British Association', *The Morning Chronicle*, 30 September 1840.
[23] 'Highland and Agricultural Society's Exhibition at Glasgow', *Caledonian Mercury*,
12 August 1844.
[24] Walton, *The Alpaca*, pp. 14–19.   [25] Walton, *A Memoir*, p. 23.
[26] Walton, *The Alpaca*, p. 19; Walton, *A Memoir*, p. 23.
[27] On the tradition of animal husbandry and the fashion for fat cattle in nineteenth-century
Britain, see 'Barons of Beef' in Ritvo, *The Animal Estate*, pp. 45–81; Woods, *The Herds
Shot around the World*.

Despite such upbeat testimonies, however, the overall picture was much less rosy, and a large number of alpacas succumbed to mismanagement and accident. Of the dozen alpacas shipped to Thomas Stevenson by his son, only two, a male and a female, survived the voyage. Another flock owned by Charles Tayleure was administered 'too strong medicine' by a shepherd, killing 'the greater part of them', while one of Bell's alpacas perished from eating a poisonous weed.[28] Most disastrously, what should have been the largest single importation of alpacas into Britain ended in tragedy in 1842 when the captain of the *Sir Charles Napier* stowed 274 of the animals above a cargo of guano, the 'effluvia' from the manure suffocating all but four of them.[29] Negligent servants and ignorant sailors were blamed for these mistakes, reflecting ingrained class prejudices.[30] Contemporaries also accused indigenous Peruvians of cheating British merchants by selling them old or diseased stock; according to General John O'Brien, native Peruvians were reluctant to part with their animals, so the ones they offered to merchants were 'almost always picked out of inferior, if not refuse stock, and, by an intelligent person, would scarcely have been deemed worth the shipping risk'.[31] Such setbacks highlighted the practical difficulties of animal acclimatisation and pointed to the need for careful supervision and local knowledge. They did not, however, extinguish the hope that, with better planning and faster transportation, alpacas might be successfully introduced to the moors and highlands.

From the Andes to the Outback

While early efforts to acclimatise the alpaca focused on Britain, attention shifted in the 1850s to British possessions overseas. This was in part a response to the limited success of programmes such as Walton's to rear the alpaca in Britain itself. It also reflected the growing importance of the colonies in this period, and an increasing desire to ensure their economic viability. Both Cape Colony and the Indian province of Sindh were considered as suitable sites for the experiment, but it was in Australia that the project came to fruition.[32] Already a major exporter of wool, thanks to the successful introduction of the merino sheep, Australia was seen as a prime location for the rearing of alpacas, which could inhabit

[28] Walton, *The Alpaca*, p. 16, 17, 21.   [29] Ibid., p. 15.
[30] Walton remarked that 'the treatment which they [alpacas] have experienced from some owners has been cruel, if not murderous in the extreme ... for all depends upon the whim of a servant – often with a wet and filthy bed under them, and not infrequently eating the offals of a green-grocer's shop'. Ibid., pp. 24–5.
[31] Ibid., p. 27.   [32] 'The Supply of Alpaca Wool', *Leeds Mercury*, 22 March 1859.

land unsuitable for other livestock.[33] Colonial acclimatisation societies listed the alpaca as one of their most promising targets, anticipating significant economic benefits for their respective colonies. The *South Australian Register* confidently asserted that 'hundreds of thousands of alpacas might roam among the mountain chains, or cover the elevated plains of Australia, neither trenching upon the sheep nor depasturing upon cultivated lands'.[34] Another newspaper, the *Sydney Morning Herald*, predicted that alpacas would 'prove more profitable than the gold mines of Australia, and certainly more durable'.[35]

Australian alpaca enthusiasts faced an additional hurdle, however, for in 1845 the Peruvian Government had enacted a decree banning the export of alpacas from the country.[36] The ban was a response to pressure from indigenous herders who, following the deaths of the alpacas aboard the *Sir Charles Napier*, had 'clamorously petitioned' congress to prevent further exports, holding meetings in the regions of Puno, Arequipa and Cusco (the prime alpaca farming areas) and refusing to sell alpaca wool until action had been taken. Reacting to their concerns, the government put in place legislation making it illegal to transport camelids 'within a distance of ... 120 miles of the sea' and threatening any who violated the ban with the 'confiscation of the animals ... [and] ten years' hard labour on the Guano Islands'.[37] Any subsequent attempts to extract alpacas from Peru would therefore require either subterfuge or a special dispensation, converting alpaca export into a form of biopiracy.

While the export ban posed a major obstacle to alpaca acclimatisation, Australians remained interested in the project, and two states, Victoria and New South Wales, managed to obtain significant numbers of the animals. In the first of these, Victoria, the local acclimatisation society took the initiative. Led by Thomas Embling, a member of the colonial assembly, and Edward Wilson, a journalist, the society promoted the introduction of a range of new and useful animals to Australia, from camels to songbirds. Convinced that the alpaca was a worthy candidate for naturalisation, Embling used his position to lobby the colonial government for financial support, claiming that alpaca farming was 'four times as profitable and at least only half the trouble of sheep farming'.[38]

---

[33] On the introduction of merino sheep to Australia, see 'Much Ado about Mutton' in Woods, *The Herds Shot around the World*, pp. 52–77.
[34] 'The Alpaca in Australia', *South Australian Register*, 10 February 1859.
[35] 'The Alpacas', *Sydney Morning Herald*, 21 August 1860.
[36] 'Cámara de Senadores, 9 de Agosto', *El Comercio*, 13 August 1845.
[37] 'The Peruvian Government and the Alpacas', *Sydney Morning Herald*, 10 September 1860.
[38] 'The Alpaca and the Camel', *Melbourne Argus*, 30 March 1858.

Wilson, meanwhile, presented papers on the alpaca at the Philosophical Institute of Victoria and expatiated at length on the potential benefits of alpaca acclimatisation in the pages of the *Melbourne Argus*. In one article, he predicted that 'within the lifetime of our present adults ... the milder portions of our continent [might] pour forth an unlimited supply of this comparatively new and most valuable material [alpaca wool]'.[39]

Early attempts to import alpacas directly to Victoria proved abortive, partly because of local scepticism and partly because of the Peruvian export ban. In July 1858, however, when Wilson, then resident in London, heard that a flock of alpacas was up for auction in the British capital, he organised a campaign to raise funds for their purchase, requesting donations from fellow expatriate Australians and British manufacturers and using the letters pages of *The Times* to publicise his cause.[40] By November, sufficient money had been collected to buy ten of the alpacas and send them to Melbourne, where they were temporarily housed in the city's Botanical Gardens.[41] The Bradford industrialist Titus Salt donated a further two alpacas from his private flock, increasing the total number of animals to twelve and providing two pure-bred males for breeding purposes.[42]

At the same time as Wilson was sourcing alpacas from Britain, another budding entrepreneur, Charles Ledger, was nearing the end of a decade-long project to introduce the species into New South Wales. A British merchant based in Tacna, Peru, Ledger had become interested in the alpaca business while employed by the firm Naylor's, which entrusted him with 'the purchase of alpaca's and sheep's wool'. His job consisted of 'receiving from the Indians the different lots as they arrived from the interior ... sorting the qualities and colours previous to packing ... and finally shipping them, principally for account of Messrs. Christopher and James Rawdon, of Liverpool'.[43] Knowing how popular alpaca wool was in Europe, Ledger conceived the idea of introducing the Peruvian animal to Britain or one of its colonies and visited Sydney in 1852 to assess the feasibility of the scheme. The trip convinced him that 'the country was most admirably adapted for the alpaca', and he proceeded to assemble a large flock of alpacas and llamas at his estate at Chulluncayani near

---

[39] Linden Gillbank, 'A Paradox of Purposes: Acclimatization Origins of the Melbourne Zoo' in R. J. Hoage and William A. Deiss (eds), *New Worlds, New Animals: From Menagerie to Zoological Park in the Nineteenth Century* (Baltimore: Johns Hopkins University Press, 1996), pp. 76–9; *Melbourne Argus*, 13 October 1858.

[40] 'The Alpacas', *The Times*, 17 July 1858.

[41] 'The Alpacas for Victoria', *Daily News*, 8 November 1858.

[42] 'Town and Country Talk', *Lloyd's Weekly Newspaper*, 22 August 1858.

[43] 'Introduction of the Alpaca into Australia', *Bradford Observer*, 29 September 1859.

Peru's southern border, smuggling the animals across the Andes into the Argentine Confederation to avoid the Peruvian Government's camelid export ban. After several months in Laguna Blanca, accustoming the alpacas to their shipboard 'rations of dry alfalfa' and bran, Ledger recrossed the Andes in perilous conditions and shipped them to Australia from the Chilean port of Caldera. Of the 322 animals stowed aboard the *Salvadora* in July 1858, 256 survived the voyage, disembarking in Sydney four months later.[44]

The story of Ledger's quest to naturalise the alpaca reads like a classic Victorian adventure, replete with heroism, tragedy and adversity. At one point, 200 of his flock perished from drinking water from a lake 'infested with leeches'.[45] On another occasion he lost half of his animals in a violent storm in the Andes; on a third, 200 alpacas died due to 'the negligence of one of the Indians'.[46] As well as enduring 'the hardships, personal danger and exhaustion suffered from cold, fatigue and privation' in the sierra, Ledger was repeatedly hounded by the Peruvian and Bolivian authorities, who arrested him on two occasions and threatened to destroy his flock. With the courage and guile typical of the plucky Victorian entrepreneur, he managed, on both occasions, to outwit his captors, the first time by 'exercising his medical skills in the cure of the wife of the detaining prefect' and the second by slipping a dose of laudanum into his gaoler's 'grog'.[47] These hardships are graphically illustrated in a series of sketches by Santiago Savage, which depict Ledger and his men crossing ravines, battling through blizzards, outwitting Bolivian officials and falling headfirst into the snow (Figure 4.3).

After such an onerous journey, Ledger doubtless expected a warm reception in New South Wales. In the event, however, the farmers who had six years earlier expressed interest in the Briton's scheme now proved cautious about the experiment, declining to buy the alpacas at auction. The colonial government was forced to step in and purchase the animals, arranging pasture for them at Sophienburg, Arthursleigh and Wingello and paying Ledger an annual salary of £300 to superintend their continued care.[48] Disappointed, Ledger nonetheless persisted with the experiment and set to work interbreeding his animals, hoping, by so doing, to obtain a superior strain of wool (he had been forced to supplement his original flock of alpacas with llamas, and intended, over several

[44] Ledger, *The Alpaca: Its Introduction into Australia*, pp. 11–13.
[45] 'Interesting Narrative Respecting Alpacas in Australia', *The Era*, 20 February 1859.
[46] 'Mr Charles Ledger and His Alpaca Contract with New South Wales', *The Era*, 25 September 1859.
[47] 'Introduction of the Alpaca into Australia', *Bradford Observer*, 29 September 1859.
[48] 'The Alpacas and Mr C. Ledger', *The Era*, 12 February 1860.

Figure 4.3 'Passage of Cordillera into Chile', from *Annotated Watercolour Sketches by Santiago Savage, 1857–1858, Being a Record of Charles Ledger's Journeys in Peru and Chile*. State Library of New South Wales MLMSS 630/1

generations, to breed them back to pure alpaca). By 1861, the flock had been shorn several times and had increased to 368 specimens – 112 more than had arrived in the *Salvadora*.[49] To gain further recognition, Ledger decided to cull seven non-breeding animals and send them to the 1862 International Exhibition along with samples of their wool – a move opposed by several 'ladies', who protested against 'the cruelty, absurdity and profitless folly of murdering seven of these beautiful creatures'.[50]

That, however, proved to be the high point for alpaca rearing in New South Wales. From a peak of 411 in July 1862, the size of Ledger's flock began to decline, and with it interest in the naturalisation scheme.[51] A severe drought in 1862–3 caused significant mortality in the flock, particularly among the nursing females, while an outbreak of mange further diminished their numbers. Ledger also confessed to breeding from the females at too young an age, thereby weakening their constitutions.[52] With rumblings of discontent at the mounting cost of the venture, the state government dismissed Ledger as alpaca superintendent and acceded to calls to sell off the remaining animals to private buyers. An attempted auction in 1864 failed to yield the prices hoped for, but in

[49] 'The Alpacas', *Sydney Morning Herald*, 26 August 1861.
[50] 'To the Editor of the Herald', *Sydney Morning Herald*, 26 August 1861.
[51] 'To the Editor of the Herald', *Sydney Morning Herald*, 8 March 1864.
[52] 'The Management of the Alpacas', *Sydney Morning Herald*, 4 March 1864.

June 1866 the surviving 111 alpacas were sold without reserve and dispersed across the state, putting an end to Ledger's ambitious breeding programme.[53] Ledger, meanwhile, returned to Peru bankrupt and embittered, finding himself 'at 48 years of age without one shilling of my own, having lost all I had in the realisation of an enterprise that I fondly hoped would have conferred great benefits on a thriving colony of my own country, and a just recompense for my capital and labour'.[54] As in Britain, alpaca acclimatisation had fallen short of expectations, an outcome blamed, by turns, on inappropriate terrain, extreme weather conditions, Ledger's personal mismanagement (which he vehemently denied) and excessive (or, according to some, insufficient) government interference.[55]

Circuits of Exchange

Recent work in the history of science has emphasised the communal and collaborative nature of natural knowledge and the crucial role played by go-betweens or intermediaries.[56] Though often portrayed as the product of individual genius and heroism, the collation, analysis and transmission of information and objects often relied on a complex set of relationships and networks that facilitated their transfer. Friendship, sociability and personal contacts were vital to this process of exchange, the latter often created and sustained through letters of introduction and written correspondence.[57] Context and place were also important in explaining how specific forms of knowledge arose and how they were communicated to other localities.[58]

---

[53] 'The Alpacas', *Sydney Morning Herald*, 23 June 1866.

[54] 'Mr Ledger's Alpacas', *Sydney Morning Herald*, 14 April 1875.

[55] 'The Alpacas', *Sydney Morning Herald*, 29 February 1864.

[56] Important contributions in this area include Simon Schaffer, Lissa Roberts, Kapil Raj and James Delbourgo (eds), *The Brokered World: Go-Betweens and Global Intelligence, 1770–1820* (Sagamore Beach: Watson Publishing International, 2009); Jane Camerini, 'Wallace in the Field', *Osiris* [2nd Series] 11 (1996), pp. 44–65; Fa-ti Fan, 'Victorian Naturalists in China: Science and Informal Empire', *British Journal for the History of Science* 36:1 (2003), pp. 1–26; Sujit Sivasundaram, 'Trading Knowledge: The East India Company's Elephants in India and Britain', *Historical Journal* 48:1 (2005), pp. 27–63.

[57] On the importance of friendships and correspondence in science, see Patience Schell, *The Sociable Sciences: Darwin and His Contemporaries in Chile* (Basingstoke: Palgrave Macmillan, 2013).

[58] See, for instance, David Livingstone, *Putting Science in Its Place: Geographies of Scientific Knowledge* (Chicago: University of Chicago Press, 2003); Kapil Raj, *Relocating Modern Science: Circulation and the Construction of Knowledge in South Asia and Europe, 1650–1900* (Basingstoke: Palgrave Macmillan, 2007).

The alpaca naturalisation scheme clearly reflects the social and spatial dimensions of nineteenth-century science and illustrates some of the diplomatic, commercial and scholarly networks that connected Britain and Latin America. Although the newly independent states of South America were not part of Britain's formal empire, they were closely integrated into British trade routes and were soon staffed with a regiment of British consular officials, many of whom furthered the study of natural history by shipping native plants and animals to British institutions.[59] Former soldiers who had gone to Spanish America to fight in the Wars of Independence often remained in the region for some time, while British merchants travelled to the continent to sell manufactured goods and purchase raw materials, establishing important connections with local people and sometimes marrying into creole families.[60] A host of naturalists also descended on South America in the years after independence, some taking up positions at local museums and universities, others conducting research on behalf of scientific institutions back in Europe. Individuals from all of these backgrounds played a role in the alpaca naturalisation project – some willingly, others less so – offering up their zoological, social and technological expertise on both sides of the Atlantic (and, indeed, the Pacific). The drive to naturalise alpacas in Britain and its colonies thus elucidates the complexity of British connections with South America in the post-independence period, and the varieties of knowledge necessary to transport a valuable zoological commodity across the globe.

Among the groups who assisted in alpaca acclimatisation, one of the most important was indigenous Peruvians, whose long contact with the animals made them experts on their needs and behaviour. Although native shepherds were often accused of failing to exploit the full potential of the alpaca and of duping gullible foreigners into buying old or diseased beasts, their expertise in farming alpacas was grudgingly acknowledged. William Walton advocated enlisting Peruvian keepers to bring alpacas to Britain, on the grounds that they knew best how to handle them – and would themselves be 'improved' by absorption into British culture (a

---

[59] Henry Southern, HM Minister at Rio, presented the ZSL with a tapir in 1853. W. D. Christie, HM Minister to the Argentine Confederation, presented 'a pair of pumas' in 1857. See Report of the Council and Auditors of the Zoological Gardens of London (London: Taylor and Francis, 1853), p. 18; Report of the Council and Auditors of the Zoological Gardens of London (London: Taylor and Francis, 1857), p. 19.

[60] For a detailed study of British and Irish ex-soldiers in Colombia, Venezuela and Ecuador, see Matthew Brown, Adventuring through Spanish Colonies: Simón Bolívar, Foreign Mercenaries and the Birth of New Nations (Liverpool: Liverpool University Press, 2006).

view that reflected the racial assumptions of the era).[61] Charles Ledger likewise relied on the expertise of Native Americans, hiring twelve Bolivian shepherds to care for his flock en route to Australia. The men remained with the alpacas until 1860, when they returned to South America, and they were assigned the task of shearing the Australian flock for the first time in 1859 due to their intimate connection with the animals. As the *Sydney Morning Herald* explained:

It was thought desirable that these men, though rather clumsy manipulators, should do the work in preference to regular [sheep] shearers, as their long familiarity with the animals has imparted to them a degree of docility and quietness while in the hands of their own keepers which they would not preserve in the presence of strangers; in addition to which the long connection of these men with the flock inclines them to use the shears more carefully, though less rapidly, than ordinary shearers.[62]

Peruvian Indians thus proved essential to alpaca relocation, although their knowledge was not always freely given or treated with respect. Ledger, for example, complained that his indigenous helpers would not shear pregnant female alpacas in the belief that they would miscarry – a view he attributed to superstition. In 1856, meanwhile, indigenous Bolivians murdered a young Frenchman named Monsieur Ibarnégaray when he attempted to purchase some alpacas for export – a clear sign that many local people actively opposed the removal of their native livestock.[63]

If Native American expertise helped with the day-to-day management of the alpacas, the collaboration of sailors in the British navy and merchant marine also proved critical in their successful relocation. This was recognised by proponents of acclimatisation, who offered detailed advice on how and where to ship alpacas with the most profitable results. Walton, for instance, conscious that a shorter crossing would improve the odds of keeping the alpacas alive, advised transporting the animals to Panama on a new line about to be set up by the Pacific Steam Navigation Company; this would allow the animals to recuperate on the isthmus for six weeks before being shipped to England from the Caribbean port of Chagres.[64] Another alpaca enthusiast, in this case from Tasmania, suggested that 'it might be possible for some of the vessels trading to San Francisco to procure a few [alpacas], by means of vessels from Chili [sic],

[61] Walton, *The Alpaca*, p. 28.
[62] 'Shearing the Alpacas', *Sydney Morning Herald*, 7 November 1859.
[63] 'The Alpacas', *Sydney Morning Herald*, 21 August 1860; 'Peruvian Export of Alpaca', *Sydney Morning Herald*, 22 September 1860.
[64] Walton, *The Alpaca*, p. 6.

there being a considerable communication between San Francisco and Valparaiso'.[65] The extension of British shipping to the Pacific in the wake of Spanish American independence and the increasing sophistication of steam-powered vessels in the mid-nineteenth century cut down journey times across the world's oceans, making it more likely that animals would survive the long crossing to new lands. Some sailors also went the extra mile to look after their live cargo, further improving survival rates. In 1841, for example, Captain Bottomley of the *Highlander* took great care to convey nine alpacas from Valparaíso to Liverpool, feeding them on lucerne and even 'washing the mouths of the animals before eating and drinking' to keep them healthy.[66]

While indigenous Peruvians and British mariners played an important role in caring for transient alpacas, the impetus behind the acclimatisation schemes came largely from British subjects based in Spanish America. Commercial links were particularly important. Charles Ledger, as we have seen, was introduced to the alpaca through his job as a wool merchant, working first as a clerk for the house of Naylor's and later operating his own business. He formed close relationships with Indian farmers, and in this way became aware of the potential value of alpaca wool. He also formed connections with fellow merchants, two of whom, Messrs Waddington and Templeton and Co., bankrolled his acclimatisation scheme.[67] Living in Peru for twenty-four years, Ledger developed an intimate knowledge of how alpacas were farmed and how their fleece was processed by the Indians. He further ensconced himself in Peruvian society by 'marrying into an influential family in Tacna', forging important links with the local community.[68] These connections – part social, part economic – enabled Ledger to carry out his alpaca-smuggling scheme.

The contribution of another alpaca advocate, General John O'Brien, illustrates even more clearly the role of itinerant Britons in appropriating the alpaca. An Irishman by birth, from Baltinglass, County Wicklow, O'Brien travelled to South America in 1812 to open a merchant house in Buenos Aires, but ended up fighting at the battles of Chacabuco, Cancha Rayada and Maipú as General José de San Martín's aide de camp. After the conflict concluded, O'Brien settled first in Peru, where he engaged in mining ventures, and later in Argentina, where he worked

---

[65] 'The Alpaca', *Maitland Mercury*, 20 March 1850.
[66] 'The Alpaca', *Liverpool Mercury*, 1 October 1841.
[67] Ledger, *The Alpaca: Its Introduction into Australia*, p. 12.
[68] 'Introduction of the Alpaca into Australia', *Bradford Observer*, 29 September 1859.

to encourage Irish immigration.[69] Upon his return to Britain some years later, he became increasingly enthusiastic about the prospect of naturalising the alpaca in his native Ireland and initiated a correspondence with British alpaca enthusiasts Danson and Walton. Possessing not only direct personal knowledge of the alpaca but also many useful contacts in Peru, O'Brien assisted the alpaca acclimatisation drive by writing a series of letters to his Peruvian friends, urging them to cooperate with Danson's naturalisation scheme. One of O'Brien's correspondents, Peter Murphy, HM Consul at Arica, was entreated to offer his 'aid and assistance' to 'whatever person [Danson] may send out to this country' to collect alpacas, with a view to conferring 'a national gift' upon 'your own dear mountains of Wicklow'. Another correspondent, Michael Crawley, prefect of the department of Lampa, was requested, for the sake of 'old friendship', to help with selecting good-quality alpacas and 'conducting them to the coast'; a third, Don Mariano Toledo, was asked, in dubious Spanish, to give Danson's agent *'todo servicio en procurer los mejores animales'* and to do *'cuento puede en facilitar sus proyectos'*. The content of these letters, with its emphasis on friendship and service to one's country, highlights the importance of social and professional networks in the study and exchange of zoological specimens and the value of personal contacts in furthering scientific and economic plans.[70]

Back in Britain, the successful rearing and exploitation of the alpaca depended on input from three other communities of 'experts': naturalists, engineers and zoo professionals. The first of these, naturalists, conducted observations on alpacas and studied their anatomy and physiology. The famous comparative anatomist Richard Owen delivered a lecture on the 'peculiar properties' of alpaca wool at the Society of Arts in 1851, in which he noted its 'glossy ... silky' quality.[71] Another scientist, Alfred Higginson of the Natural History Society of Liverpool, dissected two alpacas, observing that 'the water cells' in their stomachs 'were either empty or partly filled with masticated food in a semi-fluid state'.[72] Higginson subsequently dissected three more alpacas belonging to British owners and concluded that all 'had been over-fed', highlighting the need for a less rich diet.[73] Anatomical findings thus helped acclimatisers better understand the needs of the animal and learn useful lessons from previous mistakes.

---

[69] Clements Markham, *Travels in Peru while Superintending the Collection of Cinchona Plants and Seeds in South America, and Their Introduction into India* (London: John Murray, 1862), p. 527.

[70] Danson, *Alpaca*, pp. 13–14.   [71] Ibid., p. 20.   [72] Walton, *A Memoir*, p. 20.

[73] Walton, *The Alpaca*, p. 25.

While zoologists debated the pros and cons of transporting and farming the alpaca, the use of alpaca wool for textile production was made possible by a series of technological developments, most of them the work of enterprising Yorkshire artisans. Since it was finer and longer than sheep's wool, the fleece of the alpaca could not be spun using traditional machinery but required specially adapted spinning apparatus. Initially, no such apparatus was available. In the 1830s, however, Benjamin Outram, 'a scientific manufacturer of Greetland near Halifax', designed a machine that could spin alpaca wool economically and effectively, giving rise to a new industry.[74] Titus Salt installed Outram's machinery in his worsted factories in Bradford in 1836 and was soon producing alpaca goods on an industrial scale. Further techno-logical breakthroughs in subsequent years increased the quality and speed of the output, making the business yet more profitable; in 1847, for instance, Edward Waud 'of Bradford, Yorkshire, spinner', received a patent for 'certain improvements in the construction of machinery for preparing and spinning alpaca, mohair, wool, flax and other fibrous materials'.[75] The technical expertise of British artisans, who had perhaps never seen a living alpaca, thus played a crucial role in stimulating the demand for their wool, and, in time, the desire for their naturalisation.

Finally, in highlighting the communities that facilitated alpaca intro-duction, we ought to mention one last group: menagerists and zoo directors. When seeking advice on alpaca acclimatisation, alpaca advo-cates frequently invoked the practical knowledge of these individuals, whose direct experience of rearing alpacas in Britain made them the next best thing to native Peruvians when it came to learning about the animals' diet and habits. Danson, for instance, collaborated with Thomas Atkins of Liverpool Zoological Gardens, compiling a circular to be issued to ships' captains 'for their guide in treatment of the animals during the voyage'.[76] The Australian Edward Wilson consulted John Thompson, superintendent of London Zoo, on llama–alpaca crosses, while Walton quizzed Edward Cross, director of the menagerie at Exeter 'Change on the Strand in London, about the alpaca in his posses-sion, citing the latter's testimony that his alpaca subsisted on 'dry food, such as hay, beans and oats', that 'it never drank anything the whole time I had it', and that he cured it of a skin complaint (the 'itch') by 'rubbing a little mercurial ointment on the spine'.[77] Although Cross was a showman

[74] Ibid., p. 11.   [75] 'List of Patents', Liverpool Mercury, 7 December 1847.
[76] Danson, Alpaca, p. 19.
[77] 'The Llama and Alpaca', Melbourne Argus, 13 October 1858; Walton, The Alpaca, p. 15; Walton, A Memoir, pp. 21–3.

rather than a professional naturalist, Walton seems to have valued his opinions highly and was happy to rely on his expertise; '[F]ew men,' he remarked, 'could be found more intelligent or more observant than Mr Cross.'[78]

Even travelling menageries, usually perceived as mere sources of entertainment, could assist the alpaca acclimatisation project. When show-woman Mrs Wombwell exhibited a 'jet black' alpaca in Liverpool in 1853, the animal attracted considerable attention from local farmers, generating a series of letters to the *Liverpool Mercury*. The *Mercury* itself initiated the conversation, observing that the animal was 'entitled to much attention, not merely from motives of curiosity, but from the immense mercantile advantages which would accrue to the agriculturalist as well as the manufacturer by its naturalisation in this country'.[79] In the following weeks, letters appeared from readers concurring with this view. One correspondent, G.G., who had seen the menagerie specimen, noted that the alpaca's fleece typically weighed 'from 12 to 14 lbs', although 'on the one before us at Wombwell's (which, however, is singularly fine), we should imagine upwards of 20 lbs'. He went on to express his hope that the *Mercury*'s article would 'induce the owners or holders of hilly or mountainous districts at once to consider' the 'practicability' of domesticating the alpaca there.[80] A second correspondent, T.F., broadly agreed with these sentiments, requesting information on 'where the breed may be obtained at a price commensurate with the great risk which must be run before any return can be expected', and where he might find an 'able treatise' explaining how to avoid 'accidents arising from ignorance of the habits of the animal'.[81] While sailors, merchants and naturalists thus brought the first alpacas to British shores, it was often travelling entertainers who introduced them to people in the provinces and who, through long experience, understood best how to manage them.

Alpacas and Empire

If alpaca naturalisation illustrated the complex networks of exchange and expertise at work within and beyond the British Empire, it also exposed certain tensions in these relationships: alpacas, it turned out, meant

---

[78] Walton, *A Memoir*, p. 23. For an interesting discussion of the different forms of knowledge in play at the zoo, see Oliver Hochadel, 'Watching Exotic Animals Next Door: "Scientific" Observations at the Zoo (ca. 1870–1910)', *Science in Context* 24:2 (2011), pp. 183–214.

[79] 'Latest News', *Liverpool Mercury*, 4 February 1853.

[80] 'The Alpaca', *Liverpool Mercury*, 11 February 1853.

[81] 'The Alpaca', *Liverpool Mercury*, 22 March 1853.

different things to different people. The British, the Australians and the Peruvians all invested these valuable animals with their own specific hopes and expectations, anticipating distinct commercial benefits. Even within Britain and its antipodean possessions there were varying regional and local interests at stake. These different aspirations were not necessarily mutually exclusive; what was good for farmers in New South Wales might also be beneficial to textile workers in Bradford. They did, nonetheless, betray different priorities and emphases, and, in the case of the Peruvians and Bolivians, opposing views as to where alpacas should be farmed.

Viewed from a British perspective, alpaca farming was seen as a way to promote the nation's commerce and revitalise its agriculture by transforming barren and uncultivated regions into useful pastures. Supporters of acclimatisation believed that the introduction of alpacas would permit more effective use of Britain's farmland, already exploited as far as possible by native species, and would enable the British to produce a fine and delicate fabric capable of competing with the best French silks. They hoped it would reduce Britain's dependence on Peruvian imports of alpaca wool, which contemporaries considered insufficient to meet growing demand, and they even suggested that it would facilitate British expansion overseas by providing a material suitable for clothing colonists in the tropics. To this extent, the acclimatisation project was a tangible reflection of Britain's naval dominance and imperial reach, a demonstration of its technological and agricultural expertise and a testament to its commercial penetration of post-independence South America. By the same token, however, the desire to naturalise the alpaca was also an expression of British anxieties and economic vulnerability, at a time when people were worrying about foreign competition (Germany and the USA were beginning to industrialise) and the prospect of civil war in the USA threatened cotton supplies.[82] Some Britons feared that, if they did not acclimatise the alpaca, European rivals would take the initiative, threatening the United Kingdom's dominance in the woollen trade and putting manufacturers out of work.[83] As Walton expressed it:

---

[82] William Haines, chairing the meeting at which George Ledger advocated alpaca introduction into Australia, emphasised 'the importance of promoting increased production of wool when our supply of cotton might be in danger'. See Ledger, *The Alpaca: Its Introduction into Australia*, p. 24. On the rise of cotton and its close links with imperial expansion, see Sven Beckert, *Empire of Cotton: A New History of Global Capitalism* (London: Penguin, 2014).

[83] The French were, in fact, taking steps to acclimatise the alpaca, which they hoped would thrive in the mountains of the Pyrenees, the Alps and the Vosges, as well as in their new colony of Algeria. See M. E. Deville, *Considérations sur les Avantages de la Naturalisation en France de l'Alpaca* (Paris: Imprimerie de L. Martinet, 1851), p. 15; Michael Osborne,

In the stirring age in which we live, nations, like individuals, must compete with those who seek to outstrip them; and … in order to keep his ground in the foreign market, the manufacturer must vary his goods and adapt them to the prevailing taste, besides increasing the number of articles which he sends thither for sale.[84]

British writers thus perceived alpaca naturalisation as a possible antidote to actual and potential national problems, as well as a marker of imperial dominance. The difficulties British subjects experienced in smuggling alpacas out of Peru, however, point to Britain's limited influence over the region in the mid-nineteenth century and the vulnerability of would-be acclimatisers to shifts in local politics.

While alpaca acclimatisation was intended to benefit Britain as a whole, it is worth noting that specific regions and cities took particular interest in the project, developing their own local connections with the alpaca and sometimes pursuing distinctive regional agendas. General O'Brien, for example, though happy to collaborate with British colleagues in bringing alpacas to the British Isles, appears to have been particularly interested in the benefits the scheme would confer upon his native Ireland. He certainly emphasised the latter in correspondence with fellow expatriate Irishman Peter Murphy, HM Consul at Arica, when he envisioned the animals gracing the mountains of Wicklow.[85] Similar sentiments were at work in Liverpool, where Danson and Atkins seem to have perceived alpaca importation as a source of local pride for one of Britain's main trading hubs with South America, and as another way of advertising the port city's global reach and entrepreneurial spirit. In Bradford, meanwhile, alpaca wool contributed significantly to the city's textile prosperity and earned the West Yorkshire town royal patronage in 1845 after local artisans converted the fleece of one of Prince Albert's alpacas into an apron and 'a striped and figured dress'.[86] In 1851, when Salt constructed a special village for his employees, alpaca emblems were chiselled into several of the buildings, including the schoolhouse (Figure 4.4); by 1859, Bradford even boasted an 'Alpaca Beer-house' – further testimony to the animal's local significance and its incorporation into popular culture.[87] These cases suggest that there was a regional as well as a national dimension to alpaca naturalisation, with different counties and cities seeking specific benefits as farmers, importers and manufacturers of alpaca products.

Nature, the Exotic and the Science of French Colonialism (Bloomington: Indiana University Press, 1994), pp. 132–8.
[84] Walton, *The Alpaca*, p. 12.   [85] Danson, *Alpaca*, pp. 12–14.
[86] 'Her Majesty's Alpaca Textures', *The Morning Chronicle*, 14 December 1844.
[87] 'Bradford', *Leeds Mercury*, 21 May 1859.

Figure 4.4 Alpaca motif on the schoolhouse in Saltaire, Bradford.

Shifting our focus to Australia, we find a further difference in emphasis, but also an awareness of the wider imperial and local circumstances noted above. On the one hand, acclimatisers viewed rearing the animals in the outback as good for the British Empire as a whole and beneficial in particular to textile-producing cities such as Bradford. Wilson therefore appealed specifically to the manufacturers of Bradford to assist in raising money to purchase alpacas for shipping to Victoria, convinced that 'the enormous advantages already derived by your town from alpaca wool will doubtless induce nearly the whole of your manufacturers and spinners to contribute to this interesting experiment'.[88] On the other hand, the Australian alpaca programme had a specifically colonial dimension and was deeply inflected with elements of local pride – sometimes broadly 'Australian', but in other instances confined to a single Australian state. Seeking contributions to his alpaca fund in *The Times*, Wilson expressed his hope that 'Australians now in England' would supply the money to buy the animals and recommended that the flock be sent to 'Victoria, if possible', thus benefiting his native Melbourne.[89] Commenting on Ledger's achievements, meanwhile, *The Era* concluded that 'if there is a man who has *done well* for Australia it is Charles Ledger, and we trust the colony will mark its sense of his merit in a manner befitting a Government to bestow, and a public benefactor to receive'.[90] When the Australians let Ledger down by failing to reimburse him adequately for his services, the same paper published a further article, with the following warning:

[88] 'Llamas for Australia', *Bradford Observer*, 29 July 1858.
[89] 'The Alpaca', *The Times*, 17 July 1858.
[90] 'Mr Charles Ledger and His Alpaca Contract with New South Wales', *The Era*, 25 September 1859.

It would indeed be a disgrace to the country if Mr C. Ledger, after so far completing his hazardous and arduous enterprise, commenced on the faith of promises of handsome remuneration, should be allowed to depart to South America, and relate there … that an enterprise which had been highly extolled in England as likely to open a new era in British commerce, has been unrequited by the country which would chiefly benefit by it.[91]

This prompted the opening of a public subscription in New South Wales, partly to reward Ledger for his efforts and partly to redeem the inhabitants of Sydney in British eyes by proving that 'the high merit of Mr C. Ledger in opening up a new source of productive industry' had been 'properly appreciated'.[92] Colonial and imperial pride were thus at stake in the alpaca exchange project, as British newspapers upbraided Australian subjects for failing to show due gratitude towards a metropolitan benefactor and the inhabitants of rival settlements competed for regional pride and economic advantage – all within the wider context of the British Empire.[93] At the 1862 International Exhibition, the Government of New South Wales received a medal for 'the first alpaca wool grown in the colony', while Ledger himself 'obtained honourable mention for excellence of quality of alpaca tallow and pomade' – a source of pride for the nascent colony.[94]

Finally, viewed through Peruvian and Bolivian eyes, alpaca acclimatisation had a very different complexion, though one still heavily imbued with national significance. For these two countries, the alpaca was an intrinsic part of the landscape and its fleece an important national export. Alpaca wool earned Peru and Bolivia significant sums of money in the post-independence period, while alpacas themselves were part of the region's pre-Columbian heritage, appearing regularly on flags, coins and stamps. Indigenous people also maintained a close spiritual relationship with camelids, stretching back to before the Spanish conquest. According to Ledger (not, admittedly, an impartial witness), when news arrived in Peru of the death of the alpacas on the *Sir Charles Napier*, 'the Indians flocked from all parts of the country to the capital of the department, the city of Puno, and from thence petitioned the Government to prohibit the exportation of Alpacas from the country', attributing 'every misfortune that had happened in the district for the year past' to the

[91] 'The Alpacas and Mr C. Ledger', *The Era*, 12 February 1860.   [92] Ibid.
[93] Dane Kennedy makes a similar argument for competing civic and imperial interests in relation to mid-nineteenth-century Australian exploration. See Dane Kennedy, *The Last Blank Places: Exploring Africa and Australia* (Cambridge: Harvard University Press, 2013), p. 241.
[94] 'Mr Ledger and the Alpacas', *The Era*, 23 November 1862.

animals' 'untimely death'.[95] For both cultural and economic reasons, therefore, the Peruvian and Bolivian governments opposed the naturalisation of alpacas in other countries and issued legislation to protect this part of their national patrimony. The export of alpacas was prohibited in 1845 and further decrees in 1851 and 1868 reinforced the export ban, outlawing 'the removal from Peruvian territory of alpacas and vicuñas, and of any species of animal that proceeds from the crossing of the two races'.[96] British critics, frustrated by these impediments to free trade, repeatedly condemned this policy, referring disparagingly to the 'intense jealousy' and 'absurd decrees' of the Peruvian and Bolivian states.[97] Ledger's brother, George, accused Peru of supreme selfishness, noting that the country had benefited from the introduction of Old World animals such as the horse and the pig, and had repaid 'these gifts by the positive prohibition of its most valuable animal product ... appropriating exclusively to itself a ... blessing intended for the common benefit of mankind'.[98] To Peruvians, however, the measures made eminent sense in the face of what we might see as a nineteenth-century form of biopiracy. Moreover, as Peruvian legislators were quick to point out, the export ban 'cannot be considered in the present case as an odious restriction of trade, since the export of alpaca and llama wool is free' – an interesting distinction that demonstrated their awareness of prevailing British critiques.[99]

[95] George Bennett, *The Third Annual Report of the Acclimatisation Society of New South Wales* (Sydney: Joseph Cook, 1864), p. 95.
[96] 'Lima, Abril 1° de 1851', *El Comercio*, 8 April 1851; *Decreto Estableciendo la Prohibición de Extraer del Territorio Peruano las Alpacas, Vicuñas y Animales que Proceden del Cruzamiento de Ambas Razas*, 8 October 1868, Archivo Digital de la Legislación del Perú.
[97] 'The Alpaca', *The Times*, 17 July 1858; 'Interesting Narrative Respecting Alpacas in Australia', *The Era*, 20 February 1859. On the continued importance of camelids in Andean folklore and culture, see Jorge Flores Ochoa, *Pastoralists of the Andes: The Alpaca Herds of Paratía* (Philadelphia: Institute for the Study of Human Issues, 1979), pp. 71–85.
[98] Ledger, *The Alpaca: Its Introduction into Australia*, p. 14. This is a reference to what historian Alfred Crosby has called 'The Columbian Exchange' – the transcontinental interchange of plants, animals and microbes that occurred in the sixteenth century following the Spanish conquest of America. See Alfred Crosby, *The Columbian Exchange: Biological and Cultural Consequences of 1492* (Westport: Greenwood Press, 1972).
[99] 'Lima, Abril 1° de 1851', *El Comercio*, 8 April 1851. The export ban on alpacas paralleled similar South American legislation prohibiting the removal of antiquities and prized natural history specimens from American soil. See Robert Aguirre, *Informal Empire: Mexico and Central America in Victorian Culture* (Minneapolis: University of Minnesota Press, 2005), p. 31, 98; Rebecca Earle, *The Return of the Native: Indians and Myth-Making in Spanish America, 1810–1930* (Durham: Duke University Press, 2007), pp. 134–8; Helen Cowie, *Conquering Nature in Spain and Its Empire 1750–1850* (Manchester: Manchester University Press, 2011), pp. 165–70.

A Failed Experiment

Britons, Australians and South Americans thus all had high hopes for the alpaca and anticipated multiple benefits from its exploitation. For the former, however, as we have seen, these benefits proved elusive and the large-scale farming of alpacas never materialised. Peruvian camelids made it alive to Britain and Australia, but they did not multiply in significant numbers. Alpacas never populated the Scottish Highlands or the Australian outback and, despite repeated claims of improvement, New South Wales never usurped Peru as Bradford's main supplier of alpaca fibre. A post-mortem on alpaca naturalisation is therefore in order. Why did acclimatisation schemes fail? Was the project badly conceived, or simply poorly executed? What factors complicated or undermined naturalisation efforts?

Among the various impediments to alpaca acclimatisation, the first that stands out is the sheer difficulty of conveying live camelids across the ocean – whether the Atlantic or the Pacific. Accustomed to the frigid temperatures of the altiplano, alpacas were ill-adapted to the heat of the tropics and often succumbed to illness. Dietary changes made the problem worse, particularly if introduced rapidly, while mishandling and serious blunders added to the animals' discomfort – the latter demonstrated most graphically by the tragedy aboard the *Sir Charles Napier*. Many alpacas thus died before they reached their destination, and some of those that survived the journey arrived in poor condition. A young alpaca disembarked at the docks in Adelaide in 1858 appeared 'very low spirited' after its long voyage, sitting alone in its box and 'utter[ing] low plaintive moans as if in melancholy remembrance of its two companions, which died on the passage'.[100]

The physical challenges posed by transporting live animals over long distances are exemplified by the calamitous expedition of Alexander Duffield. Contracted by authorities in Melbourne to ship a proposed 1,500 alpacas to Victoria, Duffield purchased 600 of the animals in Carangas, courtesy of a special dispensation from the Bolivian Government, and made his way towards the sea for shipping to Australia. Although the initial part of the journey passed without incident, 285 of the flock perished in May 1863 while crossing the notoriously dry Atacama Desert and a further 120 expired on the voyage to Australia aboard the ship *Julia Farmer*.[101] Six months later, only 'a solitary white lamb' remained alive – a 'very pretty and lively specimen

[100] 'Importation of Stock', *The Adelaide Observer*, 25 September 1858.
[101] 'The Alpacas in Victoria', *Sydney Morning Herald*, 2 March 1864.

of alpaca juvenility'.[102] While several factors contributed to the high attrition rate, Duffield's sponsor, Henry de Boos, ascribed the heavy losses to embarking the alpacas too abruptly – a move necessitated by the lack of food and water on the Bolivian coast – and to changing their diet suddenly 'from sweet, moist swamp grass to English hay with crushed oats, oatmeal, bran and a short allowance of water'. He also suspected that 'the sudden change from the rarefied air of the Andes to the denser atmosphere of the sea' may have effected 'constitutional changes' on the animals, almost all of those that died aboard ship presenting 'symptoms of diseased lungs, sometimes complicated with some affection of the heart or liver, generally in the shape of enlargement or softening of those organs'.[103] Moving large numbers of animals over long distances was clearly a perilous business, and the heavy death toll had a direct impact on naturalisation efforts.

The challenge of successfully naturalising alpacas outside South America was compounded by two further factors: the Peruvian export ban and the errors of individual acclimatisers. The former was significant because it limited the number of alpacas that could be acquired outside Peru and had an effect on the quality of animals exported. Ledger, for instance, was forced to supplement his flock of alpacas with llamas in order to make up the numbers, which made it necessary for him to engage in a lengthy programme of breeding back in order to increase the number of pure alpacas. This was one of the reasons why he was reluctant to break up the flock after its arrival, since splitting the animals would vitiate any chance of restoring their purity.[104] While a few individuals (including Duffield) were subsequently granted permission to export alpacas, these were one-off dispensations, not long-term policy shifts. Introduced flocks consequently lacked regular injections of new blood from South America, and disasters like the one that afflicted Duffield could prove fatal to the success of naturalisation schemes.

Individual mistakes exacerbated the problem, sparking mutual anger and recriminations. Robert Bell's veterinarian misdiagnosed his female alpaca's prolapsed womb, leading to the animal's death.[105] The authorities in Melbourne were tardy in providing Duffield's surviving alpacas with accommodation, leaving their owner to 'shift for himself' and – in the view of some – contributing to their subsequent demise.[106] Ledger,

---

[102] 'Australian Items', *Bradford Observer*, 28 July 1864.
[103] 'The Alpacas in Victoria', *Sydney Morning Herald*, 2 March 1864.
[104] 'The Management of the Alpacas', *Sydney Morning Herald*, 4 March 1864.
[105] Walton, *The Alpaca*, p. 21.   [106] 'Australian Items', *Bradford Observer*, 28 July 1864.

meanwhile, ignored Peruvian precedent and bred from his female alpacas at 'too early an age', with the result that, when drought struck in 1862–3, the young mothers 'succumbed to the extra strain on their constitutions'. This last error was inspired by a widespread belief that 'in this colony nature is more precocious than in Europe', reflecting, perhaps, a broader overconfidence among would-be acclimatisers and an overly positive assessment of Australia's climate.[107] Mistakes of all hues diminished the already small number of alpacas outside South America, fuelling the doubts of sceptics and critics. Contrary to British claims of inertia, moreover, there is evidence that the Peruvians took measures to increase the domestic output of alpaca wool, rendering overseas naturalisation projects less urgent; in 1845, for instance, the government offered a reward of 250 pesos to anyone who introduced alpacas to the department of Junín, with a further 25 pesos for every fifty *crias* (baby alpacas) bred in the region.[108] In 1864, the Government of New South Wales accused Ledger of seriously mismanaging his flock and suspended him from his salaried position, effectively terminating the acclimatisation experiment in the colony.[109]

The failure to naturalise alpacas in Britain and Australia also reflected competing conceptions of political economy. Who should farm alpacas? How and where should they do so? How should naturalisation be financed? In Britain, the majority of alpaca owners were elite landowners who acquired the animals to beautify their estates. Acclimatiser William Walton, however, was highly critical of this practice, suggesting that such individuals were not sufficiently serious in their endeavours and had failed to give alpacas a fair trial. This was partly because they had bred their stock indiscriminately with llamas, reducing their purity and (Walton believed) their fertility; partly because they entrusted their care to servants, who failed to look after them properly; and partly because they kept the animals in inappropriate terrain, pasturing them on rich lawns rather than mountain grasses. As Walton remarked in his 1845 book, 'The great error committed by our early breeders was that they considered the few alpacas which arrived safe as mere ornament to the lawn, never imagining that, by care and attention, they might be converted into profitable stock for farms, situated on ground resembling that on which they were born.'[110]

---

[107] 'The Management of the Alpacas', *Sydney Morning Herald*, 4 March 1864.
[108] 'Proyecto de Lei', *Suplemento al Peruano*, 7 October 1845.
[109] 'The Management of the Alpacas', *Sydney Morning Herald*, 4 March 1864.
[110] Walton, *The Alpaca*, p. 20.

In Australia, the fate of Ledger's alpacas engendered a wider debate about the relative merits of public and private enterprise. When the Government of New South Wales initially stepped in to purchase the animals it was praised for its foresight, and specifically for preventing the camelids from leaving the colony. Several years later, however, when the flock succumbed to drought and disease, critics began to take a less benevolent view of the government's involvement, suggesting that the alpacas would have done better if placed in private hands. Addressing the Legislative Assembly, one commentator speculated that 'a wealthy and intelligent squatter would soon have discovered if the locality where the flock was placed was unfavourable to their health or increase, and if so he would not have rested until he found some place better adapted to their requirements. Self interest would have come to his assistance'.[111] Another contemporary – 'A Well-Wisher to Alpacas' – advised the government to put the remaining camelids 'under the charge of some hardy bushman who would live in the bush along with his charges, [and] allow them uncontrolled freedom of range, merely keeping his eye on them for security', thus ensuring their swift propagation. He also advocated selling some of the animals to rival colonies if necessary – especially Tasmania and New Zealand – arguing that '[t]o restrict the sale of the alpacas to parties who would engage to keep them within the boundaries of this colony would be a dog-in-the-manger proceeding and would defeat its own object'.[112] Contemporaries thus tussled over the pros and cons of public and private ownership, pitting the advantages of national control against the benefits of a free-market economy.

Lastly, the failure of the alpaca project may be attributed in part to two more abstract issues: technological constraints and questions of knowledge transfer. In the case of the former, the main problem related to the processing of alpaca fibre. In Peru, alpacas were typically shorn once every two years, allegedly because 'Indians could not be induced to shear [them] oftener'. Ledger saw this as a potential area of improvement and planned to outperform Peruvian wool suppliers by shearing his alpacas annually. In the event, however, he found that manufacturers rejected his samples because they had already 'fitted their machinery for a length of staple of 2 years growth' and could not cope with the shorter variant.[113] Rather than undercutting Peruvian standards, therefore, Ledger was

---

[111] 'Australian Items', *Bradford Observer*, 28 July 1864.
[112] 'The Pastures and Localities Best Suited for the Alpaca', *Sydney Morning Herald*, 4 March 1864.
[113] Ledger, *The Alpaca: Its Introduction into Australia*, p. 19.

forced to adhere to the established shearing schedule because the technology for combing shorter fibres did not exist.

Naturalisation programmes also suffered badly from the loss of crucial indigenous knowledge and expertise, which was either consciously withheld or not fully appreciated. Ledger's Bolivian shepherds lacked the skill to shear the alpacas efficiently, 'as in South America the shearing, or rather the cutting off the wool ... is done entirely by women' – a gendered element to knowledge transfer that the former wool merchant had not considered.[114] Ledger's alpacas also suffered severely from a disease called 'the scab' (or '*sarna*' in Spanish), which, according to Henry de Boos, was treated easily in Peru with an ointment of 'sulphur and lard well boiled ... applied whilst warm with a piece of flannel'.[115] This disease proved a particularly serious blow to the naturalisation project, because it triggered fears among Australian farmers that the pathogen might infect their sheep, leading to calls for the alpacas to be culled or quarantined.[116] Whether these issues stemmed from a refusal to listen to indigenous knowledge (often branded as superstition) or the indigenous people's reluctance to supply it is uncertain; writing in 1864, Ledger claimed that 'almost all that was formerly told me [by the Indians] was false', suggesting that he may have been duped.[117] Either way, the result was the same: profits evaporated and alpacas perished. Alpaca naturalisation thus suffered from the non-transfer, or only partial transfer, of particular forms of knowledge and the limitations of existing technology.[118]

Conclusion

The story of alpaca acclimatisation is one of commercial ambition, regional rivalry and scientific exchange. The initial popularity of the wool in Britain resulted from increased trade with South America in the wake of independence, coupled with the emergence of new technologies that made its processing viable. The decision to move from importation to acclimatisation was inspired by a belief that the animal would thrive in

---

[114] 'Shearing the Alpacas', *Sydney Morning Herald*, 7 November 1859.

[115] 'The Alpacas in Victoria', *Sydney Morning Herald*, 2 March 1864.

[116] Ledger denied that his alpacas posed a risk to sheep. He reported that 'I had 700 sheep badly "bottled" in two flocks, that for a year ... slept in the yards with the alpacas, and I never saw or heard of any sign of scab among them'. 'To the Editor of the Herald', *Sydney Morning Herald*, 8 April 1863.

[117] Bennett, *The Third Annual Report of the Acclimatisation Society of New South Wales*, p. 100.

[118] On the non-transmission of knowledge, and the reasons for it, see 'Conclusion: Agnotology' in Schiebinger, *Plants and Empire*, pp. 226–42.

Britain and Australia, and the conviction that the quantity and quality of wool produced would be significantly improved as a result of its trans-plantation. The actual process of naturalisation was attempted through the activation of multiple networks of exchange and expertise, with contributions from former soldiers, expatriate merchants and travelling showmen and women. It was also attempted in direct contravention of Peruvian law, which, from 1845, prohibited the export of this precious natural resource.

Efforts to naturalise the alpaca formed part of a much wider pro-gramme of biopiracy. The British wanted to control valuable natural resources, such as cinchona (quinine), rubber and tea. They believed that the best way to do so was to extract them from their native lands (Peru, Brazil, China) and cultivate them in suitable regions of the British Empire.[119] They justified their actions on the grounds that the current owners of these resources were not exploiting them to their fullest poten-tial – a process that Fa-ti Fan refers to as 'paternal imperialism'.[120] In the case of the alpaca, it was alleged that Peruvian Indians wasted much of the wool through superstition and slovenly collecting practices and that they could not meet growing foreign demand for the product. Naturalising the animal in Britain and Australia to remedy a potential shortage of raw material would, through superior animal husbandry, improve the quality of the wool produced. The Acclimatisation Society of Melbourne, for example, accused 'the South American Indian, one of the most un-improving of all the races of mankind', of having failed to maximise the alpaca's potential. It anticipated great improvements in the species now that the animal was 'subjected for the first time to the same treatment that has effected such wonders with the Leicester, Lincoln or South Down sheep, the short-horn ox [and] the thorough-bred horse'.[121]

Although it shared many features with contemporary botanical accli-matisation programmes, alpaca naturalisation differed from many of these in three key respects. First – and most obviously – it involved an animal rather than a plant, which added to the difficulties of relocation. Second, it was not managed centrally from a single scientific institution, such as Kew Gardens, but promoted independently by a number of private individuals (the ZSL, although involved in the scheme, did not assume a comparable leading role). Third, the alpacas were relocated not to a tropical colony, like most botanical seizures, but to Britain's Celtic fringe and a white settler colony in the southern hemisphere, where they

[119] See Brockway, *Science and Colonial Expansion*.
[120] Fan, 'Victorian Naturalists in China', p. 25.
[121] Ledger, *The Alpaca: Its Introduction into Australia*, p. ii.

were to be farmed by British subjects or expatriates, not coerced indigenous labour. These factors gave the alpaca project a decentred complexion and meant that it was actively encouraged – rather than passively accepted – in the receiving countries, which anticipated concrete local benefits as well as wider imperial ones. In this way, British subjects in places as diverse as Bradford, Liverpool, Sydney and Hobart took a keen interest in the scheme, publishing papers on the subject, meeting locally to discuss it and often undertaking private initiatives to bring it to fruition.

The other key thing about alpaca acclimatisation, of course, was that it failed – or at least it did not achieve the glorious results its supporters expected. Alpacas perished from ignorance or mismanagement. Indigenous Peruvians refused to share their knowledge with European acclimatisers or deliberately misled them. Australian farmers got cold feet and declined to purchase alpacas at auction, while shifting economic and political conditions on both sides of the Pacific disrupted naturalisation efforts. As a result, Ledger ended up disillusioned and bankrupt and the surviving alpacas were dispersed across New South Wales, precluding the process of selective breeding he had envisaged. One alpaca owner, Thomas Holt, donated two of his remaining animals to 'the Lunatic Asylums at Parramatta and the Cook's River' in 1873 – a rather sorry end to an ambitious acclimatisation project.[122]

Despite its limited success, however, alpaca acclimatisation retained a certain romantic appeal and was couched in the language of the military crusade or daring adventure. One contemporary newspaper characterised Ledger's quest as a 'pilgrimage', the details of which 'would compile a romance of the most extraordinary adventure'. Henry Swinglehurst, a Briton resident in Valparaíso, was even more effusive. Writing to a friend in Britain in 1858, shortly after he had dined with Ledger, Swinglehurst eulogised the merchant's achievements and likened him to the most famous explorer of his generation:

He is Livingstone No. 2, and in other times will be looked upon as a hero of trials that few have known and hardly any equalled ... Sometimes all but buried in mountain snows, with hungry men and flocks to protect; now chased by the police and again hunted by the angry natives in pursuit of the flock, and with all the privations of six years' absence from his home and family and over a journey of 6,000 leagues with animals and men.[123]

---

[122] State Library of New South Wales, Letter from Thomas Holt to Sir Henry Parkes, 23 May 1873, *Sir Henry Parkes Papers*, 1833–96, p. 366.
[123] 'A Flock of Alpacas', *The Era*, 31 October 1858.

This was very much the stereotype of the courageous Victorian explorer, assailed by hazards on all sides but emerging triumphant through ingenuity and perseverance. It was also the reincarnation of the sixteenth-century privateer who raided South American ports in defiance of the Spanish Government and exemplified British pluck and daring. Three hundred years after Sir Francis Drake preyed on Spanish ships in the Pacific, British merchants and naturalists were committing another, more modern form of piracy on Peruvian coasts, with camelids taking the place of silver as stolen treasure.

Coda: The Vicuña, Silk of the Andes

Before we leave the alpaca, it is worth spending a few moments to look at the parallel story of its close relative, the vicuña, which was also highly prized for its wool. A wild animal, the vicuña was hunted rather than farmed to obtain its precious fleece, described by one eighteenth-century naturalist as 'the silk' of America.[124] Threatened by overhunting, it was also the subject of some of the earliest colonial conservation legislation, as well as a target of several failed domestication schemes. As such, its history forms an interesting counterpoint to that of the alpaca and provides an illuminating prequel to late nineteenth-century conservation efforts.

Like the alpaca, the vicuña is native to the Peruvian *puna*, living at altitudes of 3,700 metres and above. It is supremely well adapted to life at high altitude, its oval-shaped blood cells maximising oxygen absorption into the blood stream and its large heart (almost 50 per cent bigger than the average heart weight for mammals of a similar size) pumping oxygen around its body. From a human perspective, the vicuña's most valuable asset is its thick coat, which keeps the animal warm during the chilly Andean nights. The finest of all camelid wools, at fourteen microns, vicuña wool insulates its owner from the cold and protects it from the high ultraviolet radiation experienced at high altitudes.[125] It has been both the saviour and the downfall of the species, preserving it from the elements but making it vulnerable to human exploitation.

Vicuña wool was known to indigenous Americans prior to the Spanish conquest and especially valued by the Incas, who used it to make their most exquisite textiles. According to Jesuit missionary Bernabé Cobo,

---

[124] 'Memoria sobre la Importancia de Connaturalizar en el Reino la Vicuña del Perú y Chile' in Francisco José de Caldas, *Obras Completas de Francisco José de Caldas* (Bogotá: Imprenta Nacional, 1966), p. 324.
[125] Helen Cowie, *Llama* (London: Reaktion Books, 2017), p. 18.

the clothes of the Inca emperor 'were made entirely or partially of vicuña wool' mixed with 'viscacha wool, which is very thin and soft, and bat fur ... which is the most delicate of all'.[126] The Incas did not farm vicuñas, as they did alpacas, but hunted them intermittently in drives known as *chakkus*. Pedro Cieza de León estimated that 'as many as thirty thousand head' of vicuñas might be killed during a single *chakku*.[127] Mestizo chronicler Garcilaso de la Vega, however, claimed that only the males perished in these hunts, for the Incas would release female vicuñas 'after they had been sheared'. He also stated that the Incas kept records of the number of animals caught and killed, and would only hunt in a particular region one year in every four, partly to conserve stock and partly because 'the Indians say that in this space of time the wool of the vicuña grows to its full extent'.[128] The Inca *chakku* thus appears to have been a sustainable activity 'regulated by political, religious, and other local social and cultural mechanisms'.[129]

This changed following the arrival of the Spanish in 1532. Where the Incas had hunted the graceful camelids sporadically, in carefully controlled drives, the Spanish pursued them without remission, shooting them with guns and running them down with dogs and horses. Spanish botanist Hipólito Ruíz, who witnessed a vicuña hunt in Peru in the 1780s, described how local people trapped vicuñas in an enclosure, 'shouting, beating drums, blowing whistles, and snapping whips' to intimidate the animals and running them down on horseback.[130] Half a century later, the Swiss naturalist Johann Jakob von Tschudi recorded details of a vicuña hunt he observed in the Altos of Huyhuay, in which 122 vicuñas were herded into a corral and killed (Figure 4.5).[131] While the Spanish initially exploited vicuñas for the bezoar stones often found in their intestines (believed to act as an antidote to poison), by the eighteenth century attention had shifted to the camelid's exquisite fleece, used to make scarves, gloves, stockings, hats, sheets and

[126] Bernabé Cobo, *Historia del Nuevo Mundo* (Seville, Imprenta de E. Rasco, 1895), Vol. IV, p. 205.

[127] Clements R. Markham (ed. and trans.), *The Travels of Pedro Cieza de León, A.D. 1532–50, Contained in the First Part of His Chronicle of Peru* (London: Hakluyt Society, 1864), Vol. II, pp. 45–6.

[128] Garcilaso de la Vega, *Primera Parte de los Comentarios Reales* (Madrid: Imprenta de Doña Catalina Piñuela, 1829), pp. 447–9.

[129] Glynn Custred, 'Hunting Technologies in Andean Cultures', *Journal de la Société des Americanistes* LXVI (1979), p. 14.

[130] Hipólito Ruíz, *The Journals of Hipólito Ruíz, Spanish Botanist in Peru and Chile 1777–1788*, translated by Richard Evans Schultes María José Nemry von Thenen de Jaramillo-Arango (Portland: Timber Press, 1998), p. 104.

[131] Johann von Tschudi, *Travels in Peru, during the Years 1838–1842*, translated by Thomasina Ross (London: David Bogue, 1857), pp. 313–14.

Figure 4.5 'The Mêlée: Scene in "Aparoma", April 1857', from *Annotated Watercolour Sketches by Santiago Savage, 1857–1858, Being a Record of Charles Ledger's Journeys in Peru and Chile*. State Library of New South Wales MLMSS 630/1. Note the use of the bolas to snare the vicuñas.

handkerchiefs.[132] Thousands of vicuña skins were sent to Spain to meet the growing demand, and thousands of vicuñas perished in the process. One contemporary calculated that around 100 *arrobas* (c.1,150 kilograms) of vicuña wool were imported into Spain every year by the turn of the nineteenth century, which, at 12 ounces of wool per vicuña, equated to 8,000 dead vicuñas.[133]

---

[132] Marcia Stephenson, 'From Marvelous Antidote to the Poison of Idolatry: The Transatlantic Role of Andean Bezoar Stones during the Late Sixteenth and Early Seventeenth Centuries', *Hispanic American Historical Review* 90:1 (2010), pp. 3–39; 'De las Vicuñas', *Semanario de Agricultura y Artes Dirigido a los Parrócos* X (1801), p. 263. The Spanish Crown enjoyed a monopoly over the fabric and established the Real Fábrica de Paños (Royal Cloth Factory) in Guadalajara from the 1760s to process the luxury wool. See Carlos Gómez-Centurión Jiménez, *Alhajas para Soberanos: Los Animales Reales en el Siglo XVIII* (Madrid: Junta de Castilla y León, 2011), p. 208.

[133] 'De las Vicuñas', *Semanario de Agricultura y Artes Dirigido a los Parrócos* X (1801), pp. 263–4. Hugo Yacobaccio estimates that an average of 20,410 vicuña skins per year were exported from the port of Buenos Aires during the eighteenth century. See Hugo Yacobaccio, 'The Historical Relationship between People and the Vicuña' in Iain Gordon (ed.), *The Vicuña: The Theory and Practice of Community-Based Wildlife Management* (New York: Springer, 2009), p. 12.

With such large numbers of vicuñas now being killed, it became apparent to the Spanish Crown that reform was needed to ensure the long-term viability of the trade in their wool. This gave rise to two key conservation policies that would resurface repeatedly in the nineteenth century: domestication and hunting legislation. The former, reflecting prevailing notions of agricultural improvement, centred on the idea that the vicuña, a wild animal, might be effectively tamed and farmed for its wool in a similar way to its close relative, the alpaca. The latter took the form of hunting bans, which imposed limits on the number of vicuñas that could be caught, advocated the shearing and release of those vicuñas taken, and intermittently prohibited the hunting of vicuñas completely to allow populations to recover. Both approaches represented a growing desire in the Spanish Empire to better manage its natural resources and to prevent their premature exhaustion. Where bezoar stones formed part of the post-conquest plunder economy, vicuña fleece would be increasingly subject to Enlightenment notions of improvement, as royal officials, reforming societies and – increasingly – naturalists sought to maximise Spain's agricultural and industrial production.

The first moves to domesticate the vicuña occurred in the 1760s and can be traced through surviving correspondence in the General Archive of the Indies (AGI) in Seville. On 22 February 1768, the Minister of the Indies issued an order requesting that some of the animals be sent to Spain from Peru. This elicited a reply from the Viceroy of Peru the following June, reporting that he had 'succeeded in recent days in collecting three "carneros" that are called "de la tierra" [llamas] and a vicuña ... which will be sent [to Spain] on the ship *San Miguel*'. A further letter from the Viceroy, dated 1770, expressed doubts about the vicuñas' ability to withstand 'the heat that is experienced in sailing through the tropics', while assuring the Minister of the Indies that the utmost effort was being made to ensure their successful delivery. A subsequent letter from 1772 reported the dispatch of 'three vicuñas and a guanaco' – the only survivors among the 'many that have perished, notwithstanding the care that has been taken of them'. The last reference to the vicuñas records their arrival in Cádiz aboard the ship *San Lorenzo* and the frigate *Santa Rosalia*. One of the animals 'died en route [to Madrid]'. Another was placed in the Casa del Campo, one of Charles III's royal palaces.[134]

At the same time as the Spanish were attempting – not very successfully – to introduce vicuñas to Iberia, Spanish Americans were also taking

[134] AGI Lima 651; AGI Lima 652; AGI Indiferente 1549.

a renewed interest in Peruvian camelids. One creole, the Jesuit Juan Ignacio Molina, advocated domesticating the vicuña in his native Chile, where the terrain and climate were similar to those of Peru.[135] Another, a contributor to the Argentine periodical *El Semanario de Agricultura, Industria y Comercio*, proposed farming vicuñas in the province of Salta, using local Indians as shepherds and breeding the wild camelids with llamas, alpacas and even domestic sheep to create a superior blend of wool. He attributed previous failures to domesticate vicuñas to rearing the animals in an overly hot climate and feeding them too much bread, cake and chocolate![136] Most enthusiastic of all was the polymath Francisco José de Caldas, who addressed a memoir to the Real Consulado de Comercio de Cartagena in May 1810 on the subject of naturalising the vicuña in his native New Granada (modern-day Colombia). Convinced that the vicuña – a zoological 'treasure' – would add materially to the prosperity of his homeland, Caldas urged the Consulado to purchase 1,000 of the animals (at an estimated cost of 2,500 pesos) and sketched out an itinerary for their importation from Potosí, with a sea passage from Ilo or Arica to Guayaquil, a transfer by boat to Babahoyo, a therapeutic stopover of several weeks in the foothills of Mount Chimborazo and a final overland push to Quito, Popayán and Bogotá. Caldas envisaged the majority of the camelids living wild in the mountains of the Sierra Nevada, where they could be periodically rounded up for shearing, but he also advocated forming 'flocks of domesticated vicuñas, as we have done with the common sheep'. In the interests of conservation, he proposed shearing the vicuñas alive and then releasing them, thus preventing the 'cruel butcheries' inflicted on the animals in their native Peru. As he expressed it: 'I am persuaded that the Patria [New Granada] would watch with pleasure the multiplication of the vicuña, adding one more species to the domesticated animals of this family, which is today wild in its country of birth, and New Granada would give an example of industry and economy to [the people of] Peru.' Here, then, we see an example of inter-colonial rivalry, as a creole from one colony proposed appropriating a species native to another and managing it in a more 'rational' way.[137]

The upheavals of independence put a temporary halt to vicuña domestication plans, but the project experienced a revival in the

---

[135] Juan Ignacio Molina, *Compendio de la Historia Geográfica, Natural y Civil del Reyno de Chile* (Madrid: Sancha, 1788), pp. 350–64.

[136] 'Sobre la Posibilidad de Domesticar a la Vicuña', *Semanario de Agricultura, Industria y Comercio* III:138 (1805), pp. 283–7.

[137] 'Memoria sobre la Importancia de Connaturalizar en el Reino la Vicuña del Perú y Chile' in Caldas, *Obras Completas de Francisco José de Caldas*, pp. 323–33.

Figure 4.6 Llama nurses Burra, Sarea, Cacho and Chucara suckle Ledger's nine surviving '*vicuñitas*', from *Annotated Watercolour Sketches by Santiago Savage, 1857–1858, Being a Record of Charles Ledger's Journeys in Peru and Chile*. State Library of New South Wales MLMSS 630/1

post-independence period when both foreign merchants and Peruvian entrepreneurs took a renewed interest in the species. Charles Ledger, as a sideline to his alpaca-smuggling enterprise, secured fourteen vicuña calves from local vicuña hunters in March 1857 and reared nine of them, using four llamas – Burra, Sarea, Cacho and Chucara – as foster mothers (Figure 4.6).[138] Some years earlier, in the 1830s and 1840s, a Peruvian citizen, the parish priest Juan Pablo Cabrera, started an experiment to breed vicuñas with alpacas in the remote village of Macusani. After twenty-one years of patient work, the clergyman succeeded in breeding fourteen vicuña–alpaca hybrids, each of which sported the chestnut-coloured 'snouts, ears and neck' of their vicuña mothers and the white bodies of their alpaca father – a better colour for dyeing. He kept the animals in a corral, to prevent them breeding with wild male vicuñas, and fed them – against earlier advice – on 'bread, cake, coca leaves, sugar

---

[138] 'The Llamas – Mr Ledger', *Sydney Morning Herald*, 30 November 1858; *Annotated Watercolour Sketches by Santiago Savage, 1857–1858, Being a Record of Charles Ledger's Journeys in Peru and Chile*. State Library of New South Wales MLMSS 630/1.

[and] corn on the cob'. The Peruvian Government expressed great hopes that the experiment would constitute 'the fertile germ of an incalculable treasure for our backward country', rewarding Cabrera for his efforts by giving him a special medal and paying for his portrait to be placed in the National Museum in Lima. To consolidate the domestication process, the government offered lifetime exemption from the *contribución indígena* (a form of head tax) to 'any Indian who presents to the Governor of his District ten perfectly tame female vicuñas along with a male of the same species or a male alpaca'.[139]

Unfortunately, none of these plans came to fruition, and the results of vicuña domestication were limited. Caldas was executed by royalist troops during the wars of Hispanic American independence, while Cabrera was discovered by French explorer Paul Marcoy in 1871 living in poverty, his vicuña domestication scheme abandoned.[140] Of the eighty-three vicuñas purchased by Ledger between1856 and 1858, only five survived the journey to Australia, some refusing to suckle their llama foster mothers, others perishing en route.[141] Despite these failures, however, the various vicuña domestication schemes offer an insight into a broader transatlantic programme of agricultural 'improvement' and reveal differing opinions as to where, and by whom, the vicuña should be farmed. For the Spanish Crown, the emphasis was on bringing vicuñas to Spain and domesticating them there for the benefit of Iberian artisans. For creoles such as Caldas the priority was making camelids serve the needs of South America (in Caldas's case, specifically New Granada) and taking advantage of more localised expertise. For Cabrera it was about improving the economic performance of the newly established Peruvian Republic by creating a novel, lucrative and hopefully sustainable agricultural export. And for Ledger and his supporters it was about preserving the 'innocent vicuña' from the 'blind and wilful destruction' inflicted upon it by 'ignorant' hunters and giving New South Wales the honour of being 'the only country providing this non plus ultra in wool'.[142] Vicuñas – like alpacas – thus meant different things to different people and were treated by turns as imperial, national and regional resources.

The vicuña itself, of course, did not benefit from all of this economic interest, and, with domestication a failure, was in severe need of

[139] 'Noticia exacta del Ingerto de Paco Vicuñas', *El Peruano*, 9 September 1846.
[140] C. Cumberland, *The Guinea Pig, or Domestic Cavy, for Food, Fur and Fancy* (London: L. Upcott Gill, 1886), pp. 37–8.
[141] 'Acclimatisation Society', *Sydney Morning Herald*, 2 February 1864.
[142] *Annotated Watercolour Sketches by Santiago Savage*, MLMSS 630/1.

protection. This prompted a move towards conservation, as successive regimes attempted to curb the unrestrained killing of a precious natural resource.

Conservation efforts on behalf of the vicuña began early in the colonial period and were initiated by the Spanish authorities. The first formal conservation measure came in 1557, when the Spanish Crown imposed a five-year moratorium on vicuña hunts to allow the population to recover.[143] A more intensive burst of legislation came in the mid-eighteenth century, as the Spanish authorities began to view natural resources as potentially renewable, if subjected to proper management.[144] In 1768, for instance, an order was put in place in Upper Peru to prohibit the slaughter of vicuñas for their fleeces. In 1777, a second order was passed, advocating the introduction of shear and release; in 1789, a third order reiterated the earlier instructions. The 1768 order emphasised both the economic wastefulness of the current practice and the threat it posed to the vicuña, complaining that '[t]he Indians kill the vicuñas to tear the wool after being killed [and] occasion two injuries of consequence, one that it may become extinct or diminish and the other that a lot of Wool is lost when torn'. To avoid 'these inconveniences', it specified that vicuñas should be 'shorn and after released, as it is practiced with our livestock, if at all possible'.[145] The 1777 order stated that *corregidors* (royal officials) in the Audiencia of Charcas (modern-day Bolivia) must prohibit the Indians from 'killing the vicuñas in those hunts that of their own accord or at the bidding of their priests or corregidors they often engage in'.[146] Sadly, none of these orders appear to have been enforced effectively – a fact underlined by the repeated reissuing of the same measures – and the vicuña remained in peril.

Following independence, efforts to protect the vicuña continued, reflecting both the renewed threats to the animal from a growing global market for its wool and the esteem in which it was held by Peruvian legislators. In 1825, the Liberator Simón Bolívar took action on the vicuña's behalf, outlawing the killing of vicuñas for their wool and

---

[143] Yacobaccio, 'The Historical Relationship between People and the Vicuña', p. 14.
[144] For a discussion of this change in approach, see 'Imperial Reform, Local Knowledge and the Limits of Botany in the Andean World' in Matthew Crawford, *The Andean Wonder Drug: Cinchona Bark and Imperial Science in the Spanish Atlantic, 1630–1800* (Pittsburgh: University of Pittsburgh Press, 2016), pp. 130–50.
[145] Yacobaccio, 'The Historical Relationship between People and the Vicuña', p. 16.
[146] 'Real Cédula a la Audiencia de Charcas, para que cuide de que no se hagan cazerías en que se maten las vicuñas (30 de Agosto de 1777)' in *Documentos para la Historia del Río de la Plata*, Vol. III (Buenos Aires: Compañía Sud-Americana de Billetes de Banco, 1913), pp. 40–1.

advocating a form of shear and release.[147] Twenty years later, in 1845, the Peruvian Government replaced this order with a more comprehensive framework of legal protection, designed not only to prevent the overhunting of vicuñas but to maintain the nation's monopoly over alpacas. The new law stipulated that 'the exportation of live vicuñas and alpacas to foreign countries' was 'absolutely prohibited'; that earlier Spanish decrees banning the killing of vicuñas would be reinstated; and that the authorities would prosecute anyone found continuing the 'barbarous custom' of killing vicuñas in 'traps' or 'with dogs'. To ensure that no offenders could 'plead ignorance' of the law, the government ordered that its provisions be read out by priests to their parishioners on 'four consecutive Sundays', reaching even the illiterate masses of the sierra.[148] Further decrees in 1851 and 1868 reinforced the export ban, outlawing 'the removal from Peruvian territory of alpacas and vicuñas, and of any species of animal that proceeds from the crossing of the two races'.[149] In 1907, Peru banned the export of vicuña skins completely, and in 1918 neighbouring Bolivia followed suit.[150]

Despite these measures, the vicuña was critically endangered by the mid-twentieth century, with a mere 6,000 left in the wild in 1974. This prompted a final and ultimately highly effective effort to save the species from extinction that saw vicuñas placed on Appendix I of the Convention on International Trade in Endangered Species of Wild Fauna and Flora (CITES) and an absolute export ban imposed on their fleeces. Aware of the need for international cooperation in conservation, the governments of Peru, Bolivia, Chile and Argentina issued legislation to protect vicuñas within their territories and signed an international agreement in 1979 (along with Ecuador) to coordinate conservation efforts. In 1967, the Peruvian Government established the first of several special reserves for vicuñas at Pampa Galeras, in Ayacucho province, where the animals could be monitored by scientists and protected from poachers.[151]

Thanks to these conservation measures, the decline in the vicuña population was reversed and their numbers rapidly increased – even

---

[147] Daniel Florencio O'Leary, *Memorias del General O'Leary* (Caracas: Imprenta de El Monitor, 1883), p. 363.

[148] *El Comercio*, 13 August 1845; 'Mr Charles Ledger and his Alpaca Contract with New South Wales', *The Era*, 25 September 1859.

[149] 'Lima, Abril 1° de 1851', *El Comercio*, 8 April 1851; *Decreto Estableciendo la Prohibición de Extraer del Territorio Peruano las Alpacas, Vicuñas y Animales que Proceden del Cruzamiento de Ambas Razas*, 8 October 1868, Archivo Digital de la Legislación del Perú.

[150] Yacobaccio, 'The Historical Relationship between People and the Vicuña', p. 17.

[151] Hernán Torres (ed.), *South American Camelids: An Action Plan for Their Conservation* (Gland: IUCN Species Survival Commission, 1992), p. 3.

leading to debates at the end of the 1970s over whether some of the animals should be culled or translocated. The guerrilla war unleashed by *Sendero Luminoso* (Shining Path) in the 1980s constituted a temporary setback for vicuña conservation, with poaching once again unchecked and the reserves unpoliced, but by the early 1990s the species was again in recovery, and vicuñas were downgraded in Appendix II of CITES.[152] Classed as a Peruvian heritage species, the vicuña still cannot be hunted, but it can be shorn and released by indigenous people using the traditional Inca *chakku* technique. With vicuña numbers reaching 340,000 in 2015 and continuing to rise, there is cautious optimism about the animal's future and hope that its controlled exploitation may benefit local communities.[153] In recent years, indeed, vicuña conservation has become a model for the sustainable use paradigm of wildlife management, attracting international attention; South Africans advocating the farming of rhinos for their horns cite the management of the vicuña as an example of successful conservation through sustainable use, although Chilean vicuña expert Christian Bonacic is cautious about the extrapolation of this practice to another, very different, species.[154]

---

[152] For a detailed analysis of twentieth-century efforts to save the vicuña – and their persistence across very different political regimes – see Emily Walkid, 'Saving the Vicuña: The Political, Biophysical and Cultural History of Wild Animal Conservation in Peru, 1964–2000', *American Historical Review* 125:1 (2020), pp. 54–88.

[153] 'Legalizing Rhino Horn Trade Won't Save Species, Ecologist Argues', *National Geographic*, 8 January 2015.

[154] Ibid.

5   Bitter Perfumes

In August 1824, two hairdressers, Mr Macalpine and Mr Money, were summoned before the Lord Mayor of London for keeping live bears on their premises in Threadneedle Street. The animals, both Russian bears, were sources of bear's grease, widely used as a hair product in the early nineteenth century, and were exhibited by their owners 'for the purpose of demonstrating ... that it is not scented suet, or hog's lard, or anything but genuine bear's grease which they ... sell'. Over recent weeks, however, the bears had severely tested the patience of local residents, 'disturbing the whole street by their noise' and attracting large crowds of onlookers to the vicinity, which 'blocked up the thoroughfare'. One of the animals, it was reported, 'could put his leg or arm out to its full extent and seize any passengers with its claws'. The other was said to be 'almost entirely at liberty, and might, if it so pleased him, vent his displeasure on any of his Majesty's subjects who came near him'. Both bears emitted unwanted levels of noise, making 'the place resound [during the night] with hideous howls'.

Responding to the charges against them, the hairdressers protested their innocence. Called to the witness stand, Mr Macalpine 'declaimed with fury in defence of his bear, and endeavoured to make the Lord Mayor believe that it was as harmless as a lamb'. He stated that '[h]e had killed one bear already to appease the prejudice of the place, but he would not immolate the present bear to gratify anyone'. Mr Money was unable to attend court in person, being in Paris at the time, but his underling – 'a smart young man with his hair cut and curled with mathematical precision' – testified that the bear on his premises was exceedingly tame. 'Even if it was inclined to do mischief it could not, as its cage was so far from the window as to prevent the possibility of its reaching.' The man did suggest, however, that 'whatever he had said in defence of his master's bear did not apply to any other *hanimal*', and '[i]t would doubtless promote the peace of the neighbourhood if the *opposition* bear were confined'. Having listened to the arguments of the defendants, the Lord Mayor informed both parties that he would suspend judgment for now, but that should either of their animals 'remain loose, or ... create any further annoyance', they would 'certainly be indicted as a nuisance'. 'Upon this understanding, the parties retired.'[1]

The curious incident of the bears in the night highlights the surprising source of one of Georgian Britain's most popular beauty products: bear's grease. Used widely as a hair restorative, bear's grease was made from the fat of dead bears and imported in large quantities from Russia. Commonly counterfeited, it soon became a byword for fraudulent practices and dodgy dealings, leading some vendors – Mr Macalpine and Mr Money among them – to take drastic steps to proclaim the authenticity of their merchandise. The residents of Threadneedle Street clearly did not appreciate the presence of living bears in their midst, complaining repeatedly to their owners and ultimately reporting the hairdressers to the authorities. For many customers, however, knowing where their cosmetic products came from was a major bonus, offering valuable proof that their lotions, pomades and perfumes were the genuine article.

This chapter focuses on animal substances – including bear's grease – used in perfumes and lotions. Most perfumes in the eighteenth and nineteenth centuries contained animal bodily fluids – fat, scent and even vomit – so this encompasses a wide range of products. Here, I concentrate on four widely used commodities: bear's grease, obtained from Russia and employed as a hair restorative; musk, an extremely pungent substance extracted from a 'pod' belonging to a species of Himalayan deer; civet, secreted by the animal of the same name and imported from Indonesia and later Ethiopia; and ambergris, 'the morbid secretion of the spermaceti whale'.[2] I examine how all of these products were obtained, processed and advertised and consider the ecological and humanitarian concerns associated with their acquisition.

One particularly contentious issue surrounding almost all perfumes was adulteration. As the case of Messrs Macalpine and Money demonstrates, it was an open secret that bear's grease was frequently adulterated and the original scent replaced with other, cheaper, substitutes. The same was true of musk, civet and ambergris, which, as we shall see, were also subject to repeated counterfeiting. Addressing this persistent

---

[1] 'Police', *The Morning Chronicle*, 4 August 1824.
[2] 'Ambergris', *Our Young Folk's Weekly Budget*, 19 November 1881.

concern, the chapter examines how traders and vendors in Britain and beyond diluted or fabricated animal substances and how legitimate sellers attempted to detect frauds and authenticate their own wares. I situate their efforts within a wider discourse on food adulteration, a major concern throughout the Victorian era.[3]

Bear's Grease

Bear's grease was the Georgian equivalent of hair conditioner. Made from the fat of dead bears, it was applied to the scalp to counter dryness and was believed to moisten and nourish the hair. Hairdressers and perfumers routinely marketed bear's grease to their clients, claiming that it had the power to fortify hair weakened by illness, especially fevers, and even that it could encourage hair growth on the head and face – a notion that stemmed from the belief that bears themselves were hairy and that their fat could therefore stimulate the growth of hair in humans. One 1820 advertisement for Marsh's bear's grease claimed that it 'prevents the falling off or turning grey' of the hair, countering 'the Dryness of the Head and the Debility of the Bulbs of the Hair'.[4] A later advertisement for James Lipscombe's bear's grease (1838) stated that his product, 'by bracing the pores of the head, strengthens the weakest Hair', providing nourishment after 'sickness, accouchement, alarm, trouble, study ... sea bathing and violent exercise'.[5]

According to the perfumer Alexander Ross, bear's grease was invented by a London apothecary named Mr Townsend in the late seventeenth century.[6] Initially, its use was limited, but by the mid-eighteenth century it was widely available and in high demand across Britain. The product enjoyed a substantial boost in popularity in 1795 when William Pitt's government imposed a controversial tax on hair powder to fund the war effort against revolutionary France, forcing many financially challenged Britons to abandon powdered wigs in favour of their natural hair.[7] Generally imported from Russia, bear's grease could be found in the shops of most perfumers, druggists and hairdressers and typically retailed

[3] See Jack Goody, 'Industrial Food' in Carole Counihan and Penny Van Esterik (eds), *Food and Culture: A Reader* (London: Routledge, 1997), pp. 351–3.
[4] 'London Hat Warehouse', *Leeds Mercury*, 2 December 1820.
[5] 'James Lipscombe, Hair Cutter and Dresser', *Brighton Patriot*, 27 October 1835.
[6] Alexander Ross, *A Treatise on Bear's Grease* (London: Printed for the Author, 1795), p. 10.
[7] See 'Hair Powder' in John Barrell, *The Spirit of Despotism: Invasions of Privacy in the 1790s* (Oxford: Oxford University Press, 2006), pp. 145–209; 'Powder' in Susan J. Vincent, *Hair: An Illustrated History* (London: Bloomsbury, 2018), pp. 173–89.

at around 2s 6d per pot.[8] In London, the principal vendors of bear's grease were Vickery's Fashionable Repository of Inimitable Head Dresses, at No. 6 Tavistock Street, Covent Garden; Messrs Ross and Sons, No. 119 and 120 Bishopsgate Street; and J. Atkinson, Perfumer, 44 Gerrard Street, Soho Square.[9] In the provinces, bear's grease could be purchased at Messrs Piper and Son, Perfumers, 238 High Street, Exeter; Mr Ward, Perfumer and Patent Medicine Vendor, Civet Cat, 33 College Green, Bristol; W. Payne, Hair Cutter, 13 Queen Street, Portsea; E. Parsons, Ladies and Gentlemen's Hair Cutter, No. 10, Mosley Street, Newcastle; J. Lipscombe, Hair Cutter and Dresser, 59 Grand Parade, Brighton; and J. and W. Marshall, Chemists, No. 8 High Street, Belfast, among others.[10]

Of all these vendors, the two most prominent were Alexander Ross and James Atkinson. Ross, a hairdresser, owned a shop on Bishopsgate Street, 'three Doors from the London Tavern' in Spitalfields. His rival, Atkinson, a perfumer, operated from Soho Square but distributed his stocks of bear's grease to a wide range of establishments outside London (including many of those listed above). Both invested heavily in marketing their respective products, posting regular advertisements in the London newspapers and, in Ross's case, publishing a sixty-nine-page *Treatise on Bear's Grease* in 1795.[11] Both also employed more artistic means of publicising their offerings, developing eye-catching logos for their shops and branded containers for their wares. Figure 5.1 shows a bear's grease pot produced by Ross, which features a painted scene of a bear being hunted in the wild. It has the vendor's name and address engraved around the circumference.[12] Figure 5.2 depicts a polar bear perched on an ice sheet and was used by Atkinson to attract customers into his establishment.[13]

While they both sold the same product, Ross and Atkinson obtained their bear's grease in different ways, each claiming his own practice as

---

[8] 'Vickery's Magazine for Imperial Head Dresses, Real Bear's Grease, &c.', *The Morning Chronicle*, 11 January 1805; 'Advertisement', *The Morning Post*, 17 January 1824; 'Advertisement. Caution – Bear's Grease', *The Morning Post*, 7 January 1824.

[9] 'Genuine Bear's Grease', *Trewman's Exeter Flying Post*, 31 May 1821; 'J. Atkinson, Perfumer, Gerrard Street', *Bristol Mercury*, 17 November 1823; 'W. Payne, Hair Cutter', *Hampshire Telegraph*, 11 April 1825; 'E. Parsons, Ladies and Gentlemen's Hair Cutter', *The Newcastle Courant*, 6 May 1826; 'James Lipscombe, Hair Cutter and Dresser', *Brighton Patriot*, 27 October 1835; 'The Winter Season', *Belfast News-Letter*, 8 November 1839.

[10] 'Imitation Bear's Grease', *The Morning Chronicle*, 9 December 1824.

[11] Ross, *A Treatise on Bear's Grease*.

[12] Royal Pharmaceutical Society Museum, Collection Reference MA3.

[13] Royal Pharmaceutical Society Museum, Collection Reference YBC3.

Figure 5.1 Ceramic pot lid for 'Ross & Sons' Genuine Bear's Grease Perfumed', mid-1800s. Royal Pharmaceutical Society Museum, Collection Reference MA3

Figure 5.2 Staffordshire ceramic creamware model advertising James Atkinson's bear's grease, 1799–1818. Royal Pharmaceutical Society Museum, Collection Reference YBC3

superior. Atkinson, on the one hand, imported his bear's grease from Russia, having 'entered into correspondence with a mercantile house' in that country. He insisted that this approach was best, bear's grease, when 'procured from the animal in its native climate' being 'considered more lively and penetrating than that from the animal domesticated in a foreign climate, which in that state is observed to lose much of its vivifying properties'.[14] Ross, by contrast, fattened live bears on his premises in London and slaughtered them at intervals before the public. In January 1824, he announced that he and his sons

have fatted one of the largest of those animals, which they intend killing on Tuesday, the 27th of January inst. Any Lady or Gentleman wishing to have the Grease in its original state, may see it cut from the animal on the day specified, and as long after as it will keep, which, of course, must depend entirely on the weather.[15]

In November the same year he informed the public that he had 'at present five bears, which will be slaughtered between this and February next – due notice of which will be given when those who prefer having the fat before it is rendered down may be accommodated'.[16] Ross claimed that he and his sons were 'the only parties who really fat and kill these animals' and emphasised the value of this for ensuring freshness and avoiding counterfeits (a major concern, as we shall see later). Customers could thus choose between the produce of wild and domesticated bears, depending on whether they preferred their bear's grease freshly killed or Russian-reared.

Musk

Musk is produced by the male musk deer (*Moschus moschiferus*), a small species of deer found in the Himalayas of India, Tibet and China and in the Altai Mountains near Lake Baikal. According to Jesuit Athanasius Kircher, writing in 1667, 'The provinces of Suchuen [Sichuan] and Junnan [Yunnan] greatly abound with this species of animal, and though it may be observed in all the countries of China none have it in such numbers as those which approach nearest the west.' Known as 'Xe', in Chinese, 'which means *odour*', the musk deer is a delicate little animal, well adapted to life in a cold, dry climate.[17] An 1862 article in

---

[14] 'Advertisement', *The Times*, 14 April 1823.
[15] 'Advertisement', *The Morning Post*, 17 January 1824.
[16] 'Advertisement', *The Times*, 9 November 1824.
[17] 'Musk-Deer', *Penny Magazine of the Society for the Diffusion of Useful Knowledge*, 15 June 1839.

The Illustrated London News described the species as having 'the general outline, head, legs and feet … of the common deer', the dimensions of a goat and a coat that 'is neither hair nor bristles, but is like very fine porcupine quills'.[18] Colonel Frederick Markham, who hunted the musk deer in Nepal in the 1850s, reported that it 'stands nearly two feet high at the shoulder', is 'dark speckled brownish gray' in colour, 'deepening to nearly black on the hind quarters', and is covered in 'thick spiral hairs, not unlike miniature porcupine quills … so thickly set, that numbers may be pulled out without altering the outward appearance of the fur'. Male musk deer are also noted for their long canine teeth, which are 'about three inches long' and 'the thickness of a goose-quill; sharp pointed, and curving slightly backwards'.[19]

Musk itself emanates from 'a pocket … under the belly of the musk deer', referred to by the Chinese as a 'pod'.[20] It was once thought to have medicinal virtues as a stimulant and antispasmodic, but by the nineteenth century it was prized mainly for its scent.[21] According to Markham, 'The musk … pod, which is placed near the navel, and between the flesh and the skin, in which the musk is confined … has much the appearance of the craw or stomach of a partridge, or other small gallinaceous bird, when full of food.' The musk itself takes the form of 'grains, from the size of a small bullet to small shot, of irregular shape, but generally round or oblong, together with more or less in coarse powder'. When fresh, it is of 'a dark reddish brown colour', but once removed from the pod it becomes 'nearly black'. In autumn and winter the grains are 'firm, hard and nearly dry', but in summer they grow 'damp and soft, probably from the green food the animals then eat'. Markham concluded that the musk was 'formed with the animal, as the pod of a young one, taken out of the womb, is plainly distinguishable, and indeed is much larger than in the grown up animals'.[22] Only the male deer possesses the musk pod, and each deer produces only a small amount: 'An ounce may be considered as the average from a full grown animal; but as many of the deer are killed young, the pods in the market do not perhaps contain, on an average, more than half an ounce'.[23]

---

[18] 'The Musk Deer', *The Illustrated London News*, 6 September 1862.
[19] Colonel Frederick Markham, *Shooting in the Himalayas: A Journal of Sporting Adventures and Travel* (London: Richard Bentley, 1854), pp. 85–7.
[20] Eugene Rimmel, *The Book of Perfumes* (London: Chapman and Hall, 1865), p. 245.
[21] In the medieval Islamic world, musk was used to treat a wide range of conditions, including 'heart palpitations, bad winds, fevers, indigestion, coughing up blood, shortness of breath, hiccups, liver and spleen pain, stomach complaint, delayed menses, headache, poison and insect and reptile bites'. See Anya H. King, *Scent from the Garden of Paradise: Musk and the Medieval Islamic World* (Leiden: Brill, 2017), p. 314.
[22] Markham, *Shooting in the Himalayas*, pp. 87–90.   [23] Ibid., pp. 87–90.

For centuries musk had been collected by merchants in the Far East and imported to Europe and the Middle East. Jean-Baptiste Tavernier, a seventeenth-century French gem merchant who travelled widely in Asia, stated that the musk deer was present 'in vast numbers' in the 60th degree in forested areas, whence 'in the months of *February* and *March*, after these creatures have endur'd a sharp hunger, by reason of the great Snows that fall where they breed, ten or twelve foot deep, they will come to 44 or 45 degrees to fill themselves with Corn and new Rice'. During this season, 'the Natives lay gins and snares for them to catch them as they go back: shooting some with Bows, and knocking others on the head'.[24] Markham gave a similar account of the collection of musk in northern India in the mid-nineteenth century, describing how musk deer, considered the property of the rajahs in Gurwhal, were 'hunted down with dogs' or snared in traps.

In snaring, a fence about three feet high, composed of bushes and branches of trees, is made in the forest, generally along some ridges, and often upwards of a mile in length. Openings for the deer to pass through are left every ten or fifteen yards, and in each a strong hempen snare is placed, tied to a long stick, the thick end of which is firmly fixed in the ground, and the smaller, to which the snare is fastened bent forwards to the opening; so that the deer, when passing through, treads upon some small sticks which hold it down, the catch is set free, the stick springs back and tightens the snare round the animal's leg.[25]

Extracting musk thus entailed killing large numbers of deer – not just full-grown males, but females and juveniles as well.

Once collected, musk was 'packed in peculiar silk-covered boxes or caddies, containing about twenty-five pods each', and shipped to Europe, initially by Chinese merchants and later by British traders, who began to penetrate the market in the 1860s.[26] The quantities of musk imported in this way were small, because its scent was so strong, but its value was high. In 1862, perfume manufacturer Septimus Piesse calculated that 'the average importation of musk per annum for the past five years is 9388 oz, of the value of £10,688; of which we export to France and other places 1578 oz'.[27] In May 1883, in a single sale, London brokers Barber Brothers auctioned off just two tins of Yunnan musk for a total of £52 6s.[28] Buyers could choose from four different

---

[24] Jean-Baptiste Tavernier, *The Six Voyages of John Baptista Tavernier, Baron of Aubonne through Turky, into Persia and the East-Indies, for the Space of Forty Years*, translated by Daniel Cox (London: William Godbid, Robert Littlebury and Moses Pitt, 1677), 'Travels in India', Second Book, p. 153.

[25] Markham, *Shooting in the Himalayas*, pp. 94–6.

[26] 'The Musk Deer', *The Illustrated London News*, 6 September 1862.   [27] Ibid.

[28] LMA 4605/02/002.

kinds of musk, categorised according to the geographical area from which they came: Tonquin musk (from central China), Yunnan musk (from southern China), Assam musk (from north-eastern India) and Carbadine musk (from Siberia).[29] Tonquin musk was considered the best quality, followed by musk from Assam. Carbadine musk, 'obtained from a variety of the species called Kubaya (*Moschus sibiricus*)', was regarded as 'inferior'.[30]

Civet

Civet comes from the civet cat, a small carnivorous animal of the Viverridae family. One species of civet, the Malayan civet (*Viverra tangalunga*), lives in Malaysia, Indonesia and the Philippines, while a second, the small Indian civet (*Viverricula indica*), lives in India. Another species, the African civet (*Civettictis civetta*), inhabits sub-Saharan Africa. Solitary and nocturnal, civets live in tropical forest and savannah and subsist on a diet of fruit and small mammals. The substance civet, 'a greasy and intensely strong secretion', is stored in the animal's perineal gland, in a pouch below the tail.[31]

Traditionally, civet was extracted from live animals, hunted and caged for the purpose. The creatures were confined in small cages and the scent scooped out of their pouches at intervals to supply the burgeoning market for perfumes. To accomplish this, the civets were first secured in the civet equivalent of a cattle crush and their precious scent extracted with a spoon. As the naturalist Reverend J. G. Wood explained:

The animals which belong to this group are very quick and active in their movements, and, being furnished with sharp teeth and strong jaws, are dangerous beasts to handle. As may be imagined, the civet resents the rough treatment that must be used in order to effect the desired purpose, and snaps and twists about with such lithe and elastic vigour that no-one could venture to lay a hand on it without sufficient precaution. So, when the time arrives for the removal of the perfume, the civet is put into a long and very narrow cage, so that it cannot turn itself round. A bone or horn spoon is then introduced through an opening, and the odiferous secretion is scraped from its pouch with perfect impunity. This end achieved, the plundered animal is released from its strait durance, and is permitted a respite until the supply of perfume shall be re-formed.[32]

[29] 'About Musk', *Essex Standard*, 26 January 1889.
[30] Rimmel, *The Book of Perfumes*, pp. 242–3.
[31] 'Sweets to the Sweet: Something about Perfume', *Hampshire Telegraph*, 28 January 1893.
[32] J. G. Wood, *Illustrated Natural History* (London: Routledge, 1876), p. 229.

In the seventeenth century, large numbers of civets were kept in captivity by the Dutch and, in effect, 'milked' for their scent. *Cassell's Natural History* recorded: 'The Dutch used to keep numbers of civets alive at Amsterdam, for the purpose of collecting the perfume when secreted,' removing the substance with 'a small spatula'.[33] Naturalist William Bingley reported that civets were 'kept in great number, and with a commercial view, at Amsterdam', where they were fed with 'boiled meat, eggs, birds, small quadrupeds and fish' and voided of their scent 'twice or thrice a week'.[34] Several London merchants also participated in the trade, rearing civets in their properties. An advertisement in the *Collection for Improvement of Husbandry and Trade* from 1694 stated: 'At the Civet house in Newington Green are Threescore Civet Cats, with a considerable quantity of Civet to be sold.'[35] In 1722, John Lloyd advertised the sale of 'Five CIVET CATS', kept on his premises 'at Canberry-House in Islington Parish'.[36]

In the late nineteenth century, the location of civet farming shifted from East Asia and Europe to East Africa. Writing in 1896, Prince Henri d'Orléans reported that civets were 'caught wild' there 'and tamed', with a single civet yielding up to '80 grammes of matter a week'.[37] A subsequent article in *The Cornishman* stated that 'About 20,000 ounces of civet (the perfume scraped from the pouches of the civet cat) are yearly sent to London. Of pure Jeddah civet the value is 8s 6d per ounce, of other kinds 7s.'[38] The town of Eufres in Abyssinia became the prime site for civet production and presided over a sizeable industry. According to *The Times of India*, 'Civet cats are largely domesticated there, being fed when young on farinaceous food with a little fish or flesh, and on raw flesh after they have grown up, this diet being believed to have the property of increasing the secretion, which is taken from the bag twice a week with an iron spatula.' A male cat 'in good condition' yielded 'about a dram each time, while the females secrete very much less as a rule'.[39] Once extracted, civet was poured into buffalo horns for shipment to Europe and carried by traders to the coast. 'A kilogramme of this

---

[33] 'Cassell's Founder Natural History', *The Morning Chronicle*, 12 March 1860.
[34] William Bingley, *Useful Knowledge: or A Familiar Account of the Various Productions of Nature, Mineral, Vegetable and Animal, which Are Employed for the Use of Man* (London: Baldwin, Craddock and Joy, 1821), Vol. III, pp. 41–2.
[35] 'Advertisements', *Collection for Improvement of Husbandry and Trade*, 4 May 1694.
[36] 'Advertisements', *Post Boy*, 9 August 1722.
[37] 'Events in Abyssinia', *Pall Mall Gazette*, 5 August 1896.
[38] 'African War News', *The Cornishman*, 9 January 1902.
[39] 'Civet Cats and Civet', *The Times of India*, 18 May 1907.

matter, which comes here in buffaloes' horns, is sold at from 1.100f to 1.200f in France, and is used in the manufacture of perfumes'.[40]

Ambergris

Our final scent, ambergris, originates in the intestines of the sperm whale, hunted for its spermaceti. It is sometimes found inside the animal after harpooning, but often discovered floating on the surface of the sea or washed up on beaches. Used, like musk and civet, to make perfume, ambergris was most often added to other scents to accentuate their pungency and was employed to flavour 'scented pillars, candles, balls or bottles, gloves ... hair powder' and 'pomatums for the face and hands'.[41] *Trewman's Exeter Flying Post* likened ambergris to 'dirty brown cheese, old dry and corky', noting that it was 'valuable in blending perfumes together, though it has but little perfume of its own'.[42] The *Western Times* claimed that '[e]very bottle of perfume sold in Bond-street has tiny bit of ambergris in it'.[43]

In the nineteenth century, the origins of ambergris were disputed. One article in the *Manchester Courier* hypothesised that it was 'the result of some disease in the sperm whale, analogous to gall stones'.[44] Another, in *The Manchester Guardian*, described it as 'a fragrant gummy substance containing cuttlefish beaks'.[45] Naturalist Frank Buckland concurred with the latter view, speculating that when whales swallowed cuttlefish their beaks became sites of disease in the cetaceans, triggering the production of ambergris. Following a detailed examination of the substance, Buckland concluded: 'There can ... be no doubt whatever that ambergris is the refuse of the whale's food collected in a morbid form.' 'When the whale swallows the cuttlefish the soft parts are digested but the hard beaks remain intact ... [T]hese cuttlefish beaks act as nuclei for the formation of a diseased mass, which, to use Mr Dewhurst's own words, "produces an obstipation, which ends either in an abscess, as has been frequently observed, or terminates the life of the animal".'[46] Modern

---

[40] 'Events in Abyssinia', *Pall Mall Gazette*, 5 August 1896.

[41] Franz Xavier Schwediawer, 'An Account of Ambergris', *Philosophical Transactions* LXXIII (1783), pp. 394–5.

[42] 'London Docks Museum', *Trewman's Exeter Flying Post*, 10 December 1898.

[43] 'Value of the Whale', *Western Times*, 14 January 1919.

[44] 'London Docks Museum', *Trewman's Exeter Flying Post*, 10 December 1898; 'Concerning Ambergris', *Manchester Courier*, 9 April 1909.

[45] 'British Toothed Whales', *The Manchester Guardian*, 1 September 1911.

[46] 'The Origin of Ambergris', *The Manchester Guardian*, 30 August 1871.

science broadly supports Buckland's interpretation, suggesting that ambergris is indeed formed around the undigested beaks of cuttlefish. Although usually passed naturally by the animal, these sometimes occlude the gut completely, triggering a fatal intestinal rupture.[47]

Geographically, ambergris was widespread, found in coastal regions around the globe. *The Graphic* reported that 'It is mostly found upon the coasts of Greenland, Brazil and China.'[48] Buckland claimed that it could be found 'floating in the sea, or cast up on the seashore in the neighbourhood of Madagascar, Jamaica, Bermuda, Maldives, Brazil, Molucca [Maluku Islands], Japan, China, [and the] coast of Africa', with 'the greatest supply now coming into the market ... from the Bahamas and outside Morocco'. Ambergris was also sometimes found on the west coast of Ireland, on the coast of Norfolk and in the Orkneys.[49]

Because it was so widely dispersed and so rare, ambergris was neither farmed nor hunted but usually stumbled upon as a chance discovery – a fact that made it all the more valuable. Writing in 1871, Buckland stated that 'the price varied from 10s. to 50s. per ounce'.[50] In January 1874, nineteen bags of ambergris sold in London for 50 shillings, while in November 1883 ambergris sold at 75–89 shillings per ounce.[51] Stories abounded of individuals growing rich from the discovery of a single lump of ambergris, or, more commonly, finding large quantities of ambergris and not knowing what it was. In 1889, for instance, an American whaler found 'a lump [of ambergris] worth £60,000 floating off the Cape Marie van Diemen, which the New Zealand Maoris, unaware of the value of the substance, were using to kindle fire and cook fish'.[52] In 1895, another whaler, the *Walk Helm*, returned to Hobart with the carcass of a spermaceti whale and presented it to some poor fishermen as a gift. The fishermen subsequently discovered a lump of ambergris worth £10,000 inside the animal, leading to a lawsuit over who owned this valuable find (the fishermen won).[53] Ambergris was thus a curious, elusive and extremely precious commodity – at least in European culture – worth collecting, stealing or fighting a lawsuit over.

[47] Christopher Kemp, *A Natural (& Unnatural) History of Ambergris* (Chicago: University of Chicago Press, 2012), p. 13.
[48] 'Perfumes', *The Graphic*, 22 June 1878.
[49] 'The Origin of Ambergris', *The Manchester Guardian*, 30 August 1871.   [50] Ibid.
[51] 'London Produce Market', *The Manchester Guardian*, 23 January 1874; 'London Produce Markets', *The Manchester Guardian*, 2 November 1883.
[52] 'Notes on Trade of the World', *Sheffield Daily Telegraph*, 1889.
[53] 'A Fortune for Fishermen', *Aberdeen Journal*, 16 January 1895.

The Whiff of Deception

The high prices commanded by animal perfumes made them lucrative commodities for collectors, merchants and perfumers. They also made them ripe for fraud and adulteration. With money to be made from defrauding consumers, individuals all along the supply chain found innovative ways to maximise their cut, either diluting the genuine article or replacing it entirely with some cheaper substance. Like bread, milk and tea – all commonly adulterated in the mid-nineteenth century – musk, civet and bear's grease were frequently doctored, leaving customers uncertain as to their origin and purity.[54]

Musk was routinely adulterated. Vendors added soil and blood to the musk pods to artificially increase their weight or fabricated the pods completely with skin from elsewhere on the deer's body. Writing in 1889, the *Essex Standard* reported that 'musk is often found mixed with lead, iron, copper, sand, dried blood, or even paper or rags, to increase the volume and weight. After the introduction of the foreign bodies, the pouches are closed up again in so ingenious a manner, that only the eye of an expert can discover the fraud'.[55] Markham, who visited Nepal in the 1850s, offered a detailed description of adulteration methods and noted that they were difficult to detect. 'The musk received from the Puharries,' he alleged, '[was] greatly adulterated, and pods are often made altogether counterfeit; and as they are generally sold without being cut open, it is scarcely possible to detect the imposture at the time.' In some cases pods were brought to market 'which were merely a piece of musk-deer skin filled with some substance, and tied up to resemble a musk-pod, with a little musk rubbed over to make it smell'. In other cases, the musk was removed from genuine pods 'and its place supplied by some other substance'. The 'substances commonly used for adulteration' were 'blood boiled, or baked on the fire, then dried, beaten to a powder, kneaded into a paste, and made into grains and coarse powder to resemble genuine musk; a piece of the liver or spleen prepared in the same manner; dried gall, and a particular part of the bark of the apricot tree, pounded and kneaded as above'. One gentleman was sold a musk pod 'filled with hookah tobacco'.[56]

---

[54] Food adulteration was a major problem in the nineteenth century, reaching its peak in the 1850s. Tea was often laced with plum and ash leaves, coffee adulterated with chicory and bread whitened with alum. See E. J. T. Collins, 'Food Adulteration and Food Safety in Britain in the 19th and Early 20th Centuries', *Food Policy* 18: 2 (1993), pp. 95–109.

[55] 'About Musk', *Essex Standard*, 26 January 1889.

[56] Markham, *Shooting in the Himalayas*, pp. 98–100.

Civet and ambergris were likewise susceptible to adulteration, or, in some cases, complete substitution, a practice that served to diminish their market value. According to naturalist William Bingley, writing in 1821, civet farmers in the Levant adulterated their product by 'mixing it with storax and other balsamic and odiferous substances'.[57] Nearly a century later, the same practices were still being employed in the civet industry of Abyssinia, where civet was frequently adulterated 'with butter, lard and other greasy substances, so as to increase its weight'. On the west coast of India, meanwhile, where 'tame civet cats' were occasionally brought to the towns for sale, their valuable secretion was adulterated with 'a mixture of plantain and ghee'.[58] As for ambergris, it was widely contaminated, sometimes at the point of collection and sometimes by merchants and retailers. Dr Franz Xavier Schwediawer, who wrote a chemical analysis of ambergris in 1783, reported, 'The great price of ambergris, an ounce of it being now sold in London for 1l sterling, has been hitherto the cause of its being so often adulterated' – either with 'flower of rice, or with styrax or other resins'.[59] An 1870 report from the India Home Office recorded similar practices, in this case performed by traders in the Nicobar Islands. In this instance, the ambergris was said to have been adulterated with 'the wax of a small bee', rendering it 'of a very inferior quality'.[60]

One reason why musk, civet and ambergris were so easy to adulterate was that the original scents smelled so strongly and could thus tolerate a considerable amount of dilution. Tavernier claimed that the odour of musk was so powerful that it 'would cause the blood to gush out of the nose, so that it must be qualifi'd to render it acceptable, or rather less hurtful to the brain'. He brought the skin of a dead musk deer back to Paris for closer examination, but found that its scent was 'so strong, that I could not keep it in my Chamber; for it made all peoples heads ake that came neer it'.[61] A second seventeenth-century French traveller to China, Jean-Baptiste Chardin, asserted:

It is commonly believed, that when the musk-sac is cut from the animal, so powerful is the odour it exhales, that the hunter is obliged to have his mouth and nose stopped with folds of linen; and that often, in spite of this precaution, the pungency of the odour is such as to produce so violent a haemorrhage as to end in death.[62]

---

[57] Bingley, *Useful Knowledge*, Vol. III, p. 42.
[58] 'Civet Cats and Civet', *The Times of India*, 18 May 1907.
[59] Schwediawer, 'An Account of Ambergris', pp. 394–5.
[60] Kemp, *A Natural (& Unnatural) History of Ambergris*, p. 13.
[61] Tavernier, *The Six Voyages of John Baptista Tavernier*, 'Travels in India', Second Book, p. 153.
[62] 'Perfumes and Where They Come From', *The Ladies' Cabinet of Fashion, Music and Romance*, 1 October 1855.

Just a little authentic perfume could thus cover the presence of other less desirable substances, making the work of the fraudster that much easier.

Another reason why adulteration was feasible was because the animals that produced these scents were comparatively unknown in Europe and first-hand knowledge of their excretions limited. The sperm whale, like many marine creatures, was relatively little studied, and the origin of ambergris was not widely understood. The musk deer was equally alien to European consumers, and remained so well into the nineteenth century. The first living musk deer reached Britain only in 1869, and, as a female, did not 'give forth the slightest musky odour'.[63] Two male musk deer arrived in London Zoo in 1877, courtesy of Sir Richard Pollock, commissioner at Peshawar, but they 'still show[ed] traces of the spots in the fur which are found in the young of almost all species of deer' and may not yet have developed adult musk pods.[64] Having never seen living examples or smelled the scent of a real animal, British perfume buyers would thus have found it difficult to distinguish the authentic perfume from the fake.

As for bear's grease, it perhaps exceeded all other products in its reputation for skulduggery, with only a fraction of the substance emanating from actual bears. Theoretically comprising the fat from a bear's intestines and kidneys, much bear's grease was in fact often a compound of almond oil, hog's lard or mutton suet, which were cheaper to procure and easy to prepare. James Rennie, author of *The Art of Preserving the Hair* (1826), alleged that bear's grease was frequently substituted with 'the grease of dogs or goats: or in the case of those buyers who pretend to be judges of the true bearish odour, old, rancid, yellow hog's lard, which has acquired, by being rusty, a proper shade of yellow and a sufficient perfume'.[65] The persistent adulteration of bear's grease made the product a ripe subject for satire and even resulted in a string of prosecutions. In 1838, 'a person of foreign appearance named William Sherwood' was sentenced to one month's detention in the House of Correction in Preston for 'hawking some boxes of nasty stuff, which he sold as Bear's grease at 9d. per box'.[66] Two years later, a London hairdresser, Jeremiah Riggs, went to court in Tower Hamlets to claim unpaid bills from a client, a solicitor's clerk named Wilkinson, who had refused to reimburse Riggs for a product he claimed was bogus. Perhaps surprisingly, Riggs

---

[63] 'The Musk Deer at the Zoological Society's Garden', *The Illustrated London News*, 24 April 1869.

[64] 'Zoological Society's Gardens', *The Times*, 18 December 1877.

[65] James Rennie, *The Art of Preserving the Hair on Popular Principles: Including an Account of the Diseases to which It Is Liable* (London: Septimus Prowett, 1826), p. 175.

[66] 'A Vendor of "Bear's Grease"', *Preston Chronicle*, 8 December 1838.

confessed to the deception, but felt that he still deserved payment, since it was widely known that 'every pot o' bear's grease in London vos sometime or other hinside a pig'.[67]

While the Riggs case largely elicited laughter, the fake bear's grease industry generated real scandal in 1831 when it emerged that its main ingredient in fact came from stolen pet dogs. In one particularly disturbing case, police in Marylebone searched a property belonging to Samuel Province (alias Wedgebury) and Thomas Jackson and discovered the carcasses of between 200 and 300 dogs, which they believed had 'fallen victims to "the growth of the hair"', their fat having been boiled down and 'metamorphosed into bear's grease of the finest quality'. The officers found six living dogs inside a nearby cottage, including 'two fine bull-dogs ... fastened to a peg in the ground in a field adjoining', and, 'in the immediate neighbourhood', 'about 20 fine Newfoundland dogs, and some of these with persons who did not appear able to support themselves' – animals they suspected were being 'kept there until a reward should be offered for their recovery'. They also found 'a hammer ... with a quantity of hair adhering, which was supposed to be the instrument with which the work of death was carried on', and 'a very curious instrument, hooked at one end ... supposed to be used for catching dogs by the legs'. Armed with these macabre paraphernalia, Sergeant Stedman of T Division managed to secure the conviction of both defendants, both of whom were fined the hefty sum of £20, or, in the absence of payment, 'to hard labour for two periods of six months each'. The case, however, left a sour taste in the mouth for consumers of bear's grease, many of whom had unwittingly been party to the gruesome fraud. As *Jackson's Oxford Journal* pondered, how many unsuspecting beaus had been anointing their bald heads with 'this "invaluable discovery" manufactured at the dog pit'?[68]

Stamps, Signatures and Science

With so much fake perfume on the market, the onus was on legitimate vendors to verify the authenticity of their product. If public trust in all perfumes was not to be undermined, some way of distinguishing

---

[67] 'Tower Hamlets', *Northern Star*, 4 July 1840. Even more surprisingly, Riggs won his case, the magistrate ruling that 'if people are fools enough to use bear's grease and trash of that description they ought to be made to pay for their folly'.

[68] 'The Dog Stealers', *Jackson's Oxford Journal*, 19 February 1831. On the practice of dog stealing in nineteenth-century London, see 'Flush and the Banditti: Dog-Stealing in Victorian London' in Philip Howell, *At Home and Astray: The Domestic Dog in Victorian Britain* (Charlottesville and London: University of Virginia Press, 2015), pp. 50–72.

between the genuine and the counterfeit needed to be found, and customers had to be able to tell the difference. To this end, perfumers and merchants deployed three key weapons, all calculated to certify the origin of their wares: careful branding, scientific tests and, where possible, direct evidence of the source of the scent.

Branding was employed heavily in the sale of bear's grease in the form of verbal pledges and distinctive packaging. In the former case, the impetus appears to have come from the authorities, who, in the late eighteenth century, countered a surge in adulteration claims by forcing sellers of bear's grease to swear an oath of authenticity before the Lord Mayor of London. One such vendor, William Vickery, transcribed this oath into one of his advertisements, proclaiming that:

I, William Vickery, of Bishopsgate Street, near Cornhill, London, maketh Oath and saith that I do not sell, or will cause to be sold, at any Time or Times, any other than the real Bear's Grease, and that only from the Animal or Animals I kept, or are kept and killed by one of my Order. Sworn at Guildhall, the 2nd of February 1785, before R. Clark, Mayor.[69]

In the case of the latter, the emphasis shifted from pledges of personal integrity to assiduous branding, with perfumers warning customers to pay careful attention to the price, scent and appearance of bear's grease and the container in which it was sold. Market leader James Atkinson stipulated that 'the genuine [bear's grease] of his importation has a bear printed on the top of the pot, and a printed bill round the pot with his signature and address'.[70] If the signature was not present or the image not burned onto the pot, the product inside was a fake and should be avoided.

Branding also played a role in the Chinese musk industry, where vendors of Sichuan musk used distinctive labelling to identify their product. In a move reminiscent of Atkinson's signed bear's grease pots, one Chinese firm inserted a paper circular, or 'chop-paper', into its caddies that bore witness to the authenticity of the musk inside. The text engraved on the circular read:

Our firm itself selects the best kind of superior Sze-chuen musk at Ta-tseen-loo, in that province and in Thibet, whence we send it, without any admixture, to Sco-chow, Nanking, Hwae-chow, Yang-chow and Kwangtung, for sale. Our wares are genuine, our price true, and neither old nor young are deceived in them. We beg honourable merchants who may favour us with their custom to remember our firm seal, certain shameless scoundrels having fraudulently issued notices in

[69] 'Real Bear's Grease', *The Times*, 25 May 1876.
[70] 'Advertisement – Caution – Bear's Grease', *The Morning Post*, 17 January 1824.

Figure 5.3 Print from Chinese chop-paper depicting the hunting of the musk deer, from Charles H. Piesse, *Piesse's Art of Perfumery* (London: Piesse and Lubin, 1891), p. 270

order to deceive merchants. Fearing that it may be difficult to distinguish in this confusion, we now, in Kwangtung, notify the selected designation of our firm, as a rule for guidance – The Kwang-shum-se-ke firm of Sze-chuen.[71]

The Sichuan circular thus functioned as both a testament to authenticity and a critique of disingenuous rivals. Some chop-papers also featured pictorial representations of the hunting of the musk deer, showing the animals being shot at with bows and arrows and chased down by dogs (Figure 5.3).

Where special packaging was lacking, or when consumers continued to distrust the promises made on the product's exterior, science could sometimes serve as a counter to adulteration. In the case of musk, for instance, some simple chemical tests could determine authenticity.

[71] Charles H. Piesse, *Piesse's Art of Perfumery*, fifth edition (London: Piesse and Lubin, 1895), p. 273.

According to *The Ladies' Cabinet of Fashion*, 'Boiling water dissolves ninety parts of genuine Tonquin musk, alcohol only fifty parts,' while genuine musk was also 'soluble in ether, acetic acid, and yolk of egg'.[72] In the case of bear's grease, the clue, apparently, lay in the colour, which buyers were advised to check before purchasing. London perfumer Mr Winter observed that it was not 'generally known that the colour of real bear's grease, whether procured from the animal in a wild or domesticated state, is quite white'.[73]

Ambergris was more difficult to authenticate, given its highly elusive nature, but as knowledge of chemistry deepened, several simple tests were devised to help consumers expose frauds. One 1891 article, aimed at a juvenile readership, stated: 'The true ambergris, which is the morbid secretion of the spermaceti whale, gives out a fragrant smell when a hot needle is thrust into it, and it also melts like fat, but the counterfeit often sold instead of the real thing does not present these features.'[74] Naturalist Frank Buckland, who made his own independent study of ambergris, commented: 'When melted and placed upon a glass, it is the colour and consistency of light glue ... when held up to the light lovely bands and shades of a beautiful green are seen in the fluid. This is a sure test for the presence of ambergris.'[75] A few comparatively simple scientific tests could thus be deployed to determine whether a purported animal product was genuine or fraudulent.

Finally, in perhaps the most dramatic assertion of authenticity, some perfumers kept live animals on their premises to certify the origin of their product. In the seventeenth century, as noted above, the Dutch sent 'numbers of civet cats to Holland from their Eastern possessions so as to obtain the secretion in an unadulterated form'.[76] The British perfumer at 'the Civet house in Newington Green' presumably also kept his 'Threescore Civet Cats' so that buyers could see where their civet was coming from.[77] This practice added a gloss of authenticity to the process of civet extraction, although it appears to have receded from fashion from the late eighteenth century – probably because it was more economical, albeit more risky, to import the product in large quantities from overseas.

Sperm whales clearly could not be reared in situ, while musk deer also resisted domestication, reportedly going blind and dying 'soon after they

---

[72] 'Perfumes and Where They Come From', *The Ladies' Cabinet of Fashion, Music and Romance*, 1 October 1855.

[73] 'Winter's Real Bear's Grease', *The Morning Chronicle*, 7 August 1837.

[74] 'Ambergris', *Our Young Folk's Weekly Budget*, 19 November 1891.

[75] 'The Origin of Ambergris', *The Manchester Guardian*, 30 August 1871.

[76] 'Civet Cats and Civet', *The Times of India*, 18 May 1907.

[77] 'Advertisements', *Collection for Improvement of Husbandry and Trade*, 4 May 1694.

are caught'.[78] Bears, however, did survive in captivity, and a few perfumers and hairdressers vouched for the authenticity of their bear's
grease by keeping live bears on their premises. Some of these, like
Messrs Macalpine and Money, did so primarily for show and were
reluctant to cull their own animals. Others, however, staged theatrical
killings of the beasts to verify the origin of their product, usually publicising the event in advance in the local press, as we have seen. In 1834,
for instance, Ross and Sons announced: 'An immense bear from the
Zoological Gardens, Brighton, which has been fattened on bread only,
will be killed on Monday next, the 6th of January ... when anyone wishing
to have the fat from the animal can be accommodated.'[79] In 1842,
hairdresser Mr Langfield from Oxford Road, Manchester, slaughtered
'a fine bear of extraordinary size' purchased at auction from the defunct
Manchester Zoological Gardens 'for the purpose of obtaining real bear's
grease', an event witnessed by 'a party of eight or ten gentlemen' who
'shortly afterwards, sat down to supper at the Three Horse Shoes, kept by
Mrs Sarah Bradley, and partook of the animal's heart, liver and some
bacon'.[80] Public killings such as these served to authenticate the bear's
grease sold by supposedly reputable vendors and boosted confidence in a
much adulterated substance. Not all were what they seemed, however,
for some retailers found ingenious ways of extending the shelf lives of
their bears and passing off pretend killings as the real thing. Writing in
the 1820s, social commentator Charles Edwards chronicled the case of
one dealer in fake bear's grease who 'had but one bear in all the world,
which he privately led out of his house after dark, every night, and
brought him back (to seem like a new supply going in) in the morning ...
writing in his window "Our fresh bear will be killed tomorrow."'[81] Five
decades later, a journalist for the *Daily News* recounted a similar story
about a London barber who regularly faked the slaughter of a live bear
that he kept on the premises by intermittently removing the animal from
view and getting a local coal-heaver named Leather-mouthed Jemmy to
mimic the groans and yelps of a dying animal. The ruse apparently
worked for years, with the same bear growling at customers, being taken
once more for slaughter, 'with the assistance of Leather-mouthed
Jemmy', and then being surreptitiously returned to the barber's shop to
be 'killed' again the following week.[82] The eye, the nose and even the ear

[78] Piesse, *Piesse's Art of Perfumery*, p. 261.   [79] *The Morning Chronicle*, 4 January 1834.
[80] 'A Bear (Not a Bare) Treat', *Manchester Times and Gazette*, 26 November 1842.
[81] 'Posthumous Letters of Charles Edwards, Esq', *Blackwood's Edinburgh Magazine* XIX
     (January 1826), p. 22.
[82] 'Tame Bears', *Daily News*, 18 October 1873.

could thus all deceive when it came to determining the authenticity of animal perfumes.

Fading Scents

From around the middle of the nineteenth century, bear's grease and animal perfumes started to decrease in popularity and the use of these substances declined. The causes of this change were various. On the one hand, concerns about the sustainability of the supply and about animal welfare played their part, although in this case only to a limited degree. On the other hand, a change in taste among consumers and the advent of synthetic perfumes gradually rendered animal substances less desirable and less necessary, reducing – though not eliminating – the demand for zoological fragrances.

Conservation concerns were certainly raised in relation to animal perfumes, though not as extensively as they were for sealskin, ivory or birds' feathers. While no concerted movement appears to have emerged to protect musk deer, civets or whales, individuals did express anxieties about the long-term viability of some perfume-bearing animals and the suitability of the methods used to catch them. This was particularly true of two species: the musk deer and the sperm whale.

In the case of musk deer, Markham explicitly commented on the wastefulness of current hunting techniques – at least in northern India – and their potential impact on the species' survival. As detailed earlier, the typical way of catching musk deer was to trap them by placing snares within a fenced enclosure. This was an indiscriminate form of hunting, for it ensnared not only the male deer – the only sex that produced the coveted perfume – but also any female or juvenile deer that happened to pass through the area, thus diminishing the reproductive fitness of the species. Moreover, as Markham noted:

polecats often find out the snares, and after tasting the feast, if not destroyed soon, become a terrible annoyance, tracing the fence almost daily from end to end, and seizing on everything caught … Musk-deer are frequently lost to the snarers in this manner, for when one is eaten by the polecats the pod is torn to pieces, and the contents scattered on the ground.[83]

It has been calculated that 140 musk deer had to be killed to obtain just one kilogram of the coveted perfume – a huge collateral cost.[84]

---

[83] Markham, *Shooting in the Himalayas*, p. 96.
[84] Jonathan Reinarz, *Past Scents: Historical Perspectives on Smell* (Urbana: University of Illinois Press, 2014), p. 128.

While no serious studies appear to have been done to determine whether musk deer numbers were decreasing as a result of this continued attrition, it seemed clear – at least to Markham – that any decline was entirely down to the global demand for musk. As he expressed it:

> This little persecuted animal would probably have been left undisturbed to pass a life of peace and quietness in its native forests, but for the celebrated perfume with which it is provided. Its skin being worthless from its small size, the flesh alone would hold out no inducement for the villagers to hunt it while larger game was more easily procurable; and its comparative insignificance would alike have protected it from the pursuit of European sportsmen. As the musk, however, renders it to the Himalayan mountaineers the most valuable of all game, no animal is so universally sought after in every place that it is known to inhabit.[85]

If the musk deer did become extinct, therefore, it would be the fault of the perfumer and his customers.

Another species that appeared to be in jeopardy by the late nineteenth century was the sperm whale. In an article published in 1893, the *Hampshire Telegraph* observed: 'The whale that secretes ambergris has become so scarce now that little of the perfume is found, and it is hard to obtain in the market at any price.'[86] Six years later, the journal *Humanity* asserted that 'the wholesale slaughter of whales carried out by Norwegian and other steamers equipped with harpoon guns is so great that the animals seldom reach maturity'. The journal also claimed that whaling was 'still more horrible' than sealing, for 'whales frequently take several hours to kill' and had been seen by witnesses to '*shudder* with pain' when harpooned.[87] Although the decline in whale numbers – and the cruelty exposed by *Humanity* – was not directly caused by the trade in ambergris, the increasing scarcity of ambergris was perceived as symptomatic of a wider problem and was cited as evidence that the animal that produced it was facing extinction. While whales were killed primarily for oil (for gas lighting), spermaceti (for candles), meat (for consumption) and baleen (for making women's corsets) (Figure 5.4), a drop in the amount of ambergris constituted proof, for some, of sperm whale decimation and reinforced the view that the species was in trouble.[88]

Not everyone, it should be said, agreed with this analysis. Writing in 1891, the *Portsmouth Evening News* remarked that '[a] good deal has been said as to the virtual extinction of whales in the South Pacific', but 'recent

---

[85] Markham, *Shooting in the Himalayas*, pp. 84–5.
[86] 'Sweets to the Sweet: Something about Perfume', *Hampshire Telegraph*, 28 January 1893.
[87] 'The Whale-Fishery', *Humanity: The Journal of the Humanitarian League*, November 1898, p. 85.
[88] On the multiple uses of whale products, see Edwin Lankester, *The Uses of Animals in Relation to the Industry of Man* (London: Robert Hardwicke, 1860), pp. 134–6.

Figure 5.4 Benjamin Waterhouse Hawkins, 'Graphic Illustrations of Animals: The Whale', c.1850. Courtesy of Oxford Science Archive/ Heritage Images

events' have 'falsified this statement, two whales having been taken off the Tasmanian coast in August 1891 yielding ambergris worth £4,000'. Indeed, the paper concluded that there was scope to exploit the sperm whale more extensively, and that 'steam whalers' should be employed 'in those regions in substitution for the old sailing vessels that now do the work perfunctorily'.[89] But while some denied that sperm whale numbers were in decline, the general consensus was that the species was receding and that ambergris would soon grow even scarcer than it already was. A lack of knowledge about the lifecycle, population and migration routes of whales, moreover, made it difficult to assess the true impact of persistent hunting or to formulate viable conservation strategies. As a 1914 report in *The Manchester Guardian* admitted, scientific knowledge

[89] 'Whaling in the South Pacific', *Portsmouth Evening News*, 9 October 1891.

of the whale was sketchy, and '[o]f what we may term the life-cycle of the whale we know very little; the periods of gestation, of suckling, of growth to adult age, and of longevity are unknown to us for any of the cetacean'.[90] This would remain a problem into the mid-twentieth century, when the first international efforts at whale conservation foundered on a lack of basic information about whale biology, behaviour and population size.[91]

Neither civets nor Russian bears were considered at risk from the trades in their respective body parts, but in both cases the question of cruelty may well have come into play. Civets, as indicated above, were not killed to obtain their scent – at least not in most cases – but concerns were raised about the pain and discomfort caused by the extraction of civet from living animals. Writing in 1745, the British maritime explorer George Anson had made the somewhat fantastical claim that civets in fact became deeply uncomfortable if the fluid in their pouches were not removed, implying that humans were actually doing them a favour. 'There are a great number of Civet Cats: and if their Civet is not taken away every Month, they suffer so much Uneasiness from it that they tumble about the Ground till the Bladder breaks, which eases them.'[92] A century later, however, Reverend J. G. Wood contradicted this view, suggesting that the removal of scent was, at the very least, uncomfortable, and possibly painful, hence the need to disable the animal while it underwent this 'rough treatment'.[93] Whether or not the actual extraction process was cruel, the cramped conditions in which farmed civets were typically kept were certainly a cause for concern, raising a number of objections. An 1898 article in *The Nottinghamshire Guardian* observed that the animals were often 'confined ... in close cages in which there is hardly room for them to move'.[94] More emotively, a correspondent in the RSPCA's *The Animal World* magazine expressed concern for 'a beautiful civet' on show at the Crystal Palace, which could be seen 'pacing restlessly up and down' in its tiny cage, 'its semi-retractile claws clinking on the unnaturally smooth, hard, cold, zinc-covered floor'.[95] This particular civet was kept for show, not for its scent, but its suffering graphically highlighted the plight of Abyssinian civets housed in similar

---

[90] 'To Study the Whale', *The Manchester Guardian*, 10 April 1914.
[91] Kurk Dorsey, *Whales and Nations: Environmental Diplomacy on the High Seas* (Seattle: University of Washington Press, 2013), pp. 165–70.
[92] 'Continuation of a Voyage to the South Seas under the Command of Commodore Anson', *Penny London Post*, 21 January 1745.
[93] Wood, *Illustrated Natural History*, p. 229.
[94] 'The Civet Cat', *The Nottinghamshire Guardian*, 9 July 1898.
[95] 'A Caged Civet at the Crystal Palace', *The Animal World*, November 1871, p. 31.

conditions. The fact that some civets were kept as pets in the late nineteenth century may have further increased sympathy for the species, making the public more aware of its needs and habits. In 1883, for instance, a man from Bolton advertised a 'Genet, from Abyssinia, spotted like [a] leopard, smells always of musk, rare and beautiful little animal, allowed freedom in houses in the East'.[96]

As for bears, their slaughter appears to have evoked little sympathy in the eighteenth century and certainly does not seem to have deterred anyone from purchasing bear's grease. This was perhaps unsurprising in an era when bear baiting was still a popular sport and people took pleasure in seeing the animals mauled by dogs in the bear pit. By the middle of the nineteenth century, however, attitudes towards bears had softened somewhat, and there is evidence that at least some forms of abuse towards them had become unacceptable. In 1835, bear baiting was formally outlawed by the Cruelty to Animals Act as part of a wider drive to extirpate blood sports.[97] A few decades later, the 1870s and 1880s witnessed a growing concern for the mistreatment of dancing bears on British streets – usually by foreign showmen – and a spate of prosecutions for cruelty. In 1878, for instance, magistrates in Exeter fined Jean Sergant and Jacques Four £1 3s for putting a ring through the upper lip of a bear, which 'must cause the animal pain'.[98] In 1882, magistrates at Greenwich Police Court fined 'Jean Baique, 33, and Francis Fant, 50, Frenchmen', 40 shillings each for hitting a performing bear with 'a stick thicker than a broom handle', 'evidently causing it great pain, as blood was flowing'.[99] Although neither of these animals was used for bear's grease, the increasing frequency of such prosecutions suggests that bears were at least credited with feeling pain and deemed worthy of humane treatment. When the RSPCA prosecuted a showman in 1912 for keeping a bear 'in a cage only 35 in. long, 24 in. wide and 22 in. high', magistrate Mr Denman recommended that 'it would be very just and proper punishment to put the defendant in a precisely similar cage to that in which the unfortunate animal had been confined, and let him stay there for the same period' – a penalty that, sadly, the law did not allow.[100]

If concerns regarding cruelty and sustainability played some role in reducing demand for animal-based perfumes, two other factors ultimately

---

[96] *The Bazaar*, 23 February 1883, p. 672.
[97] Great Britain Parliament, *An Act to Consolidate and Amend the Several Laws Relating to the Cruel and Improper Treatment of Animals, and the Mischiefs Arising from the Driving of Cattle, and to Make Other Provisions in Regard Thereto*, 1835: 5 & 6 William 4 c.59.
[98] 'Cruelty to a Bear', *Western Times*, 4 July 1878.
[99] 'Cruelty to a Performing Bear', *Manchester Courier*, 11 April 1882.
[100] 'Cruelty to a Bear', *The Animal World*, March 1912, pp. iii–iv.

proved more influential: a change in taste and the rise of synthetic perfumes. The former became noticeable around the middle of the nineteenth century as the pungent animal scents favoured by the Tudors and
the Georgians gave way to more subtle floral fragrances. Once the epitome
of good taste, musk and civet ceased to be fashionable in the Victorian era
and were gradually superseded by lavender, jasmine and rose. By 1878,
the French region of Provence was exporting 2,000,000 kilograms of
orange blossoms, 500,000 kilograms of roses, 80,000 kilograms of jasmine
and 80,000 kilograms of violets every year to meet a growing demand for
these new scents, while an extensive lavender industry was in operation at
Mitcham in Surrey. Flowers were also exported from Italy, Spain, Algeria
and India for use in the perfume trade, along with rose of attar from
Bulgaria and ylang-ylang from the Philippines.[101]

The shift from animal- to plant-based perfumes reflected changing
consumer preferences and the emergence of more sophisticated techniques for extracting the scent from flowers and fruits. It did not, however, signal a complete end to the use of animal substances in perfumes,
since even perfumes manufactured from plants often contained animal
products. According to Henry Barton-Baker, for instance, one method
for extracting the odour from plants consisted of 'steeping the flowers in
hot clarified reindeer fat', which was then made into a pomade.[102]
Writing in *The Standard* in 1857, another journalist noted that 'the
preparation of suet' was 'an important branch of the perfumery business',
'the cheapness of mutton' in Australia boding well for the establishment
of flower farms there.[103] Musk, civet and ambergris continued to be used
widely as fixatives for floral scents, and, if Charles Piesse is to be believed,
they remained more popular than contemporary parlance suggested.

It is a fashion of the present day for people to say that they 'do not like musk'; but,
nevertheless, from great experience in one of the largest manufacturing
perfumatories in Europe, there can be no doubt that the public taste for musk
is as great as any perfumer desires. Those substances containing it always take the
preference in ready sale – so long as the vendor takes care to assure his customer
'that there is no musk in it'.[104]

Animals thus continued to be caged and killed for use in perfume even
after their scents went out of fashion.

---

[101] 'Perfumes', *The Graphic*, 22 June 1878. On the emergence of a global perfume industry,
   see Reinarz, *Past Scents*, pp. 72–4.
[102] 'Perfumes', *The Graphic*, 22 June 1878.
[103] 'Perfumery', *The Standard*, 8 September 1857.
[104] Piesse, *Piesse's Art of Perfumery*, pp. 266–7.

While changing tastes alone did not terminate the use of musk and civet in perfumes, the emergence of viable alternatives to these substances did eventually have an impact on their use. Motivated in part by sustainability concerns, and – to a greater degree – by fears of adulteration, scientists sought organic and inorganic substitutes for these coveted but expensive perfumes and succeeded in formulating acceptable replacements. In the case of bear's grease, the hair restorer was superseded by alpaca pomatum after Charles Ledger sent a sample of the latter to the London Exhibition of 1862 (see Chapter 4). After that too struggled to keep up with demand, a new product made from washed pomatum (a by-product of perfume manufacturing), olive oil and 'otto' of nutmeg became popular.[105] In the case of musk, the scent of the musk deer was replaced by the leaves of the so-called musk tree, native to Australia and New Zealand, and by various animal-based alternatives, including scents from the musk ox and the musk rat.[106] Chief Justice Temple of British Honduras (now Belize) even championed an 'odiferous substance in the auxiliary glands and under the jaw of the alligator' as a potential 'substitute for musk', hoping that it might become 'a valuable article of commerce' for his Central American outpost.[107]

Finally, and most conclusively, the late nineteenth century witnessed the creation of the first synthetic perfumes, which, manufactured in the laboratory, sought to artificially replicate the smell of musk, civet and ambergris. According to a report by the Imperial German Consul at Shanghai, the first 'artificial musk ... was brought to market in Paris and New York towards the end of the year 1889', triggering a drop in demand for genuine Tonquin musk from China.[108] This was followed in 1891 by the development of 'Musk Baur', patented by a scientist of that name, and, in 1926, by Leopold Ružička's discovery 'that the essential principles of musk and civet are compounds of the large ring type', which enabled subsequent chemists to imitate their qualities more effectively.[109] In 1896, Piesse and Lubin were advertising a range of 'Synthetic Scents' whose 'exquisite odours' represented 'a triumph of the science of the Parfumeur-Chimiste over Nature' – although they

---

[105] Ibid., p. 406.
[106] J. C. Sawer, *Odorographia: A Natural History of the Raw Materials and Drugs Used in the Perfume Industry* (London: Gurney and Jackson, 1894), p. 403.
[107] 'On British Honduras, its History, Trade and Natural Resources', *Journal of the Society of Arts*, 16 January 1857, p. 125.
[108] Sawer, *Odorographia*, pp. 396–7.
[109] Ibid., pp. 397–8; 'Musk for Perfume Made Artificially', *New York Times*, 1 January 1936.

continued to use the image of a male musk deer on their merchandise.[110] The advent of authentic substitutes significantly reduced the demand for real musk, giving the musk deer something of a respite. To this day, however, some high-end perfumery businesses continue to use genuine musk and civet in their products, drawing criticism from environmentalists and animal rights campaigners.

Conclusion

In June 1891, a journalist for *The Spectator* visited London Zoo and conducted an unusual experiment on some of the animals. Wishing 'to test for himself the reported fondness of many animals for perfumes', the man entered the gardens armed with 'bottles of scent and a packet of cotton wool' and made his way around the different enclosures. He tried his subjects first with lavender water, one of the most popular scents of the era, and repeated the experiment with rose water and lilac blossom. He closely observed the beasts' reactions and recorded what he saw.

The results of the perfume experiment were interesting. One leopard, when presented with a ball of scented cotton wool, 'shut its eyes, opened its mouth and screwed up its nose'. Another 'smelt it and sneezed, then caught the wool in its claws', before fetching a third leopard, which took the ball between its teeth and inhaled 'the delightful perfume with half-shut eyes'. The lion 'laid his broad head on the scented cotton', snuggled up to it and 'purred', while the ocelot, 'after inhaling the perfume, *ate* the small piece of paper on which it was placed'. The racoon, 'when the bottle was presented to it corked, with great good sense, pulled out the stopper; but this may have been due to curiosity, as it was at once thrown away'. Only the otter objected to the perfume, giving 'a snort of disgust' upon smelling it and diving straight back into the water. The journalist hypothesised, on the basis of these responses, that many animals did indeed derive pleasure from human-manufactured scents, and that their enjoyment was 'made intensely more delightful to them than ourselves by the wonderful development of their sense of smell'.[111]

The Spectator's olfactory experiment offers a counterpoint to earlier ideas about animals and perfume. In 1824, when hairdressers Macalpine and Money were indicted for housing live bears on their premises, animals were sources of scented cosmetics, kept in the British capital to authenticate their owners' wares. In 1891, by contrast, the inhabitants of London Zoo were recipients – and arguably beneficiaries – of floral

---

[110] 'A Novelty!', *The Illustrated London News*, 31 October 1896.
[111] 'Animal Esthetics; Scents and Sounds', *The Spectator*, 30 May 1891, pp. 13–14.

perfumes, given to them as a test of their taste and sense of smell. Once highly prized for their pungent odours, animal scents such as civet and musk had fallen out of fashion, to be replaced by the lavender waters, rosewaters and lilac blossom presented to the animals in the Regent's Park. The first synthetic musks were also coming onto the market – a move inspired partly by changing tastes, partly by concerns about adulteration and partly by environmental and humanitarian qualms.

While the perfume trial at London Zoo highlighted some of the changes that had occurred in the industry since the beginning of the nineteenth century, it also revealed important continuities. First, the perfumes given to the animals, though floral in name, almost certainly contained civet and musk as fixatives – probably the reason why the big cats liked them so much. In modern zoos, keepers often give caged animals aftershave and other perfumes as a form of environmental enrichment, with scents containing genuine or synthetic musk proving the most popular. The big cats at Dudley Zoo love a perfume made from hyraceum, the fossilised excrement and urine from the Cape hyrax, while jaguars apparently go wild for Calvin Klein's Obsession for Men, which contains synthetic civet.[112] On a darker note, the benign experiments in London Zoo arguably foreshadowed the more controversial practice of testing perfumes and other cosmetics on animal subjects before licensing them for human use. The testing of cosmetics on animals became mandatory from the 1930s and was not banned in Britain until 1998. Although the victims of this practice were rabbits, rats and guinea pigs, rather than lions and leopards, the growing use of such testing meant that animals continued to suffer for human vanity, having beauty products applied to their skin and their eyes to check for irritation. The move away from animal perfumes was thus partial and gradual, while a rise in concern for human safety has made animals the test subjects for new synthetic scents. Viewed in this light, the history of perfume has a distinctly bitter aftertaste.

---

[112] See perfumesociety.org/dudley-zoos-big-cats-go-crazy-salome/ (accessed 20 November 2018); 'Jaguars Obsessed with Calvin Klein Scent', *The Guardian*, 11 June 2010.

Kind home wanted for extra tame thoroughly acclimatised monkey, exceptionally funny, clever, interesting, enjoys bath every Saturday, companionable as a child, far less trouble, 40/-, worth £5 to any lady or children.   *Supplement to The Bazaar*, 3 October 1877, p. 709

In September 1886, James Noble, a carter, appeared before magistrates in Sheffield charged with 'having unlawfully and cruelly abused and tortured a parrot'. According to Frank Hopkinson, the defendant's former employee, Noble had returned home one evening 'a little the worse for drink' and 'had a dispute with his wife'. 'A parrot which they kept in the house began to talk or make a noise of some sort' and Noble, declaring 'that he would not be laughed at by a parrot, took a knife and began stabbing it through the wires of the cage'. The carter removed the parrot from its residence and 'hit it several times with the knife, nearly cutting off its wing'. He then threw the dying bird onto the fire, dousing it with gravy as it 'quickly burnt to death'. Hopkinson's brother corroborated his testimony, adding that Noble had also 'tried to screw the parrot's neck around' before he commenced stabbing it. Magistrates ruled that Noble had committed an act of 'great cruelty' and ordered him to pay a fine of £2.[1]

The tragic case of James Noble's parrot offers an unusual insight into the lives (and all too often deaths) of exotic pets in Victorian Britain. Probably imported from Africa as part of a lucrative wild beast trade, Noble's bird may have arrived at one of the several animal dealers, large and small, scattered across the United Kingdom, and was either purchased directly from the dealer's warehouse or bought second-hand via one of the many classified advertisements in local newspapers. As the testimony of Frank Hopkinson indicated, the parrot was kept in a wire cage inside Noble's house. It had evidently been taught to talk – the ultimate cause of its downfall – and presumably functioned as a source of

---

[1] 'Gross Cruelty to a Parrot in Sheffield', *Sheffield Independent*, 17 September 1886.

amusement, a luxury accessory or a cherished companion (for Mrs Noble at least). The bird's premature death, though exceptionally violent, was not atypical for members of its species, many of which perished at the hands of negligent animal dealers or brutal owners. In 1890, a writer in *The Animal World* lamented that she had recently 'lost five beautiful [parrots] in succession within a few weeks of purchase' – in this case from disease rather than abuse.[2] In 1866, in a case more closely resembling Noble's, a police constable arrested London costermonger Thomas Joseph Underwood for 'st[icking] a dinner fork into [his wife's] head' and 'thr[owing] a parrot with its cage and a dog on the fire'.[3] The lot of the Victorian parrot was not, it would seem, a very happy one.

This chapter examines the rise of a rather different type of animal commodity: the exotic pet. Unlike ivory, sealskins or feathers, which arrived in Europe as the lifeless remnants of dead animals, exotic pets reached the continent alive (or at least some of them did), adding a dash of exoticism to the Victorian home. Parrots appeared in middle-class parlours. Monkeys cavorted in kitchens and escaped from gardens. Tortoises were hawked through the streets and guinea pigs sipped tea in old ladies' drawing rooms.[4] While the majority of exotic pet owners were upper or middle class, working men and women also kept parrots, monkeys and canaries, either for companionship or for show. Robert Clough, an innkeeper from Thornley, County Durham, kept a parrot, which sadly had its wings broken by a drunken miner.[5] John Harris, 'a seaman R.N.' from Falmouth, 'was killed in London ... in falling out of the window of a railway train ... trying to recover a pet monkey'.[6]

Historians have recently paid increasing attention to the practice of pet keeping in eighteenth- and nineteenth-century Britain, highlighting its social, cultural and emotional dynamics and noting the pet's dual role as commodity and companion.[7] Compared with cats and dogs, however,

---

[2] 'The Transit of Parrots', *The Animal World*, November 1894, p. 175.
[3] 'Thames Police Court', *The Times*, 20 March 1866.
[4] 'An Interesting Pet', *The Animal World*, June 1883, p. 86.
[5] 'A Violent Pitman', *Northern Echo*, 30 August 1875.
[6] 'News', *The Cornishman*, 9 September 1897.
[7] Important studies on the history of pets include, Kathleen Kete, *The Beast in the Boudoir: Petkeeping in Nineteenth-Century Paris* (Berkeley: University of California Press, 1994); Katherine C. Grier, *Pets in America: A History* (Chapel Hill: University of North Carolina Press, 2006); Ingrid Tague, *Animal Companions: Pets and Social Change in Eighteenth-Century Britain* (Philadelphia: Penn State University Press, 2015); Sarah Amato, *Beastly Possessions: Animals in Victorian Consumer Culture* (Toronto: University of Toronto Press, 2015); Philip Howell, *At Home and Astray: The Domestic Dog in Victorian Britain* (Charlottesville and London: University of Virginia Press, 2015); Neil Pemberton, Julie-Marie Strange and Michael Worboys, *The Invention of the Modern Dog: Breed and Blood in Victorian Britain* (Baltimore: Johns Hopkins University Press, 2018).

exotic pets such as parrots, monkeys and tortoises have received comparatively little study and have more commonly been examined within
the contexts of zoological gardens and travelling menageries.[8]
Concentrating on this under-researched but substantial tranche of
animals, this chapter explores the exotic pet market in Victorian and
Edwardian Britain and considers the ethical debates generated by the
importation and ownership of wild animals. I begin by examining where
exotic pets could be bought and sold, what qualities owners most prized
in them and how they were housed, trained and cared for. I then go on to
assess two key concerns arising from the nineteenth-century pet industry:
the environmental impact of catching and importing animals, and the
cruelties associated with keeping wild creatures in captivity.

Animals Wholesale and Retail

A thriving trade in exotic animals existed in nineteenth-century Britain.
Since the late eighteenth century, foreign birds and beasts had been
imported and offered for sale by animal dealers, street vendors and
commercial menagerists.[9] From the mid-nineteenth century, this trade
significantly expanded, facilitated by improvements in steam shipping
and more extensive penetration of overseas territories. Large-scale
animal dealerships emerged in major port cities such as London,
Liverpool and Hamburg and regular advertisements for exotic beasts
began to appear in newspapers such as *The Era*. On 11 June 1881, for
instance, Liverpool dealer William Cross announced the arrival of '45
pairs of Butcherguards [budgerigars], 147 Pair Red-faced African Love
Birds, 1,600 Tortoises, 1 Monster Boa Constrictor, 1 Horned Viper,
2 Philomtambo Antelopes, 13 Caratrix Monkeys … 1,000 Pairs Mixed
Small Birds, 1 Pair Californian Sea Lions (Adult), 1 Tamandua Anteater
[and] 5 Python Snakes'.[10] While the larger animals supplied travelling

---

[8] See, for instance, Elisabeth Baratay and Eric Hardouin-Fugier, *Zoo: A History of
Zoological Gardens in the West* (London: Reaktion Books, 2002); Nigel Rothfels,
Savages and Beasts: The Birth of the Modern Zoo (Baltimore: Johns Hopkins University
Press, 2002); Elizabeth Hanson, *Animal Attractions: Nature on Display in American Zoos*
(Princeton: Princeton University Press, 2002); Helen Cowie, *Exhibiting Animals in
Nineteenth-Century Britain: Empathy, Education, Entertainment* (Basingstoke: Palgrave
Macmillan, 2014); Christopher Plumb, *The Georgian Menagerie* (London: I. B. Tauris,
2015). An exception is Louise Robbins, *Elephant Slaves and Pampered Parrots: Exotic
Animals in Eighteenth-Century Paris* (Baltimore: Johns Hopkins University Press, 2002),
who examines exotic animals in menageries and in the domestic sphere, but she focuses
on eighteenth-century France rather than Victorian Britain.
[9] See 'The Material Conditions of Pet Keeping' in Tague, *Animal Companions*, pp. 14–49.
[10] 'Arrivals This Week at Cross's Menagerie', *The Era*, 11 June 1881.

showmen and zoological gardens, many of the smaller ones would have
been destined for private homes, catering to a growing demand for
exotic pets.

Given their reliance on regular importations of animals, most of the
largest animal dealers were located close to the docks, giving them easy
access to sailors and their wares. Cross's Menagerie was situated in
Tabley Street, Liverpool, adjacent to the Wapping, Salthouse and
Albert docks. London dealer Charles Jamrach had a shop on the
Ratcliffe Highway in London's East End, while his rival John Hamlyn
operated from an address in Upper East Smithfield, close to St Katharine
Docks. Dealers received regular imports of live animals from whalers,
traders and naval vessels and also commissioned their own collecting
expeditions to secure rarer or more demanding creatures. Cross claimed
to have 'agents and agencies in London, Gravesend, Plymouth,
Southampton, Bombay [and] the Cape' and 'something like 1,000 captains of vessels sailing from numerous ports to nearly every known region
of the world who are always ready to secure anything and everything
I want'.[11]

What was it like to visit one of these wild beast emporiums? A couple of
journalists who visited Jamrach's establishment fifteen years apart give us
an insight into this unusual and exhilarating experience. Writing for the
Daily News in 1869, the first of these writers described how he initially
entered Jamrach's retail shop, where he was greeted by a cacophony of
shrieking macaws, parrots and parroquets. From there, he proceeded 'a
little further down the … Highway, and, turning a corner', came upon
'what looks like a stable door, save that its bolts and bars are of a more
substantial order than are usually put up even for stolen steeds'. Inside
this second warehouse the writer admired 'pelicans gorged with fish-
gobbling; antelopes thrusting forward their graceful heads [and] emus
fretting against the bars', before heading up a ladder to view a mandrill
chattering 'demoniacally' and a snarling lynx 'with a fixed stare and so
decided a determination to spring, that we involuntarily receded a pace
or two, covering our crouched heads with arms'. Less threateningly, the
writer also observed a 'white terrier pup, which insists fancifully on
playing with your trousers with its teeth', and a wombat that was 'put
on the floor to tumble'. The tour concluded back in Jamrach's private
apartment, where the journalist admired a sloth 'suspended by his four
claws from a chair-back … in front of the fireplace'.[12]

[11] 'Mr William Cross, Naturalist', *The Era*, 29 June 1895.
[12] 'Lions and Tigers Wholesale and Retail', *Daily News*, 12 August 1869.

A second observer, visiting Jamrach's in 1884, recounted a similar experience, although his guide was Albert Edward Jamrach rather than his father, Charles. Arriving at the establishment by river steamer, this writer was first impressed by the distinctly 'nautical tone' of the locality, which was packed with 'shops of second hand clothing', places where sailors could have letters written for them in multiple languages and 'the studio of a professional tattooist'. On entering the business itself, the journalist, like his predecessor, was shown first into the 'bird department', which on this occasion contained 'Chinese magpies and jay thrushes, two American mocking-birds' and, upstairs, a whole room in which 'the air is literally darkened by the sight of 1,500 pairs of zebra-finches from Australia'. His guide then led him to 'the animal stables in Bett's street', where 'a double-humped camel in a corner scornfully turns his back upon us; a Barbary two-horned sheep makes a savage attempt to bruise our fingers' and an 'Afghan hound' was 'eager to make our acquaintance'. As the journalist exited the establishment via a dark passage, he saw some black swans being loaded into a case for train transit, and, more alarmingly, an alligator 'in a long box, with cross-pieces nailed over the top', which, on being roused by Jamrach with a piece of wood, 'raise[d] himself on his legs, open[ed] his mouth and wheeze[d] furiously for several consecutive minutes'.[13] His final encounter was with a stuffed elephant – 'one of four out of six who died in Jamrach's possession' – a sober reminder of the 'speculative' nature of the wild animal business, 'with its chance of death among the stock'. Like the earlier commentator, the writer conjured a colourful picture of Jamrach's shop as somewhere loud, chaotic, potentially dangerous and sensorially overpowering (Figure 6.1).

While the major animal dealers such as Jamrach and Cross inevitably captured most of the headlines, these were not the only places where Victorian and Edwardian Britons could get hold of exotic animals. On the contrary, closer analysis reveals the existence of many smaller dealers who served customers in London and in the provinces from at least the mid-nineteenth century. In 1872, for instance, Mr Broad of 4 Patna Place, Plymouth, advertised a 'beautiful little lemon crested cockatoo', a 'beautiful grey parrot, will talk like an auctioneer' and a 'large Amazon green crested parrot ... manly voice ... whistles like a locomotive'.[14] In the late 1870s, Mr Spedding, a pork butcher from Dewsbury, advertised a 'Tortoise, in good health', an 'American opossum, will swing on the finger by its tail' and a 'Beautiful little animal, the skunk, about the size of

---

[13] 'A Jaunt to Jamrach's', *The Era*, 13 September 1884.
[14] 'Country House', *The Bazaar*, 20 March 1872, p. 396.

Figure 6.1 'Teddy Bear, pet of Whitechapel Children', c.1913.
Working-class children from the East End feed 'handfuls of sugar' to a
bear cub from Jamrach's shop.

[a] half grown cat, colour black and white, can be handled as wished'.[15]
In the 1890s, Frederick Kings, a grocer from Redditch, began a lucrative
sideline in exotic pets, advertising civets, coatis and opossums,[16] while in
the 1900s, Mr Ewart from Ilminster operated a successful dealership in
monkeys, selling 'bonnet [monkey] Angelina, wears earrings, 20/- [shil-
lings] … rhesus, Old Nick, 30/-, bonnet Sairey Gamp, 30/- [and] Adonis,

[15] 'Country House', *The Bazaar*, 21 February 1877, p. 454; 31 May 1876, p. 1172;
18 December 1878, p. 1527.
[16] 'Country House', *The Bazaar*, 19 January 1891, p. 161; 18 August 1898, p. 467;
13 September 1895, p. 772. Kings moved his business to Kenilworth in 1896,
operating out of the Globe Hotel. See 'Country House', *The Bazaar*, 17 April 1896,
p. 1487.

wears blue velvet coat and trousers, 42/-'.[17] Other provincial animal
dealers included 'Mr Martin's Animal Dealer, Buchanan Street',
Glasgow, who marketed a special medicated dog soap; 'Mr Harvey,
Bird and Animal Dealer', who owned a shop 'Three doors from the
Bars Hotel, Chester'; Exeter dealer Mr Pook, who sold songbirds,
parrots and a 'small tame black-faced monkey' at his shop in Fore
Street; and Mr Thorpe, who sold 'Young leopards and bears ... Diana
and other monkeys, lemurs, &c.' from his warehouse at 75 South Parade
in Hull.[18] Though far from comprehensive, such reports give the impression of a widespread network of animal dealers, some enduring and
others more ephemeral, serving buyers in towns across the British Isles.

Nor, indeed, did one necessarily have to go to a dealer to acquire
foreign animals, for a flourishing market in second-hand beasts also
existed throughout the nineteenth century. Animals – native and exotic –
appeared regularly in the classifieds section of local newspapers and also
featured prominently in the commercial magazine *The Bazaar, Exchange
and Mart*, published several times a week from 1871. In 1877, for
example, T. Hewett from Acton advertised a 'small handsome tame
monkey ... quite tame, dress in doll's clothes and can be wheeled in
perambulator, quite a playmate for children. Price 20s.'[19] In 1878, Mrs
Garrington from Bilston advertised a 'strong and healthy African parrot,
a most beautiful bird, grey with scarlet tail, can say "Who are you",
"What do you want", "Mother", "Thank you", "Walk in", "Come,
Polly, take a walk", "Come boy, now then", "I'm the Shah", "Harry
here", "Shut up", laughs, barks, mews and cackles, a good mimic'.[20] In
1886, G. B. Bromley from Goole advertised a 'Young raccoon ... clean
in habits, will eat from the hand, likes to sit on the shoulder, has always
been made a pet', and in 1906 Mrs Cranfield from Lancaster advertised a
'Beautiful grey black striped wee marmoset monkey, black velvet like
collarette round neck, a gem'.[21] The frequency of exotic animal advertisements in *The Bazaar* and other commercial papers is evidence of both
the popularity of such pets and also, perhaps, the difficulty of looking
after them; one parrot was sold 'for bad behaviour' and another for
'making so much noise and disturbing the children'.[22] The

[17] 'Country House', *The Bazaar*, 17 August 1906, p. 646.
[18] 'Dog Soap', *Glasgow Herald*, 2 March 1871; 'Notice of Removal', *Chester Observer*,
    8 February 1879; 'Country House', *The Bazaar*, 7 February 1872, p. 169; 29 May
    1872, p. 787; 19 May 1906, p. 2458.
[19] 'Country House', *The Bazaar*, 1 September 1877, p. 456.
[20] 'Country House', *The Bazaar*, 21 August 1878, p. 343.
[21] 'Country House', *The Bazaar*, 2 October 1886, p.1339; 16 October 1906, p. 1415.
[22] 'Country House', *The Bazaar* 12 September 1877, p. 531; 17 September 1913, p. ix.

Figure 6.2 'Cocky, the Fireman's Friend', *The Animal World*, February 1907, p. 39. Cocky lived at Camden Town fire station and 'penetrat[ed] the air with loud screams' when the fire bell rang.

advertisements reveal, too, the diverse social and geographical profile of Victorian exotic pet owners, with vendors including a draper from Aberystwyth, a baker from Kettering, a postman from Whitby, a wardrobe dealer from Leeds, an urban sanitary inspector from Wellington and a mineral water manufacturer from the Isle of Man (Figure 6.2).[23] In 1903, Mr Hoole from Stud Farm in Shropshire advertised 'the tamest female monkey imaginable … very clean in the house, sole reason for selling daughter leaving home for boarding school, same was her pet', suggesting that at least some exotic creatures were owned by children.[24]

Which exotic animals were most coveted as pets, and what qualities did prospective owners most value in their companions? Based on the advertisements placed in newspapers, the most popular imported animal seems to have been the parrot, with African grey parrots arriving in the largest numbers (Cross, for instance, claimed to import 80,000 of the birds every year).[25] Amazonian parrots, macaws, parakeets and cockatoos were also widely available, while budgerigars provided a

---

[23] 'Country House', *The Bazaar*, 16 October 1872, p. 322; 26 August 1874, p. 267; 10 February 1877, p. 354; 19 June 1878, p. 1553; 19 November 1884, p. 1770; 16 September 1898, p. 833.

[24] 'Country House', *The Bazaar*, 3 December 1903, p. 2572.

[25] '80,000 Parrots Imported Annually', *The Era*, 17 March 1883.

Figure 6.3 'The Best of Friends', 'Animal Anecdotes', *The Animal World*, March 1912, p. 56.

smaller and cheaper alternative. Beyond the parrot family, songbirds attracted a large following, particularly mockingbirds, Virginia nightingales, Australian love birds, waxbills and cardinals. Canaries were a stand-out favourite, especially among working-class buyers, and were bred extensively in Germany as well as in Britain.[26] Among mammals, monkeys dominated the exotic pet market, arriving in Europe from Africa, Asia and America (Figure 6.3). The precise species of these creatures was often imperfectly defined, but an 1888 guide for monkey keepers listed capuchins, spider monkeys, howler monkeys, marmosets, Diana, rhesus, bonnet and macaque monkeys as among the most

---

[26] An article in *The Animal World* reported: 'The city of Norwich, with the surrounding villages and hamlets, counts its breeders by the thousand; while in Coventry, Derby, Northampton, Nottingham and other towns in the Midland district where labour is of a sedentary character, as well as in many towns in Yorkshire and Lancashire, the canary is the poor man's savings-bank – the family pig where sanitary laws forbid the erection of a stye.' See 'Extent of the Traffic in Canaries', *The Animal World*, 1 September 1877, p. 136.

common varieties.[27] Guinea pigs, mongooses, raccoons and opossums (both American and Australian) were all advertised with some regularity in *The Bazaar*, while dealers such as Cross sold a steady stream of non-native felines and canines, such as Persian, Siamese and Angora cats and Japanese pug dogs.[28] There was also a growing market for certain non-mammalian species, notably tortoises, of which Cross imported 35,000 in 1888, and goldfish, of which Hamlyn imported 1,200 in April 1893.[29] Finally, at the more extreme end of the spectrum, baboons, snakes and bears made unconventional, if demanding, pets and surfaced with some regularity in the second-hand market. One vendor in *The Bazaar* advertised a 'Young pet Russian bear, Edward, 15 months old, reared in captivity, beautiful condition, and perfectly tame, follows like dog, formerly regimental mascot, £10'.[30] Another advertised 'Jennie, the nicest baby baboon ever brought from Cape Colony, very tame, affectionate, most amusing, clever, healthy and hardy, write for photograph and fullest particulars. Take £2; worth £5 if not for damaged tail.'[31] Other more unusual offerings included a 'fine handsome Indian fruit bat, large size, feed from hand', a 'Young alligator, 3ft. long … been kept for 3 years in conservatory', an 'Indian python, 9ft. long … just shed skin entire from lips to tail' and a 'Chimpanzee, Dobbie … abundant long hair and long whiskers, beautiful teeth, accustomed to freedom of butler's pantry'.[32]

When it came to selecting an animal for purchase, buyers prioritised health, attractiveness and intelligence, with terms such as 'handsome', 'tricky', 'docile' and 'affectionate' predominating in advertisements. 'W.' from London, for instance, advertised 'a grey parrot, a thorough good one, will say anything or whistle anything, but does not swear or squall'.[33] Mr Grove from Caerleon advertised a 'young monkey (Rhesus), will not bite, shakes hands, puts out lighted matches, &c., child can handle freely'.[34] Mrs Major from Saffron Walden advertised a 'Splendid Australian grey male opossum, will climb a rope 20ft. high', while T. A. Thurstan from Enfield advertised a 'Pair of beautifully tame

---

[27] Arthur Patterson, *Notes on Pet Monkeys and How to Manage Them* (London: L. Upcott Gill, 1888), pp. 45–65.

[28] In May 1891, for instance, Cross advertised '1 Blue Persian Cat' and '2 Orange Persian Kittens'. See 'Arrivals at Cross's Menagerie', *The Era*, 9 May 1891.

[29] '35,000 Tortoises Just Imported', *The Era*, 23 April 1881; 'Seals', *The Era*, 22 April 1893.

[30] 'Country House', *The Bazaar*, 5 July 1912, p. ix.

[31] 'Country House', *The Bazaar*, 7 January 1903, p. 71.

[32] 'Country House', *The Bazaar*, 10 January 1902, p. 128; 22 October 1913, p. ix; 2 November 1901, p. 1752; 12 November 1915, p. vii.

[33] 'Country House', *The Bazaar*, 16 October 1872, p. 322.

[34] 'Country House', *The Bazaar*, 4 February 1910, p. 27

miniature marmosets, so small both will go in a pint pot'.[35] As these examples indicate, owners prized elegant, clever and gentle pets and looked for different qualities in different species. Parrots needed to talk, canaries to sing, monkeys to entertain and marmosets to fit in drinking glasses. One vendor from Hampshire advertised a multitalented grey parrot, whose vocabulary included:

How are you off for soap, my dear? What's up now? All's well that ends well, Do you see any green in my eye? You're a brute, you're an ugly brute, Such a good Polly ... Hip, hip hurrah, Come in, wipe your shoes, Shut the door, There's a good boy, Be quiet, Shut up, Goodbye, DOG dog, CAT cat.[36]

Another advertised a 'Lovely monkey, 2 years old, very tame, household pet, boxes with little boy, ducks and feints like a professional, pure white cat sleeps in its arms ... must sell together as they are inseparable'.[37]

Above all, animals needed to be healthy, so evidence of strength and robustness was particularly welcome. Dealer Mr Ewart, keen to alleviate buyer anxiety, asserted that his marmoset was the 'Hardiest ... living, [has] wintered in England, practically certain liver, never coddled, always lived in unheated room'.[38] F. G. Wrigley from Sheffield described his monkey as a 'Pretty little pet, perfect in every detail, black head, arms, legs and tail, back and stomach golden, sweet little face ... without any scratch whatever, perfect tail and free from vermin'.[39] Such assurances were highly valued because un-acclimatised exotic pets were a notoriously risky investment. As monkey expert Arthur Patterson put it in pragmatic, if rather crude, terms: 'You do not want to lay out money on a "dier".'[40]

Pet Monkeys and How to Manage Them

The importation and sale of exotic animals was, of course, only the first step in a longer process. Unlike an alpaca jacket or a sealskin coat, a pet parrot was still alive when it reached its final owner (albeit not always for long), so welfare issues extended beyond the point of purchase. To assess the humanitarian impact of the pet trade, therefore, we need to know more about how exotic pets were kept and treated and how their owners

---

[35] 'Country House', *The Bazaar*, 16 March 1892, p. 895; 6 February 1903, p. 584.
[36] 'Country House', *The Bazaar*, 27 September 1882, p. 914.
[37] 'Country House', *The Bazaar*, 9 March 1900, p. 858.
[38] 'Country House', *The Bazaar*, 26 August 1904, p. 838.
[39] 'Country House', *The Bazaar*, 6 November 1908, p. 2283.
[40] Patterson, *Notes on Pet Monkeys*, p. 68.

perceived them. Did owners view their parrots and monkeys as loved companions or as luxury possessions to be flaunted? What kind of life could exotic animals expect in captivity? What practical advice was available on how to house, feed, train or medicate non-native species?

One source of information on the treatment of pets in Victorian Britain is the many advice manuals that were published on the subject in the second half of the nineteenth century. While most of these focused on domestic species, such as dogs, cats and rabbits, a significant number addressed the keeping of exotic animals, which, being less familiar, often required more careful guidance. Sometimes written by dealers in rare animals, with long experience in caring for them, and sometimes penned by amateur connoisseurs, manuals typically offered advice on the accommodation, feeding, medication and breeding of pets, providing insight into the joys and tribulations of pet ownership. Their content addressed the common logistical problems associated with keeping parrots, monkeys and other exotic species, while their tone revealed something of contemporary attitudes towards such caged companions.

Take, for instance, *The Parrot Keeper's Guide, by an Experienced Dealer*, first published in 1857. Written, as the title suggests, by a dealer in exotic birds, the guide promised 'general observations on the best modes of treatment, the diseases to which they are subject and methods of cure etc.' and provided detailed information on the practical aspects of parrot keeping.[41] Addressing the important issue of accommodation, the author insisted that parrots should be given as much space as possible. A perch should be supplied, which should be 'made thicker in the centre than at the ends, thus enabling the bird to choose that part best suited to his grasp', and a large ring or hoop should also be installed to allow the parrot to 'exercise itself'.[42] When it came to cleanliness, the guide advised removing fruit stones of 'plums, cherries and the like' from the bottom of the cage, cleaning the parrot's feet at intervals to remove 'any incrustment of dirt that it may have contracted' and bathing the birds in lukewarm water to prevent them from being 'troubled with vermin'.[43] As for feeding, it recommended a diet of canary seed and water, supplemented by bread soaked in water and the occasional 'portion of loaf sugar, biscuit or hard fruit (thoroughly ripe)'.[44] Like all captive animals, parrots were subject to diseases, so the author offered advice on how to treat common conditions such as asthma, wasting, convulsions, surfeit, diarrhoea, consumption, diseased feet and vermin infestation. Plunging a

---

[41] Anon., *The Parrot Keeper's Guide, by an Experienced Dealer* (London: Thomas Dean and Son, 1857), frontispiece.
[42] Ibid., p. 32.   [43] Ibid., pp. 27–8, 34–5.   [44] Ibid., pp. 28–31.

bird into cold water could help stop a fit, spring water and milk were good for diarrhoea, while a dose of chopped chillies would 'warm the system and renew the tone of [a parrot's] stomach'.[45] Finally, if the owner wanted to teach his bird to talk, he should treat it 'with continual kindness, constantly caress and indulge it with little niceties'. He should not alarm it by approaching it with a glove on, nor should he have 'the tongue of the bird slit to enable it to talk' – a practice apparently common in the 1850s.[46] *The Parrot Keeper's Guide* thus outlined best practice in the treatment of exotic birds, while also revealing the casual cruelties sometimes inflicted on captive creatures.

A second advice manual, Arthur Patterson's *Notes on Pet Monkeys and How to Manage Them* (1888), offers insights into the treatment of another popular exotic pet. Written by a naturalist from Great Yarmouth who had kept monkeys for many years, the guide opened with a discussion of the different species of monkey available, listing their various virtues and vices. Mangabeys, for example, were 'to be recommended as comical yet well-behaved pets'; the mona monkey from West Africa was a 'hardy' monkey but in middle age developed some 'nasty, revengeful, peevish ways'; while capuchins were 'peculiarly suited to ladies'.[47] Patterson went on to address housing, issuing detailed instructions on how to construct a cage and advising owners to install a trapeze or monkey wheel for 'exercise and fun'.[48] He advocated a staple diet of 'boiled rice and milk', supplemented in cold weather by 'a little dainty bit of horse-flesh or beef', offered guidance on the treatment of broken limbs, fits and consumption, and described a novel method for winning a monkey's affection.[49]

To get into your monkey's confidence at once, let a friend go up to the cage with a stick and somewhat frighten the animal; whilst in the midst of his nonsense, rush forward and pretend to take the part of your pet, thrash your friend within an inch of his life with the very stick he has been using and put him out. Next take the monkey some savoury morsel, such as a date, or an apple, and sympathise with it. You are sworn friends from that time.[50]

The book concluded with advice on how to euthanise a dying monkey (hit it on the back of the skull with an iron bar), dispose of an old, 'crusty' monkey (give it to a passing menagerie) and stuff a dead monkey (skin it, soak the skin in arsenic and pose it in a natural posture), thus covering all stages of the animal's lifecycle.[51]

[45] Ibid., pp. 37–45.   [46] Ibid., p. 35.   [47] Patterson, *Notes on Pet Monkeys*, pp. 45–67.
[48] Ibid., pp. 26–44.   [49] Ibid., pp. 71–4, 88–94.   [50] Ibid., p. 79.
[51] Ibid., pp. 95–105.

A Parrot Difficulty

Pet-keeping manuals tell us how owners were supposed to care for their animals, but they cannot tell us how they actually treated them. Did they adhere to the feeding instructions? Did they keep their animals in the manner suggested? Did they follow the advice on veterinary care?

One place where we can find at least some of the answers to these questions is the correspondence section of the RSPCA's monthly magazine, *The Animal World*, which regularly featured letters from concerned pet owners. The content of these letters ranged from obituaries for departed companions to pleas for advice on the treatment of bald parrots and sickly monkeys, illustrating the pleasures, desires and anxieties inspired by domestic pets. Of course, these letters constitute a biased sample, for the people who subscribed and wrote to an RSPCA publication were, of necessity, self-defined animal lovers and were probably more conscientious and discriminating pet owners than the average Victorian or Edwardian. Nonetheless, a detailed examination of the correspondence reveals how at least some real owners acquired, named and fed their pets, as well as the sentimental attachments they sometimes formed with them.

The letters submitted to *The Animal World* can be divided into three broad categories: personal stories, unsolicited advice and requests for information. The first of these generally took the form of entertaining vignettes and were usually written by owners who wanted either to honour a beloved companion or simply to amuse fellow subscribers. In 1883, for instance, the Reverend C. G. Blaydes rhapsodised over the accomplishments of a parishioner's guinea pig, who would sit up on its hind legs to receive 'large pieces of orange peel, of which it is very fond when dried', and who enjoyed sipping tea 'out of a teaspoon'.[52] In 1886, Derby physician Lawrie Gentles recounted the antics of his pet canary, Jumbo, named after the famous elephant ('not on account of his size ... but on account of the difficulty we had at first getting him into his cage'), who flew around his surgery pecking at his pen nib and subsisted on a diet of 'cold custard pudding, cheese, chewed toast, salt, fig-seeds and butter'.[53] In 1892, A.M.S. from Sydney reminisced about her (now deceased) pet marmoset, who used to groom the cat, lick the yolk out of egg shells and sleep in 'a small eau-de-Cologne box', while in 1909, J. L. Bevir celebrated his pet lemur, Venus, who 'fraternised with guinea pigs', made himself a 'noisy nest of paper' behind a bookcase and enjoyed

[52] 'An Interesting Pet', *The Animal World*, June 1883, p. 86.
[53] 'Our Pet Canary', *The Animal World*, April 1886, p. 59.

his daily breakfast of coffee and marmalade.[54] Perhaps most effusive was Alexandra Peckover, who wrote with great affection about the adventures of her monkey, Jemima. In a lengthy article, Peckover recounted how the monkey 'was first caught when young in the woods near Sierra Leone, and was brought to England from Madeira'. She described Jemima's accommodation, which consisted of 'a room in the basement with a central pane opening into the front garden', and detailed her catholic diet of 'dried figs', any fruit except grapes, and 'her special favourite', 'hot potatoes'. The letter concluded with a recital of Jemima's daily antics, which encompassed teasing the dog, stealing raisins from the kitchen and digging up hyacinth bulbs in the garden.[55]

While the above descriptions all focused on the unique qualities of specific animals, a second tranche of letters appears to have been written with the intention of sharing knowledge and experience among fellow owners. More didactic in tone, these contributions explained what contemporaries had done to improve the lives of their own pets and which good practices might be adopted more widely. Parrot owner, H.J., for instance, recounted how she had 'added much to the happiness' of her Polly by hanging a looking glass 'outside the cage at the end of her perch', a practice she hoped 'other possessors of parrots' would copy.[56] A second correspondent, E. Harridge, described how he had cured a parrot of bronchitis with a diet of gruel, linseed tea and blackcurrant jam – the latter diluted with hot water and 'given to him in his mug when nearly cold'. A third, in this case anonymous, detailed how he had treated a cockatoo's 'swollen and inflamed' leg with daily massages.[57] A fourth contributor, Edmund Harvey, contacted *The Animal World* to relate 'the, to me, interesting sight' of a friend's 'small pet monkey enjoying its bath', in the belief that 'some of your readers might care to know that at least some monkeys do in fine weather'. Harvey stated that the monkey in question came from Borneo and that 'the water in which it bathes is always warmed to about blood heat' – important advice for those wishing to follow his example.[58] Drawing on personal experience, the writers of these letters sought to improve the lives of other pets by disseminating interesting nuggets of information, as well, perhaps, as parading their expertise and humanity.

[54] 'A Pet Opossum', *The Animal World*, December 1892, p. 179; 'Venus', *The Animal World*, April 1909, pp. 87–8.
[55] 'Jemima', *The Animal World*, May 1882, p. 75.
[56] 'Treatment of Parrots', *The Animal World*, May 1876, p. 78.
[57] 'Treatment of Parrots Affected with Bronchitis', *The Animal World*, October 1902, p. 159; 'A Cure by Massage', *The Animal World*, December 1905, p. 191.
[58] 'Monkeys Bathing', *The Animal World*, December 1888, p. 186.

The third, and most common, type of letter to appear in *The Animal World* was the pet-related query. Letters in this category came from concerned pet owners and reflected the anxieties and problems that went along with keeping an exotic pet, from hygiene and diet to what to do with a sick companion. One correspondent solicited advice on 'how to cure a parrot of screaming', while a second asked 'how it is possible to keep a white Persian kitten clean'.[59] Other queries included: what to feed an opossum during the winter (when 'blackberries, grapes, cherries plums, etc.' were out of season), 'if parsley is poisonous to parrots', how to treat a parakeet 'suffering from diarrhoea' and whether or not tortoises had teeth.[60] In most cases, these questions received responses from fellow readers, establishing a dialogue among pet owners in different parts of the country. The letters pages of *The Animal World* thus functioned in part as a kind of agony column where correspondents could solicit advice from the editor or from other subscribers.

A couple of more detailed examples of this question-and-answer genre reveal the nature of this dialogue and illuminate the trials and tribulations of owning an exotic pet. In September 1886, for instance, one reader, Georgina Weldon, contacted *The Animal World* for advice on how to treat her three pet monkeys – Karky, Paddy and Andy – who had recently been afflicted by illness. Karky, a marmoset, had been 'cruelly ill-treated and neglected' before Weldon acquired him, sporting 'matted' fur and a 'broken tail'. He had been nursed back to health by his new owner, but was still suffering from paralysis in his lower limbs. Andy, a tiny monkey of unspecified species, had initially been 'fat ... sprightly and active' but had recently been 'taken with a fit (just like a little baby), and ever since has been getting weaker and weaker'. Weldon had attempted to relieve the monkeys' ailments by 'rubbing them and saturating them in brandy', but worried that this was not a long-term solution. She insisted that all three of her monkeys were carefully 'wrapped up' at night, to prevent them from getting a chill, and that they were 'kept clean, combed and brushed daily'.[61]

Responding to Weldon's cry for help, fellow *Animal World* readers proffered a range of advice, in the process revealing their own experiences of living with simian companions. One correspondent, E.R.B., suspected that 'a warm bath and a dose of castor oil might do good to

---

[59] 'A Parrot Difficulty', *The Animal World*, February 1875; 'How Can a Persian Cat Be Kept Clean?', *The Animal World*, January 1875, p. 15.

[60] 'A Pet Opossum', *The Animal World*, February 1887, p. 24; 'Food for Parrots', *The Animal World*, September 1882, p. 143; 'Parroquet's Ailments', *The Animal World*, July 1874, p. 110; 'Tortoises', *The Animal World*, February 1882, p. 26.

[61] 'To Monkey Fanciers', *The Animal World*, September 1886, p. 158.

Mrs Weldon's monkey' and advised giving marmosets 'a hot water bottle at night' – 'a small stone ink bottle covered with flannel is the proper thing'.[62] A second writer, Frances Maitland Savill, wondered whether applying iodine to the spine would cure Karky's paralysis, having seen the technique used successfully on 'a little pet Abyssinian guinea pig which we had last year'.[63] Sadly, the advice came too late for Andy, who succumbed to a fit on 14 October, but Karky and Paddy were still alive by February 1887 and Weldon hoped they might recover. In a follow-up letter to *The Animal World*, she thanked the senders for both the published and private letters she had received on the subject and related her suspicion that the sickness had been caused by allowing her monkeys to run about on an 'oilcloth or kamptulicon' rug.[64]

A similar, though slightly less tragic, dialogue took place in the pages of *The Animal World* three years later, this time involving a balding cockatoo. On this occasion, the discussion was initiated by Leila Bishop, whose 'white, lemon-crested cockatoo' had recently 'acquired the habit of pecking out his feathers ... evidently causing himself great pain, and sometimes drawing blood'. Bishop stated that 'we have always fed him with canary seed and a little hemp ... [and] bread soaked in cocoa'. She suspected, however, that 'a former servant' may have 'given him grease or meat', causing his illness. The bird, 'a wonderful talker', was now quite 'a pitiable sight; legs and breast and part of the back are bare, and what feathers remain are dirty and draggled'.[65]

Like Weldon's, Bishop's letter generated a range of responses, some of which were published in the following edition of *The Animal World*. One writer, A. A. Huddleston, concluded that 'there is little doubt that the white, lemon-crested cockatoo spoken of ... has had improper food given to it, such as meat or butter', and advised giving the bird 'old cotton reels, pieces of hard wood and stones to bite at', as well as 'pinders (or monkey nuts ... flat maize (uncooked))' and 'a couple of picnic biscuits a day'.[66] Another correspondent, F.G.B., seconded Huddleston's 'excellent advice' but also recommended that the cockatoo be 'kept warm' and its cage 'well-covered at night'.[67] Whether either treatment had the desired result is not recorded, but we get the sense, once again, of an active community of animal lovers, familiar with exotic species and keen to share their knowledge with a less experienced owner.

[62] 'Bearing-Reins, Monkeys and Dogs', *The Animal World*, November 1886, p. 174.
[63] 'Paralysis in Monkeys', *The Animal World*, March 1887, p. 46.
[64] 'Treatment for Monkeys', *The Animal World*, February 1887, p. 30.
[65] 'Diseases of Parrots', *The Animal World*, December 1890, p. 191.
[66] 'Diseases of Parrots', *The Animal World*, January 1891, p. 14.
[67] 'Diseases of Parrots', *The Animal World*, February 1891, p. 31.

Looking at the correspondence in *The Animal World* overall, we thus gain an interesting insight into the joys and sorrows of keeping pets in late Victorian Britain. The letters written to the RSPCA's journal reveal the tenderness and affection inspired by certain companion animals and the existence of a virtual community of conscientious pet owners who exchanged information on how best to care for non-native creatures. At the same time, however, they also reveal the darker side of pet keeping – the illnesses and premature deaths suffered by exotic animals and the tendency of even some self-professed animal lovers to treat their pets as luxury toys rather than wild beasts. One writer, for instance – in this case from Burma – described how she would 'dress up' her pet gibbon, Dinah, in 'a pale silk dress, trimmed with lace', even though the ape strongly objected to the practice.[68] Many middle- and upper-class pet owners seem genuinely to have cared for their animals, showing considerable sensitivity to their needs and emotional well-being, but the range of complaints described in their correspondence indicates the difficulties of looking after exotic animals and the high death rate occasioned by ignorance and misguided affection. The impression one receives from reading these sometimes touching letters is therefore somewhat ambiguous: *Animal World* readers loved their pets, but they were consciously or unconsciously encouraging a brutal trade that accounted for the importation of tens of thousands of exotic animals each year – a subject to which we now turn.

A Cruel Trade

While keeping pets was a source of pleasure for owners, it was often a source of misery for animals, particularly exotic species. Some pets may have been loved and doted on, but others were neglected, abandoned or, worst of all, actively abused. From a conservation perspective, the impact of the wild animal trade on both native and foreign fauna was also substantial, with large numbers of animals dying during capture or in transit. What, then, was the environmental impact of keeping exotic species as pets? And what abuses did monkeys, parrots and other animals face after they arrived in Europe?

If we focus first on the exotic animal trade, we find a callous and brutal business that consistently put profit above welfare. Dealers such as Jamrach freely admitted that they lost a large proportion of their stock through death, a loss they accepted as inevitable in their line of work.

[68] 'Dinah!', *The Animal World*, November 1894, p. 173.

Losses were equally high in transit and during capture, with many animals dying for each one that reached Europe alive. While it is difficult to quantify precisely the impact of these deaths – statistics for which are often lacking – an examination of the ways in which wild animals were collected and transported gives us a sense of the wider environmental cost of the pet trade and reveals its hidden cruelties. Anecdotal accounts of attrition need, moreover, to be put in the context of the overall volume of the trade, which, particularly in the 1880s and 1890s, was substantial; Jamrach reported that it was common for him to have 'five thousand pairs of cockatoos, &c.' shipped over 'in a single vessel', while Cross claimed that 'I think nothing of having a consignment of sixty or seventy [monkeys] in one steamer, some of them nearly as big as a man'.[69] If even a small proportion of these animals died, this would have had a significant effect on wild populations – and, as we shall see, the proportion was not small.

To understand why losses were so high, we need, first, to look at how wild animals were captured. This was typically a violent and extremely costly process with a lot of collateral damage. Unlike sportsmen, who generally targeted large male animals in pursuit of an epic battle, professional animal catchers were 'rarely concerned with the ritualised confrontation of man and animal on a field of honour' and more worried about profit.[70] They consequently resorted to tactics that any self-respecting sportsman would have seen as underhand, slaughtering females, or sometimes entire families, in their efforts to secure their more tractable cubs or calves. This brutal strategy often sentenced the young animal to death if it was still dependent on its mother, and it could result in whole herds being wiped out in a single onslaught. In 1866, for example, a group of sailors aboard the Dundee whaling ship *Ravensburg* killed a mother polar bear and (accidentally) one of her cubs in order to snare the surviving cub.[71] In his 1901 book *Kamerun: Sechs Kriegsund Friedensjahre in deutschen Tropen* ('Cameroon: Six Years of War and Peace in the German Tropics'), meanwhile, German animal catcher Hans Dominik relates, with chilling sangfroid, how he slaughtered a

---

[69] 'Jamrach's', *Reynold's Newspaper*, 23 March 1879; 'The Wild Animal Trade', *Leeds Mercury*, 2 January 1886.

[70] See 'Catching Animals' in Rothfels, *Savages and Beasts*, pp. 44–80. For further discussion of contemporary hunting practices, see 'Imperial Glory' in Baratay and Hardouin-Fugier, *Zoo*, pp. 113–30; 'The Imperial Hunt in India' in John MacKenzie, *The Empire of Nature: Hunting, Conservation and British Imperialism* (Manchester: Manchester University Press, 1988), pp. 167–99.

[71] 'Capture of a White Polar Bear by a Dundee Whaler', *Liverpool Mercury*, 3 November 1866.

whole herd of elephants so that he might more easily catch his target – a single calf.[72] These examples were, sadly, not atypical; indeed, they remain the modus operandi in the modern exotic pet trade.

For those animals that survived the initial onslaught, the suffering did not end there, for they then faced a gruelling journey back to Europe in what were often horrifying conditions. The cub captured by the crew of the *Ravensburg* was stowed in an empty oil cask with an improvised grating and arrived in Dundee caged and growling.[73] He at least reached Britain alive; many other animals were not so fortunate, succumbing to starvation, injury or disease en route to their destination. A manatee transported to Jamrach from South America, for instance, perished 'from exhaustion of its food and cold and wet weather during the voyage'.[74] Of the '32 elephants, 8 giraffes, 20 antelopes, 16 buffaloes, 2 specimens of rhinoceros, 1 hippopotamus, 12 hyenas, 4 lions, 4 ostriches, 12 hornbills, 2 adjutants [and] 1 bustard' shipped from Upper Nubia to Trieste by traveller Lorenzo Casanova, meanwhile, only '[e]leven elephants, five giraffes, six antelopes, no buffaloes, one rhinoceros, no hippopotamus, twelve hyenas, no lions, seven hornbills, two adjutants and four ostriches' arrived alive – a survival rate of just 42 per cent.[75] Nor, indeed, did the attrition end with an animal's arrival in Britain, for many further deaths occurred on the dealers' premises or very soon after purchase by a private owner. Jamrach's 'death roll' from 1886 revealed that 'in May [of that year] the loss through death amounted to £169, in June £70, in July £105, in August £96, in September £112 and in October £95; or for 6 months £647'.[76] Naturalist Frank Buckland, giving this financial loss an animal face, recounted how his pet monkey, Susey, arrived at Jamrach's 'lying on her side breathing very hard, and very, very ill', surviving only through the prompt administration of 'port wine, beef-tea and hot flannels'.[77] For every animal that lived long enough to become a menagerie inmate or a domestic pet, at least two or three others probably perished in the field, at sea or in the dealer's warehouse.

To assess the impact of the exotic pet trade in particular, let us look at three popular companion animals: parrots, tortoises and monkeys. In the case of the former, the birds were imported into Britain from Africa, South America and Australia, with the African grey parrot arriving in by

[72] Rothfels, *Savages and Beasts*, pp. 60–1.
[73] 'Capture of a White Polar Bear by a Dundee Whaler', *Liverpool Mercury*, 3 November 1866.
[74] 'Trafficking in Wild Beasts', *Royal Cornwall Gazette*, 25 October 1873.
[75] 'Wild Beasts', *Pall Mall Gazette*, 27 July 1868.
[76] 'The Wild Animal Trade', *Leeds Mercury*, 2 January 1886.
[77] 'My Monkeys', *The Morning Post*, 26 December 1867.

far the greatest numbers. According to one contemporary, a keeper at London Zoo, parrots previously had a better survival rate, as they were brought over in small quantities aboard sailing ships and 'were partly acclimatized by the long weeks of travel before reaching England'. By the 1870s, however, parrots were collected in large numbers by African suppliers, 'packed in crates, and stowed in the hold of a steamer, the transit of which is too rapid to accustom them to the change of climate', with the result that 'hundreds die on the way, and almost as many more die within two or three weeks of arrival of a fever contracted in the unhealthy hold'.[78] *The Bazaar*'s bird expert, C. W. Gedney, claimed that 90 per cent of African grey parrots taken from their nests in the African interior perished from consumption (i.e. pulmonary disease) soon after they reached Britain.[79] His successor, Dr W. T. Greene, who performed post-mortems on readers' pets, informed one grieving owner that '[n]ot one [grey parrot] survives out of a thousand, and yet people will keep on wasting their money in buying them. It is not a bit of use your buying another, as it is bound to go the same way.'[80]

Fearing that such losses would 'result in the probable extinction of grey parrots', one commentator, naturalist Albert H. Waters, suggested that 'if three fourths of the imported birds die on the passage, the inference should be that only one-fourth the number should be placed on ship-board, and provided, as they might be, with roomier cages'.[81] Another writer, 'South Coaster', proposed the establishment of a 'Grey Parrot Importation Co-operative', whereby fellow parrot fanciers might source healthy, well-vetted birds, independently of the 'regular dealers', via 'a respectable trader in Africa, who might be willing to send the first consignment on Deposit System'.[82] Such humane recommendations might have stemmed the live bird trade, but they cut little ice with dealers, who ranked profit above welfare. As RSPCA secretary John Colam reflected:

The dealers buy birds at so low a price, and sell them with so much profit to thoughtless people who 'want to keep a pet', that they can afford to lose 60 or 70 per cent of these poor prisoners by death better than invest money in larger cages, pay rent for larger rooms, which the use of larger cages would involve, and find attendants to clean out the miserably little cages now used, and give the captives food and water regularly.[83]

[78] 'The Transit of Parrots', *The Animal World*, November 1894, p. 175.
[79] C. W. Gedney, 'The Grey Parrot', *The Bazaar*, 3 February 1877, p. 76.
[80] 'Cage Birds', *The Bazaar*, 27 September 1893, p. 801.
[81] 'Cruelty to Parrots', *The Animal World*, November 1894, p. 175.
[82] 'Grey Parrots', *The Bazaar*, 21 November 1890, p. 719.
[83] 'Caged Birds, etc.', *The Animal World*, September 1882, p. 131.

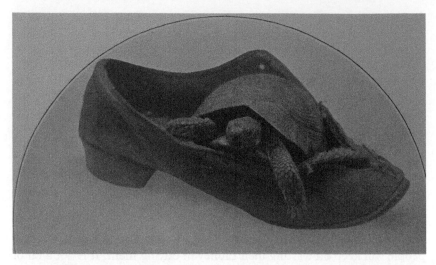

Figure 6.4 G. Bletcher, 'Quite comfortable', *The Animal World*, April 1912, p. 110.

For tortoises, the situation was, if anything, even worse, as these animals were considered to be 'of low nervous organisation' and therefore impervious to pain, as we shall see below.[84] Typically collected during their hibernation period, tortoises were packed 'one on top of another in holds of vessels, like peas in a sack', and then, 'upon the vessel reaching port', 'unloaded and submitted to further ill-treatment by being placed like slabs of stone on the quay or landing place, where they are left until convenient for their removal'. Some of the animals were sent to restaurants to satisfy the 'gourmand tastes' of elite diners, while others were 'hawk[ed] … about the streets in barrows' by costermongers, for 'the poorest people' to buy 'as toys and curiosities' (Figure 6.4).[85]

The cost of this commerce was high, with many tortoises perishing en route to Britain and many more expiring shortly after their arrival from a combination of injury and neglect. One opponent of the trade, writing in *The Animal World*, recounted a recent incident at St Katharine Docks, London, when 'upwards of 10,000 living land tortoises, all … packed like herrings in barrels', had been abandoned on the quay, many already dead and 'hundreds apparently … dying from privations arising out of a long voyage, detention at the docks and bad packing'.[86] Another critic

---

[84] 'Cruelty in the Importation of Tortoises', *The Animal World*, January 1882, p. 2.
[85] 'Tortoises', *The Animal World*, August 1872, p. 186.
[86] 'Cruelty in the Importation of Tortoises', *The Animal World*, January 1882, p. 2.

described how it was 'no uncommon thing ... to see upon a street barrow some wretched tortoise – the largest specimen of the collection generally – with a considerable dent in his upper shell or carapace', a deformity caused by cramming too many animals into a single barrel and indicative of the animal's 'broken back and fractured ribs'.[87] Tortoises thus suffered much cruelty en route from Africa and their importation had a significant (but unquantifiable) impact on wild populations.

As for monkeys, their route to Britain was in some ways less straight-forward, as they emanated from many different parts of the globe and comprised a range of different species. Bonnet and rhesus macaques came from Asia, capuchins, spider monkeys and marmosets from South America, and vervet, colobus and Diana monkeys from Africa; in 1891, a vendor in *The Bazaar* offered for sale a 'white, tame, Colobus monkey, from [the] Taita Range, Kilimanjaro, East Africa'.[88] While monkeys may have come from different places, one thing that most of them shared was a vulnerability to the British climate. Many monkeys died during the journey to Britain and many more expired soon after arrival, most commonly from pulmonary complaints brought on by the cold. So high was the attrition rate during shipping that primate expert Arthur Patterson advised anyone thinking of purchasing a monkey to 'see him [in person], and if he looks dull and mopy, or if he has a cough, or is thin and poor, refuse him as a gift', for he would soon perish from his ailments.[89]

For those monkeys that did survive the ordeal of importation, other maladies often awaited. Dysentery and diarrhoea were common, brought on by inappropriate diets, while fits were another frequent affliction, the result, most probably, of lack of exercise. A number of monkeys also developed the disturbing habit of 'nibbling their tails' – a behavioural abnormality caused by lack of company and insufficient mental stimula-tion.[90] Some owners clearly did their utmost to mitigate these problems, seeking the best possible conditions for their pets and vetting potential purchasers of their animals when they parted with them. One man posted an advertisement in *The Bazaar* asking 'if anyone who has [a] hot house

---

[87] 'Tortoises', *The Morning Post*, 3 April 1884.
[88] 'Country House', *The Bazaar*, 4 November 1891, p. 1282.
[89] 'Management of Monkeys', *The Bazaar*, 27 January 1886, p. 99.
[90] 'Some Odd Monkey Notes', *The Bazaar*, 16 February 1892, p. 237. Rhesus monkeys, for instance, form strong social bonds and can develop behavioural abnormalities in captivity if these are lacking. See Avanti Mallapur and B. C. Choudhury, 'Behavioural Abnormalities in Captive Nonhuman Primates', *Journal of Applied Animal Welfare Science* 6:4 (2003), pp. 275–84.

would take charge of [his] marmoset for the winter'.[91] Another, putting his 'Fine hardy monkey' up for sale, stipulated that the 'buyer must guarantee good home, no organ grinder need apply'.[92] Despite such efforts, however, the outlook for the average pet monkey was rather bleak, and few survived beyond a few years. When a correspondent named Fifi wrote to *The Bazaar*'s monkey expert for advice on how best to treat her beloved marmosets – one of whom had just died after 'eating some slugs' – the expert remarked, 'Three years may be regarded as a long life in captivity in this country' – a pretty poor statistic for a species that can live up to the age of twelve in its native South America.[93] Monkeys of all species thus suffered considerably at the hands of hunters, traders and pet owners, exchanging life in the trees for solitary captivity.

Given the serious casualties of the wild animal trade, the question arises as to whether any action was taken to prevent this attrition, and, if so, how effective it was. For parrots, tortoises and monkeys, as noted above, little appears to have been done, for the collection of the animals fell outside the jurisdiction of the British courts and profit trumped preservation in the eyes of dealers. For other species, however, action was taken. As in the anti-plumage and anti-sealskin movements, this took the form of both legal measures to protect wild animals and increased pressure on consumers to change their buying habits.

To begin with, from the 1870s, legislation was enacted to protect Britain's wild birds, which, like parrots, were captured by dealers for sale as pets. The first piece of legislation to cover birds, the Sea Birds Preservation Act of 1869, banned the shooting of twenty named species of sea bird.[94] This was followed by the Wild Birds' Preservation Act of 1872, which outlawed the taking of thirty-nine listed species of birds, alive or dead, during the breeding season (15 March–1 August). Further acts followed in 1876, 1880 and 1894, extending the close season, and, in the last case, giving county councils the power to add other locally threatened species to the banned list.[95] Several bird catchers and dealers were prosecuted under these acts, providing a level of respite for native bird species. In 1881, the RSPCA recorded sixty-five convictions for

[91] 'Country House', *The Bazaar*, 10 October 1898, p. 1096.
[92] 'Country House', *The Bazaar*, 22 April 1903, p. 1999.
[93] 'Marmosets in Captivity', *The Bazaar*, 29 January 1915, p. 111.
[94] Great Britain Parliament, *An Act for the Preservation of Sea Birds*, 1869: 32 & 33 Victoria c.17.
[95] 'Wild Birds Protection Act', *Western Times*, 21 August 1872; 'Protection of Wild Birds', *York Herald*, 28 April 1876; 'The Wild Birds' Protection Act', *Essex Standard*, 20 November 1880; 'Society for the Protection of Birds', *The Star*, 24 February 1894.

'attempting to take wild birds'; in 1889, it listed sixty convictions for the same offence.[96]

While the Wild Birds' Preservation Acts made some inroads into the trade in British songbirds, they had significant limitations. In some cases, magistrates were reluctant to convict, letting first-time offenders off without a fine or deliberately turning a blind eye to crimes in their locality. On other occasions, bird catchers exploited loopholes in the legislation by snaring birds not explicitly enumerated as protected or claiming that they had taken them outside the close season. In 1888, for instance, a man arrested 'for having a thrush in his possession during the close season' was acquitted of any crime 'as neither blackbirds nor thrushes are included in the schedule of the birds protected by the Act'.[97] In 1898, meanwhile, a bird catcher sent 100 live larks to a dealer in Manchester but was let off after claiming that 'he caught them before the close season, and kept them weeks in confinement' – a claim that was difficult to disprove.[98] Even when some of these loopholes were closed, the Wild Birds Acts applied only to the British Isles, meaning that parrots and other foreign birds continued to be collected and transported with impunity. Some steps were taken to address this issue – most notably in 1904, when Louisiana passed the Audubon Society's Model Law, preventing the export of 'Mockingbirds, Cardinals, Nonpareils and Indigo-Buntings' from the state – but these were piecemeal and only partially effective.[99] The importation of tortoises into Britain would not be made illegal until 1984, while a worldwide ban on the trade in African parrots only came into effect in 2016 – many years after Albert Waters and others first expressed concerns for their survival.[100]

If legislation had only limited impact, what about the other key driver of change, consumer action? Was disapproval expressed over the keeping of exotic species in captivity? And, if so, did it effect any change?

Like other abuses of animals in the late nineteenth century, the keeping of wild creatures as pets generated concern, giving rise to the by now familiar refrain that if consumers did not buy exotic pets, dealers would not catch or sell them. Parrot lover C.R.E. insisted that anyone who purchased a young grey parrot was guilty of fuelling a 'great cruelty', for 'it only leads to more being imported'; 'the sooner more publicity is given

[96] 'Analysis of Convictions, 1881', *The Animal World*, August 1882, p. 175; 'Analysis of Convictions, 1889', *The Animal World*, August 1890, p. 121.
[97] 'The Protection of Wild Birds', *The Star*, 26 June 1888.
[98] 'The Protection of Wild Birds', *The Morning Post*, 1 February 1898.
[99] 'National Committee Notes', *Bird Lore* VI (1904), p. 141.
[100] 'Global Trade in Wild African Grey Parrots Banned', *The Telegraph*, 2 October 2016.

to the great suffering the birds undergo from the moment they are "boxed" on board ships until the ninety-nine out of every hundred die, the sooner ought their importation to be stopped'.[101] This view was seconded by a journalist in the Portsmouth *Evening News*, who observed that, 'owing to a growing taste for keeping birds in confinement, thousands of birds are caught annually, and hosts of songsters are doomed to a speedy death or a short life of misery in an ill-conditioned cage, because people fancy they have a love for the feathered race, when in reality they have nothing of the sort'.[102] Both of these writers believed that supposed bird lovers, in frequenting dealers, were making those dealers' businesses profitable and encouraging the trade to continue. The blame for any cruelty should therefore lie firmly with the buyer, who, through neglect or incompetence, might make the animal's existence still worse. As an opponent of the tortoise trade reflected – with a noteworthy sideways swipe at the sealskin industry – 'The thought of the pain [these animals] are made to endure in order that silly people may amuse themselves with a plaything for a few days, cannot fail to afflict the minds of humane persons; just as the same people have been horrified by recitals of cruelty to seals.'[103]

The extent to which such views gained traction and made a tangible impact on the wild beast trade is more open to question. On the one hand, there is some anecdotal evidence that the popularity of exotic pets diminished in the 1890s and 1900s, with many dealers reporting a contraction in their business. As early as 1886, an article in the *Leeds Mercury* noted that 'buyers of feathered and four-footed pets of the smaller kind' were 'falling off', dropping 'by more than half compared with what it was only half a dozen years ago'.[104] A report in *The Standard* five years later made similar observations, remarking that '[t]he monkey is not so saleable as it was, the snake has ceased to be profitable and voyages are too short for sailors to be able to teach parrots the marketable amount of highly spiced language' to attract buyers.[105] Albert Edward Jamrach, meanwhile, informed a reporter in 1900 that the outbreak of the Boer War had reduced demand for pets, since 'fashionable women' now had 'far more serious things to think about and spend their money on'. He cited the example of a fennec fox, 'a pretty-little fawn-coloured beast', who would once have made 'an excellent pet for a lady' but who

[101] 'Grey Parrots', *The Bazaar*, 21 November 1890, p. 718.
[102] 'The Protection of Wild Birds' *The Evening News*, 23 June 1880.
[103] 'Cruelty in the Importation of Tortoises', *The Animal World*, January 1882, p. 2.
[104] 'The Wild Animal Trade', *Leeds Mercury*, 2 January 1886.
[105] 'Charles Jamrach', *The Standard*, 9 September 1891.

had been languishing in his store for several weeks.[106] This would seem to suggest that Colam and his supporters were making headway and that consumers were making a conscious decision not to buy exotic animals.

In reality, however, the picture was less clear-cut, for other, less noble causes may also have impacted on the wild beast trade, perhaps more significantly than any shift in attitude towards unusual pets. The Mahdi Rebellion in the Sudan (1881–99), for instance, was reported by some contemporaries as an obstacle to the importation of larger wild animals such as elephants and giraffes, while increased competition from European dealers and the decline of several major travelling menageries during this period undoubtedly affected British outfits such as Cross and Jamrach.[107] It is also possible that growing numbers of consumers were bypassing dealers and purchasing animals second-hand – not least because this improved the beasts' chances of survival. One vendor in *The Bazaar*, for instance, advertised '2 young Boer monkeys, beautiful, tame and amusing, very small, taken from a deserted Boer farm near Klerksdorp, Transvaal by the owner', suggesting that the Boer War provided a new source of exotic pets rather than being a deterrent to buying them.[108] The exotic pet trade may thus have experienced a dip towards the end of the nineteenth century, but this was not permanent and not necessarily an indication that the conservation message had hit home.

Battered Monkeys and Roasted Parrots

The conservation aspect of the exotic pet trade was one source of concern for Victorian animal lovers, but another, equally troubling, issue was the treatment surviving pets received at the hands of their owners. Unlike fur seals, elephants and egrets, pets were subject to abuse not only overseas but in the more visible settings of British homes and streets. This gave rise to a different form of welfare anxiety, focused, in the first instance, on the physical well-being of the pet and, in the second, on its overall quality of life. What legal rights did exotic pets possess, and what sanctions could be brought against owners who abused, deformed or wilfully neglected their domestic companions?

When it came to physical mistreatment, the scope was broad, ranging from the mutilation of pets for aesthetic reasons to the killing of animals

---

[106] 'Pets of Peace', *Courier and Argus*, 4 October 1900.
[107] Charles John Cornish, *Life at the Zoo: Notes and Traditions of the Regent's Park Gardens* (London: Seeley, 1895), p. 184.
[108] 'Country House', *The Bazaar*, 22 July 1901, p. 303.

in fits of rage. One correspondent in *The Animal World* denounced the 'cruel practice ... of breaking the noses of pug dogs, when puppies, to produce the form which is the most admired'.[109] A seller in *The Bazaar*, meanwhile, advertised a 'Pretty tame black faced she monkey' with a 'docked tail', suggesting the full or partial amputation of what was, for monkeys, effectively a fifth limb.[110] Such endemic practices were paralleled by one-off acts of violence that resulted in the injury or death of individual pets, at the hands of either their owners or other visiting parties. In 1886, as described earlier, Sheffield carter James Noble appeared before magistrates charged with 'having unlawfully and cruelly abused and tortured a parrot by stabbing it with a knife, and afterwards burning it to death on the fire'.[111] In 1897, in a similar case, Hull magistrates sentenced 'a tradesman named James Tennant' to one month's hard labour after he 'rubbed a red-hot poker' on a parrot's 'breast and feet' to make it talk.[112]

While both Noble and Tennant were punished for their offences, the position of parrots and other exotic pets was, in fact, complicated in terms of the law. This was due to the terminology of the 1835 Cruelty to Animals Act, which officially covered only 'any horse, mare, gelding, bull, ox, cow, heifer, steer and pig or any other domestic animal'.[113] The RSPCA, recognising this legal loophole, lobbied to get it closed, launching a series of test cases in the 1870s and 1880s to ascertain whether the Act could be extended to wild animals in captivity. The results of this campaign were rather mixed, however, and appeared to depend largely on the whims of individual magistrates. In 1877, for instance, when lion tamer William Edward Nicholls struck a camel with a manure fork 'because it threw some froth from its mouth on his clothes', Hull magistrate Mr Travis convicted the defendant, fining him 10 shillings and costs.[114] Three years earlier, however, when the RSPCA prosecuted menagerie owner Harriet Edmonds and keeper Frederick Hewitt for making hyenas leap through burning hoops, magistrate Mr Bruce at Leeds Borough Court reluctantly dismissed the case on the grounds that,

---

[109] 'Shameful Cruelty to Pug-Dogs – If True?', *The Animal World*, September 1872, p. 205.
[110] 'The Protection of Small Birds', *Derby Mercury*, 7 August 1872; 'Country House', *The Bazaar*, 23 August 1886, p. 548.
[111] 'Gross Cruelty to a Parrot in Sheffield', *Sheffield Independent*, 17 September 1886.
[112] 'Horrible Cruelty to a Parrot', *Hull Daily Mail*, 28 October 1897.
[113] Great Britain Parliament, *An Act to Consolidate and Amend the Several Laws Relating to the Cruel and Improper Treatment of Animals*, 1835: 5 & 6 William 4 c.59. A further statute in 1849 updated the animal cruelty legislation but simply repeated the passages defining 'animal' and banning fights. See Great Britain Parliament, *An Act for the More Effectual Prevention of Cruelty to Animals*, 1849: 12 & 13 Victoria c.92.
[114] 'Cruelty to a Camel', *Hull Packet and East Riding Times*, 19 October 1877.

'if you asked anybody who understood the English language and who was not a lawyer, what was a domestic animal, the answer would not include a lion or a panther or a hyena'.[115] It was the same story in 1884, when RSPCA inspector Eli Udden attempted (unsuccessfully) to prosecute lion tamer Hezekiah Moscow for beating four bears at the East London Aquarium with 'a drop thong whip', leaving visible 'weals about their bodies'.[116] The position of menagerie inmates – and, by implication, exotic pets – was thus rather ambiguous, with certain forms of cruelty and certain classes of perpetrator regarded as exempt from prosecution.

If we turn specifically to the cruelty cases concerning parrots, monkeys and other exotic pets, we find a complex picture in which some offenders were convicted and others released without charge. On the positive side, the RSPCA was generally rather more successful in defending pets than in defending menagerie inmates, and it managed to secure several high-profile convictions, mostly involving extreme violence perpetrated by working-class offenders. In 1858, for instance, magistrates in Portsmouth convicted sailors James Gill and George Joyce for burning the tail of a monkey, 'the property of Frederick Palmer, a convict warder'.[117] In 1877, magistrates in Birkenhead fined Edward George Carter 40 shillings for breaking the neck of his wife's parrot 'in a fit of spleen' (he was angry that his dinner was not ready when he came home from work).[118] In 1884, 'William Wood, collier, was sent to gaol for fourteen days, without the option of a fine', for 'roasting a live linnet' in a public house, while, in 1895, 'Stephen Fordham, 78, a labourer receiving parish relief', was fined '1s, 9d damage and 11s costs' for killing his neighbour's Persian cat, after it 'annoyed him by laying on his flowers'.[119] There were also several successful prosecutions of Italian organ grinders for mistreatment of their monkeys, among them Michael Angelo, fined 7s 6d for shaking a monkey 'viciously', and Amelia Pascoe, fined 2s 6d for 'bumping' his monkey's head against Blackpool pier (Figure 6.5).[120]

---

[115] 'Singular Charge of Cruelty in a Menagerie', *Leeds Mercury*, 5 December 1874.
[116] 'Charge of Cruelty to a Bear', *Manchester Times*, 9 February 1884.
[117] 'Atrocious Cruelty to a Monkey', *Southampton Herald*, 25 December 1858.
[118] 'Birkenhead', *Royal Cornwall Gazette*, 23 March 1877.
[119] 'Ferae Naturae, or Domestic?', *The Animal World*, May 1884, p. 67; 'The Ashdon Veteran and the Cat', *Chelmsford Chronicle*, 31 May 1895.
[120] 'Cruelty to a Monkey', *Devon Journal*, 13 September 1894; 'A Monkey in a Police Court', *Preston Chronicle*, 12 August 1893. On the relationship between animal cruelty prosecutions and race/nationality in Edwardian Britain, see David Wilson, 'Racial Prejudice and the Performing Animals Controversy in Early Twentieth-Century Britain', *Society and Animals* 17 (2009), pp. 149–65.

Figure 6.5 'Captive and an Alien', *The Animal World*, December 1909, p. 274. An Italian street entertainer poses with a chained rhesus monkey.

While parrots and monkeys were perhaps more likely than bears and hyenas to meet the criteria for domesticity, not all cases involving exotic pets received such sympathetic hearings. Sometimes entire species were deemed to lie outside the protection of the law, while on other occasions individual victims were judged insufficiently tame to be classed as 'domestic' animals. When the RSPCA sought to prosecute those responsible for the abandonment of a cargo of African land tortoises on St Katharine Docks, magistrate Mr Saunders of Thames Police Court 'expressed a doubt that the law forbidding cruelty could be applied to these animals' and took 'no further steps' beyond compelling 'the consignee to remove the casks and liberate the wretched captives'.[121] When 'John Smith, boat builder, William Davis, railway porter, and Joseph

---

[121] 'Cruelty in the Importation of Tortoises', *The Animal World*, January 1882, p. 2.

Gittings, builder' were brought before Wednesbury Police Court charged with cruelty to a monkey, a long discussion ensued as to whether or not the animal was 'domestic'. The RSPCA took the view that the monkey, which 'was kept tied up in the yard of Cottage Spring Inn ... Tipton', was a captive pet and therefore eligible for protection. Presiding magistrate Mr Neville, however, rejected this argument, concluding that 'this was a wild animal, not kept for any purpose subservient to the use of man, but for the purpose of giving people, who were perhaps lower in the scale of intellect than the animal itself, the opportunity to tease it'. Neville did at least express some sympathy with the monkey, which 'had given proof of its higher order of intellect' by biting Smith on the hand, but he declined to impose any additional penalty on its tormentor.[122] Whether an exotic pet was classed as domestic depended, therefore, on what it was used for (work or entertainment), on its presumed capacity for pain (tortoises and turtles were believed to be 'of low nervous organisation' and less capable of suffering) and on whether the abuse directed at it represented deliberate assault or merely ignorant neglect.[123]

This situation changed in 1900 when Parliament passed the Wild Animals in Captivity Protection Act. Intended to prevent travesties of justice like those described above, the 1900 Act clarified the position of exotic animals under British law and made it an offence to cause any suffering to a wild animal in captivity or to 'cruelly abuse, infuriate, tease or terrify it'. The new legislation enabled magistrates to sentence offenders to up to three months of hard labour or a fine of up to £5 and – significantly – it covered both the mental and physical suffering inflicted on performing animals.[124] In May 1908, for instance, in a landmark case, magistrates in Exeter fined animal trainer Richard Sedgewick 20 shillings for 'cruelly infuriating a lioness' by rattling the bars of her cage, firing a revolver and poking her with 'a long stick'.[125] The following September, magistrates at West London Police Court convicted elephant trainer William Schreida and his assistant Havaida for 'cruelly terrifying' an elephant by making it slide down a steep chute into a pool of water – a

---

[122] 'A Man and Monkey Fight', *Berrow's Worcester Journal*, 19 October 1895.

[123] 'Cruelty in the Importation of Tortoises', *The Animal World*, January 1882, p. 2. On the many different ways of classifying and categorising animals in nineteenth-century Britain, see Harriet Ritvo, *The Platypus and the Mermaid and Other Figments of the Classifying Imagination* (Cambridge: Harvard University Press, 1997).

[124] 'The Protection of Wild Animals', *Humanity: The Journal of the Humanitarian League*, August 1900, p. 59.

[125] 'Tame Lioness Stirred Up to Provide Excitement at Exeter', *Western Times*, 1 May 1908.

trick accomplished by digging 'the pointed end of [an] elephant stick six times into the right cheek of the animal' and prodding 'the hook into his back'.[126] In both cases it was the psychological as much as the bodily abuse of the animals that formed the basis of the conviction; the lioness was said to have 'jumped about', 'infuriated', and the elephant 'trumpeted loudly', giving 'vent to shrill cries of pain and terror'.[127]

Although directed primarily at menageries and circuses, the Wild Animals in Captivity Protection Act had two significant implications for the protection of exotic pets. First, it encompassed new species of animals whose mistreatment might previously have escaped punishment. In 1901, for example, in what was to become the very first conviction under the new Act, an 'Italian lad', Stephen Stromboli, found himself in the dock for 'having cruelly terrified and ill-treated a lemur' – a type of animal not previously accorded protection. According to passer-by Mr Closh, who witnessed the incident, Stromboli entered Hyde Park 'chasing a lemur, to which a chain was attached'. The boy quickly 'caught up with the animal, and stopped it with a jerk by putting his foot on the chain'. He then 'took hold of the chain and swung the lemur violently against a tree', causing it to fall 'upon some chairs'. Summoned before Marlborough Street Police Court, Stromboli admitted hitting the lemur, but claimed that it 'had misbehaved itself by biting at a lady's child' and had been taken into the park for chastisement. Magistrate Mr Fenwick, however, said that, 'if the animal were dangerous, it ought not to be taken about'. He ordered the Italian to pay a fine of 40 shillings or to serve a sentence of twenty-one days in prison – a longer sentence than had been given for comparable offences prior to 1900. This conviction was significant because, before the Act, the lemur's domesticity would most likely have been questioned, especially since it 'had a habit of biting at people'.[128]

A second, perhaps even more important, consequence of the 1900 Act was that instances of neglect could now be prosecuted alongside cases of overt physical cruelty. This is illustrated by an unusual case in 1903, in which the RSPCA charged a middle-class lady, Mrs Mary Richardson, with 'causing unnecessary suffering to a parrot by leaving it in her house for about eleven days, while she and others usually residing in the house were absent'. Appearing before magistrates in Leeds, Mrs Richardson

---

[126] 'The Elephants at the Franco-British Exhibition', *The Animal World*, November 1908, pp. 245–7.

[127] 'Tame Lioness Stirred Up to Provide Excitement at Exeter', *Western Times*, 1 May 1908; 'The Elephants at the Franco-British Exhibition', *The Animal World*, November 1908, pp. 245–7.

[128] 'The First Conviction under the New Act', *The Animal World*, January 1901, p. 16.

confessed to leaving the parrot unattended, although she explained that she had given it 'a sufficient supply of seeds to support it during her absence'. The parrot, however, 'owing to the want of distribution of the food daily', quickly became 'famished' and was discovered by RSPCA officer Inspector Clayton 'in almost a dying condition', with 'its feathers ... standing up and ... scarcely able to move'.[129] Clayton fed the bird with some sop, provided by a neighbour, and with the assistance of local vet Mr Frank Somers succeeded in reviving it. The RSPCA nonetheless decided to prosecute, and magistrate Mr Atkinson, after a week's deliberation, ordered Mrs Richardson to 'pay the costs of the prosecution, including the fee for the veterinary surgeon'.[130] Atkinson meditated that this was a highly unusual case, 'for it is only three years ago since an offence of this kind could be regarded as of any importance in this country'.[131] Interestingly, the wide reporting of the trial also caused many contemporaries to write to the Richardsons asking how the parrot was doing and offering to re-home it, eliciting a statement that the animal was now 'well and cheerful and that it is neither for sale nor for disposal otherwise'.[132]

The Wild Animals in Captivity Protection Act therefore expanded the scope of existing cruelty legislation in regard to wild animals and, by extension, exotic pets. Before 1900, blatant cruelty towards animals adjudged to be 'tame' was sometimes prosecuted successfully, especially when it was connected to other forms of violence (such as wife beating) and perpetrated by foreigners (particularly Italian organ grinders and French dancing bear keepers). After 1900, new abuses (such as neglect) and new animals (such as lemurs) were covered by the law, and both animal dealers and middle-class pet owners found themselves in court; in 1902, ex-schoolmaster and hawker William Kay from Stalybridge reportedly committed suicide after the RSPCA charged him with 'causing unnecessary cruelty to a number of goldfish'.[133] Exotic pets were thus increasingly subject to formal protection and one important legal loophole had been closed.

Animals Out of Place

Physical abuse was one definition of cruelty. For some humanitarians, however, the concept stretched further to encompass more insidious,

---

[129] 'Alleged Cruelty to a Parrot Left in a Leeds House without Food', *Yorkshire Evening Post*, 16 April 1903.
[130] 'Alleged Cruelty to a Parrot', *Leeds Mercury*, 21 April 1903.
[131] 'Alleged Cruelty to a Parrot Left in a Leeds House without Food', *Yorkshire Evening Post*, 16 April 1903.
[132] 'The Leeds Parrot', *The Animal World*, December 1903, p. 192.
[133] 'Goldfish Cause a Tragedy', *Tamworth Herald*, 3 May 1902.

chronic forms of mistreatment. While most lemurs were not beaten and most parrots were not doused with hot gravy, many exotic pets were given the wrong food, deprived of exercise and companionship or kept in close confinement. Although less visible than violence in public places, this kind of cruelty preoccupied many members of humanitarian organisations and formed a second strand of criticism against pet keeping. Should people purchase pets if they were not competent to look after them? Was it morally right to cage wild animals for human entertainment?

Neglect was one major concern for humanitarians. Not everyone who owned an exotic pet looked after it properly, and humanitarians feared that many domestic companions were the victims of improper treatment. Writing in *The Animal World* in 1893, for instance, one reader alleged that parrots were 'often placed in bell-shaped cages, so narrow that I can scarcely think that they are even able to expand their wings to their fullest extent'.[134] Ten years later, another correspondent expressed concern for goldfish, which were 'frequently placed in their globe in a window, where they are always in the full glare of the light of the day, and that of the artificial light by night'.[135] A third commentator, parrot expert Frank Finn, warned readers that '[p]eople who can only lodge a parrot where it has its own company for hours together, or in a little-used room or lonely back yard', ought not to 'think about keeping one at all, for it is positive cruelty to keep so sociable a bird in this way'.[136] Captive songbirds thus languished in dirty cages, goldfish suffocated in deoxygenated water and parrots were deprived of the space and companionship essential to their physical and psychological well-being. The result was often premature death or self-mutilation – the latter evidenced by the myriad reports of bald parrots and tail-chewing monkeys.

Ignorance was another major cause of animal suffering, particularly in the case of non-native animals. People did not know what exotic creatures ate or how best to care for them, and sometimes they made choices that proved fatal to their domestic companions. Writing in *The Bazaar* in 1900, for example, traumatised monkey owner V.E.N. recounted how he had allowed his common white-eared marmoset 'to roam about a warm room', only to catch the animal 'in the very act of devouring the head of my poor [bullfinch], whom he held suspended through the top wires of the cage, by his strong claws'. V.E.N. had thought marmosets were 'harmless, insectivorous creatures' that subsisted on nuts and

[134] 'Parrots in Captivity', *The Animal World*, March 1893, p. 47.
[135] 'Gold Fish as Pets', *The Animal World*, February 1903, p. 22.
[136] 'The Treatment of Parrots', *The Animal World*, March 1912, p. 57.

mealworms, but he learned, to his cost, that this was not the case. Disabused of this misapprehension, he now kept his surviving parrot in a separate room and gave his marmoset 'raw beef, egg, and plenty of fruit', though 'I am bound to say "Binkie" does not appreciate dead meat'.[137]

Tortoises also suffered from popular ignorance about their diet – in this case a widespread belief that they subsisted on beetles and snails. In a letter dated November 1880, one concerned *Animal World* reader asked for clarification about the proper food for a tortoise, thinking the oft-prescribed repast of 'flies and slugs, and all sorts of pupae and larvae' might be incorrect.

Naturalists tell us that land-tortoises are vegetarian feeders, and my own experience goes to prove the same, yet I am often told by people who have tortoises that they eat the slugs and caterpillars in the garden (though they have never seen them); and not long ago I heard of a gardener buying a tortoise and shutting it up in a greenhouse without any leaves or green food, because he wanted it to clear the house of beetles.[138]

Over three decades later, fellow subscriber Ethel Wyrill contacted the RSPCA with a similar concern, sparked, in this case, by the publication of 'a photograph of a large number of tortoises' in the *Daily Mirror* under the headline '14,000 tortoises to act as scavengers in London gardens'. Wyrill feared that such a claim would perpetuate 'the mistaken and popular idea that the poor animals live on snails and beetles', and that, as a result, 'most of this immense number will die of starvation, which is too common amongst them as it is'. To counter the misinformation, the humanitarian sent a letter to 'the *Daily Mirror*, *Standard* and *Daily Mail*, hoping they would print it, to give the poor things a chance'.[139]

Wyrill was a tortoise owner herself, and had no objection to keeping the animals as pets as long as they were cared for correctly. Frank Finn, meanwhile, considered parrot ownership a positive good if done in the right way, since parrots were 'the most destructive of all birds to crops and fruit' and would 'have to be killed down to a great extent if not valued for caging' – a fact he believed 'should do away with any objection to caging them from a humanitarian standpoint'.[140] For other writers and other species, however, the question was not whether pets were properly treated, but whether it was morally right to keep pets at all. Was it acceptable to keep an animal in a cage? Were there species that

[137] 'A Bird Killing Monkey', *The Bazaar*, 12 December 1900, p. 1477.
[138] 'Tortoises' Food', *The Animal World*, November 1880, p. 174.
[139] 'The Treatment of Tortoises', *The Animal World*, June 1912, p. 180.
[140] 'The Treatment of Parrots', *The Animal World*, March 1912, p. 57.

should never be kept in captivity, no matter how well their needs were catered for?

Anna Perrier exemplified this more extreme position. Prompted by the harrowing sight of a friend's caged goldfinch, which 'was literally beating itself to death against the wires of its cage', Perrier wrote to *The Animal World* in 1883 to offer a searing critique of pet keeping. Caging birds, she argued, was an act of great cruelty, and, because the victim suffered for longer, a worse abuse than pigeon shooting (then a key target of RSPCA condemnation). Keeping small dogs in stuffy drawing rooms was little better and should also be stopped: 'Are the humane ladies who drag about wheezy lap-dogs aware that scarcely any animal requires, to keep it in perfect health, more exercise than a dog?' Perrier directed much of her criticism at women, with whom caged birds and lapdogs were 'favourite pets', but she reserved some of her ire for *The Animal World* itself, which published 'many articles' on the keeping of pets without making it clear that there were some species that, 'however petted, cannot be kept without cruelty, and cruelty very often much greater than is practised in many of the most objectionable sports'. In Perrier's view, indeed, the only species that could truly be kept without suffering was the cat, which, 'if well-fed and allowed to do as it pleases … has little to make it uncomfortable'.[141]

Although certainly a minority opinion, Perrier's was not a lone voice. Other humanitarians also raised objections to the imprisonment of any wild animal, suggesting that denial of freedom was in itself an act of cruelty. When one writer contacted *The Animal World* with the question 'Are Cage Birds Happy?', fellow reader, E. St J.V. argued that it was 'cruel to deprive any animal of its liberty', particularly a bird, which derived pleasure from being able to fly. 'All birds have been given wings, which must surely be intended for use.'[142] In 1902, meanwhile, when A. Armstrong wrote to *The Animal World* soliciting information on 'the best methods of taming a squirrel; also the best diet, and if it is safe to allow it out of its cage?', he was bombarded by letters from fellow readers urging him to liberate his pet in the interest of humanity.[143] One respondent, E. Davies, protested: 'To me who has seen the pretty creatures rejoicing in the freedom of the woods, it is most painful to watch them penned up, trying in vain, by the treadmill arrangement used in their cages, to take the exercise they need.' A second, Mr Wilkins, remarked that '7 months' imprisonment – solitary confinement' was 'a

---

[141] 'The Keeping of Pets', *The Animal World*, September 1883, p. 143.
[142] 'Caged Birds', *The Animal World*, May 1893, p. 79.
[143] 'Taming a Squirrel', *The Animal World*, February 1902, p. 32.

hard punishment for the gay little fellow', and that 'a creature which God made to leap from tree to tree and call the wide world his own' should not be kept in a cage 'a foot square'.[144] For all these writers, keeping any wild animal in captivity was morally suspect and could not be mitigated by human kindness and affection.

Taken together, what these more emotive responses show is the existence – at least by the late nineteenth century – of a more nuanced response to animal cruelty centred on welfare. At the more moderate end of the scale, the RSPCA and the majority of *Animal World* readers worried about the treatment of monkeys, parrots and goldfish and urged owners to care for their pets responsibly. At the more extreme end, Anna Perrier, E. St J.V. and the members of the Humanitarian League considered captivity an evil in itself and advocated an end to the imprisonment (as they regarded it) of all wild animals. For the former group, education was the answer; prompted by Wyrill's concerns about starving tortoises, one RSPCA branch secretary 'hit on the practical plan of purchasing a supply of the Society's pamphlets on the care and treatment' of the animal and 'presenting them to the proprietors of a shop where these unfortunate creatures are sold, on condition that a copy of the pamphlet be presented with each tortoise'.[145] For the latter, all pets (with the possible exception of cats) were victims of cruelty and all owners, rich or poor, were guilty of creating a market for bird catchers and animal traders. E. St J.V., opposing both the caging of wild birds and the caging of polar bears in zoological gardens, concluded that 'we have no right to deprive any animal of its natural life of freedom, whether in the tropics or in the polar regions'.[146] Humanitarian concern for exotic pets thus fell along a spectrum, from horror at deliberate violence to anxiety about ignorant neglect to philosophical doubts about the deprivation of liberty.

Conclusion

The nineteenth century witnessed a growing demand for exotic pets. Expanded trade with Africa, Asia and the Americas facilitated the acquisition of imported species, while the advent of steam shipping made it quicker and easier to transport live animals. Parrots, songbirds and monkeys emerged as the most popular exotic animal companions, but

[144] 'Taming Wild Squirrels', *The Animal World*, March 1902, p. 47.
[145] 'Tortoises for Sale', *The Animal World*, July 1913, p. 122.
[146] 'Caged Birds', *The Animal World*, May 1893, p. 79.

there was also a market for tortoises, guinea pigs, goldfish, lemurs and raccoons.

Like other animal commodities, exotic pets presented significant ethical issues, drawing criticism from some contemporaries. First, as noted above, the high attrition rate among imported animals raised concerns about the survival of the most exploited species. Accurate statistics for the number of animals killed by the wild animal trade are not readily available, particularly for the period prior to a creature's arrival in Europe, but the figures we have suggest a high level of mortality at all stages of the process, from point of capture to point of purchase. One writer compared the shipment of grey parrots to the 'horrors of the "middle passage" in the old days of the slave trade to America and the West Indies', so extreme was the suffering and so high the level of attrition.[147] Legislation was passed to alleviate some of these horrors, with bans on the catching of certain wild birds and on the export of particular species from certain nations (notably the USA), but this was far from comprehensive in its coverage and was often circumvented. While there are anecdotal reports of a drop in demand for exotic pets in the 1890s and 1900s, the reasons for this are not entirely clear, and it did not necessarily reflect a change in consumer attitudes. Even the carnage of the First World War did not completely stem the trade in exotic pets, which continued to be bought, sold and even taken to the front; in 1916, Mr James from Durham advertised a 'Male monkey' with 'Belgian uniform to fit, has been in the trenches for 11 months'.[148]

The keeping of exotic pets also raised concerns about animal cruelty, not so much during transit, when suffering was less visible, but in Britain itself. In part, these concerns focused on the legal status of exotic animals, which could, in theory, be abused with impunity until the 1900 Wild Animals in Captivity Protection Act brought battered lemurs and famished parrots within the remit of the law. In part they centred on the more insidious cruelties associated with pet ownership, from keeping goldfish in dirty water to caging wild squirrels or inbreeding lapdogs. The RSPCA lobbied to secure legal protection for exotic pets, while the more radical Humanitarian League denounced pet keeping altogether. Despite their efforts, however, parrots, monkeys and other exotic beasts continued to suffer in captivity, experiencing deprivation, neglect and sometimes physical violence.

---

[147] 'Cruelty to Parrots', *The Animals' Friend*, 1907, p. 109.
[148] 'Country House', *The Bazaar*, 27 May 1916, p. vii.

Conclusion

In 1899, a Mrs A. C. Tracy wrote to *The Animals' Friend* requesting advice on 'Humane Dress'. As she explained, 'Last winter I had, on two or three very cold days, to wear sealskin and bear alike; this year I want to avoid that, but being very far from strong, I shall be made to wear them, unless some neck wrap is in place equally warm.' While several fur substitutes were already on the market, Mrs Tracy was yet to be convinced by the ones she had tried, rejecting ostrich feather boas, which were ruined by the 'damp and fogs', along with 'woollen "clouds", scarves, or any ugly or unfashionable article'. Indeed, she had 'a stiff neck now from want of a warm necklet, and several ladies whom I have persuaded to give up furs have asked me what to do'.[1]

Mrs Tracy's letter was published in the October edition of *The Animals' Friend*, where it elicited a wide range of responses. One writer, J.M.W., recommended 'seal plush', which was 'very warm and nice looking' and 'could be made into necklets'. A second, F. Dismore, suggested using 'muffs made of crinkled chiffon, which appear "dressy" even if made in black' and 'can be made in different colours to suit the costume'. A third, J. C. Brace, confessed, 'Before my moral sense had evolved to the Humanitarian point, I wore fur and feathers with the complacency of a Red Indian adorned with his girdle of scalps.' Since switching to 'lace, silk, or the finest Shetland wool wraps', however, she had been 'much more comfortable than when wearing furs, which frequently made me unhealthily hot where the fur touched and bitterly cold by contrast where it did not'. A fourth writer, E. M. Beeby, advised replacing fur with 'the exquisite fabric knitted by the Shetlanders' and concurred with J. C. Brace that 'fur necklets are unhealthy'. A fifth, Fanny Emily White, rejected fur boas on both health and aesthetic grounds, insisting that they were 'a prolific source of throat and chest affections and, apart from being fashionable', gave women a 'huddled up

---

[1] 'Humane Dress', *The Animals' Friend*, 1899, p. 30.

appearance' that 'would be voted ugly in the extreme, as all symmetry is destroyed, head, neck and shoulders being so buried in the fur until it is difficult to determine where one begins and the other terminates'. A sixth, Helen Buckland, submitted samples of a wool imitation of astrakhan to the magazine, expressing confidence that, 'if a few people set *any* fashion *boldly*, it must catch on'.[2] Mrs Tracy thus had a variety of viable alternatives at her disposal, some of animal origin, others man-made.

The epistolary exchange that took place in the pages of *The Animals' Friend* was only a tiny part of the broader public debate surrounding luxury animal commodities, yet it offers a fascinating snapshot of the perspectives of several contemporary consumers – in this case, all female. Like Mrs Tracy, many women wanted to banish fur and feathers from their wardrobes and to escape the censure of humanitarians. Dispensing with cherished garments was not always easy, however, and some turned to a wider community of knowledge in search of advice and solidarity in their quest. As demonstrated by the responses to Mrs Tracy's letter, this community concerned itself not only with animal welfare – though that was the prime driver for change – but also with shifting conceptions of beauty, comfort and physical health, issues that perhaps eluded male conservationists but that mattered a great deal to female consumers. While scientists and politicians wrestled with the protection of seals, elephants and egrets in the field, British consumers made individual and often very personal decisions about which animal products to buy and which to reject.

Mrs Tracy's story also illustrates in microcosm some of challenges posed by global animal commodities and the complex moral dilemmas surrounding their use and procurement. For female consumers such as Mrs Tracy, there was the question of what to wear and how to be certain that it was ethically sourced – not easy when the fruits of pelagic sealing were indistinguishable from skins obtained in the seal drive and when many hats were adorned with faux 'artificial' egrets. For conservationists, there was the question of which species to protect, and at what cost – a decision that usually rested on anthropocentric factors. For humanitarians, there was debate over which animals suffered most in the name of fashion and how that suffering should be mitigated. And for vendors of some animal commodities – especially exotic pets – there was the quandary of to whom to sell them and how to ensure their good treatment. Dr Wright from Croydon, for instance, stipulated that he would not sell

[2] 'Humane Dress', *The Animals' Friend*, 1899, pp. 43–4.

his 'Valuable crevet [grivet] monkey ... to menageries, zoos, or to [a] family where there are children, or to poor people'.[3] While birds' feathers, sealskin, ivory, alpaca wool, animal perfumes and exotic pets all presented slightly different logistical, ethical and ecological challenges, all followed similar trajectories and invited similar critiques and solutions. By way of conclusion, therefore, I bring together the common elements within these six commodity stories and chart the parallel movements to which they gave rise. What lessons did contemporaries draw from the fate of the egret, the fur seal and the African elephant? What does this tell us about shifting attitudes towards global commerce, conservation and animal protection?

Common Problems

To begin with, looking at a range of different animal commodities reveals a number of commonalities. Although they emanated from different animals and arrived from different parts of the world, birds' feathers, sealskin, ivory, alpaca wool, musk and exotic pets were often purchased by the same people and presented common environmental and ethical problems. Conservationists, humanitarians and sometimes consumers themselves were aware of these similarities and often referenced them explicitly.

First, one thing that stands out in an analysis of nineteenth-century animal commodities is their common origin. Although they arrived in Britain via distinct and often quite localised commodity chains, all the animal products examined in this book were the result of imperial expansion, new trade routes and technological innovation. Egret feathers passed from the Everglades of Florida and the Llanos of Venezuela and Colombia to auction houses in London. Alpaca wool travelled from the Andean sierra to the weavers of Bradford via merchant houses in Arequipa and Liverpool. Ivory passed through the hands of African hunters and middlemen in German East Africa to traders in Zanzibar, cutlers in Sheffield and piano-makers in Feltham, while musk came to London perfumers from Sichuan and Nepal via Chinese merchants in Shanghai. Grey parrots were collected alive in West Africa and shipped by steamer to London and Liverpool, where they were sold direct to the public by dealers such as Jamrach and Cross or forwarded to pet shops in the provinces. The penetration of European (especially British) merchants into distant territories brought new commodities within reach of

---

[3] 'Country House', *The Bazaar*, 8 July 1910, p. 121.

metropolitan consumers, while faster shipping – first by sail, later by steam – enabled delicate and perishable items (such as monkeys) to be transported quickly across oceans. New technologies, from refrigeration to the invention of chemical dyes, further facilitated the conversion of animal body parts into fashionable consumer goods. Feathers, furs, perfumes, wools and ivory travelled further and in larger volumes than they had ever done before, undergoing new manufacturing processes and reaching new, and often distant, markets.

A second factor that linked all of these animal products was the shift in their consumer base. In the early modern period, and in many cases well into the eighteenth century, the consumer of exotic animal products would have been a member of the nobility – one of the few able to afford such goods. By the mid-nineteenth century, however, the consumer profile for most of these products had changed, as commodities once confined to a privileged few filtered down to the middle and lower classes. Ivory, valued in the eighteenth century for its use in intricate carvings, became the substance of choice for knife handles and piano keys – both staples of the bourgeois home. Vicuña wool, at one time a status symbol of the Inca nobility, found its way into European luxury markets, before giving way to cheaper alpaca. Feathers, previously worn by a courtly elite, appeared on the bonnets of middle-class women, while the monkeys and parrots that had entertained early modern princesses became the companions of working-class men; in 1878, C. Freeman, a grocer from Stockport, advertised a 'Monkey, Topsy, splendid animal, sensible as a human being, quite tame, sister to Mr Ford's celebrated Pedro Senior, with cage, chain, brass collar and lock, new clothes and small fiddle'.[4] Global animal commodities were more accessible, more desirable and more fashionable than they had ever been before, and the pressure on wild animal populations exponentially greater.

Of these new, middle-class consumers, a considerable proportion were women, providing humanitarians and conservationists with a common enemy. Ladies might not go out into the field to club seals or shoot songbirds, but their fashion choices caused others to do so on their behalf and were seen as the main reason why many animals suffered cruelty or faced extinction. As one article in *The Animal World* reflected:

As long as ladies are willing, and indeed anxious, to give fifty guineas for a handsome sealskin jacket, the war of extermination against seals will be relentlessly waged. The time will come, perhaps before very long – when there will be no elephant tusks for billiard balls, no sealskins for jackets, no sable tail for

---

[4] 'Country House', *The Bazaar*, 16 November 1878, p. 1170.

trimmings and muffs, no ostrich plumes for court headdresses and no whalebone for feminine uses. The demand for billiard balls is, it is true, almost exclusively confined to the nobler sex; but in most other cases the war of extermination will be found to have been waged in the milliner's behalf.[5]

In reality, of course, as this comment in part concedes, male consumers also purchased animal products and were not above donning a sealskin vest, owning a monkey or potting an ivory billiard ball. Indeed, the evidence suggests that alpaca wool, bear's grease and ivory were used almost equally by both sexes, that gentlemen's sealskin jackets were frequently advertised in trade magazines such as *The Bazaar* (forming perhaps 25 per cent of the market), and that military officers wore egret plumes in their hats until 1899, when Queen Victoria ordered their substitution by ostrich feathers.[6] Women, however, bore the brunt of the criticism, and their apparent hypocrisy as both maternal lovers of animals and shameless wearers of musk, feathers and chinchilla-trimmed dresses was frequently used to undermine their humanitarianism in other arenas – perhaps, in some cases, with good reason (Figure 7.1). As prominent anti-vivisectionist Louise Lind af Hageby remarked: 'To think that when we go out into the world and try to stop some fearful brutality, some shameless cruelty to an animal, whether it be in cattle transport or the slaughterhouse, or in cruel sport or vivisection, to think that we should always have the answer, "look what women wear"!'[7]

Lastly, in charting commonalities between our six animal commodities, we should note that those who opposed the exploitation of animals for human adornment or amusement often made explicit comparisons between the different animal industries, invoking one cruelty to expose or elucidate another. One writer in *The Animal World*, M.S.H., a defender of the fur seal, cited an earlier letter in the publication regarding the killing of blue jays and canaries 'to trim a lady's ball dress', but reflected that 'these poor little creatures were probably killed in a moment, not skinned alive like seals', which consequently suffered more.[8] Another correspondent, noting the recent conviction of a 'brutal passer-by' for pulling out the tail feathers of a peacock belonging to the Dean of Peterborough, wondered 'whether it does not hurt an ostrich (who has no Dean to plead for him) as much as a peacock to have his feathers pulled out', suggesting that pets and farmed animals were accorded

---

[5] 'Revolting Cruelty to Seals', *The Animal World*, April 1875, p. 52.
[6] SPB, *Ninth Annual Report* (London: J. Davy and Sons, 1899), p. 12.
[7] 'Women as Humanitarians', *The Humanitarian*, June 1910, pp. 15–16.
[8] 'Cruelty to Seals', *The Animal World*, April 1875, p. 62.

Figure 7.1 'Miss Betty Hicks Collects for the RSPCA at Messrs Selfridges', *The Animal World*, June 1915, p. 71. The lady donating to the charity shamelessly sports a fur boa around her neck (likely raccoon) and appears to be exiting a department store.

different rights.[9] A third writer, a conflicted consumer, worried that her continued use of tortoiseshell combs might make her as bad as the much condemned feathered woman:

Does the wholesale production of tortoiseshell goods signify the slaughter of countless tortoises for that purpose alone? Have I any more justification in wearing tortoiseshell (*real* of course) combs in my hair than I should have if I wore feathers and fur of birds and animals slaughtered solely for the purpose of

---

[9] 'One Law for the Peacock, Another for the Ostrich', *The Animals' Friend*, 1909, p.116.

adornment? I admit that it would be a deprivation to me *not* to use real tortoiseshell (celluloid is so dangerous), but I try to be as consistent as I can.[10]

Such comments suggest that it was tactically astute, but also logical, to think of animal cruelty in comparative terms. It also, however, suggests the creation of hierarchies of suffering among animal victims, based, in part, on the visibility of the abuse (the peacock was plucked in a British street, ostriches in a distant colony).[11] in part on the manner of the killing (M.S.H. thought that canaries were killed more quickly than seals, mainly because of their smaller size), and in part on the use to which their skins, feathers or fat were ultimately put (killing whales, for instance, was cruel, but it was harder to give up spermaceti candles than hummingbird earrings).

Common Solutions

By the end of the nineteenth century, proto-conservationists recognised that a whole gamut of valuable species was under threat from the insatiable demands of the fashion industry, centred, in most cases, on the city of London. Having identified the problem, however, the question was what to do about it. How could animal populations be protected, or managed sustainably? How could suffering be reduced?

Central to almost all conservation efforts was curbing the supply of animal parts and offering some form of protection in situ for threatened species. This generally took the form of hunting quotas, licences, close seasons and moratoriums, the establishment of designated sanctuaries and export or import bans for specific animal products. For fur seals, it meant limiting the size of the land catch to sustainable levels, outlawing pelagic sealing, prohibiting the slaughter of females and, in 1897, banning the importation into the USA of sealskins from animals caught at sea. For elephants, it meant setting up game reserves, issuing hunting licences to limit the number of individual animals hunters could legally kill and imposing a minimum tusk weight for ivory imports to ensure that female and young animals were excluded from the slaughter. For wild birds, it meant imposing close seasons when they could not be killed or captured, instituting complete bans on the hunting of particular species and enacting a range of export and import bans on plumage to stop the trade in feathers. Together, these measures served to regulate the slaughter of valued species and made it harder to trade in illegally obtained

---

[10] 'The Use of Tortoiseshell', *The Animal World*, December 1905, p. 191.
[11] 'Cruelty to a Peacock', *The Animal World*, September 1909, p.205.

animal parts. In almost all cases, however, hunting and export bans failed to ensure complete protection and were undermined by the difficulty of policing large reserves against poachers, the tendency of vulnerable species to stray into unprotected areas, the failure of all nations to support conservation measures, and the practical difficulty of distinguishing between female and male sealskins or birds caught inside and outside the close season. Ultimately, as long as there was a demand for ivory, sealskin, pet monkeys and egret feathers, someone would be willing to evade or violate legal protections in order to supply them.

With this in mind, a second key strand of nineteenth-century conservation work centred on reducing consumer demand by exposing the ecological impact of using particular commodities and also, crucially, highlighting the cruelties perpetrated on innocent animals on behalf of (often female) consumers. To this end, campaigners employed several common strategies still visible in today's animal welfare literature. First, they connected impersonal dead animals to specific living examples, such as the famous elephant Jumbo. Second, they published gruesome descriptions of skinning seals alive, starving egret chicks and the 'cruel butcheries' inflicted on the vulnerable vicuña. Third, they attempted to make explicit the role of the consumer in causing these massacres, insisting that it was the US or European buyer – not the Aleut seal skinner or Venezuelan egret killer – who was ultimately to blame for causing such brutality. As illustrated by the sartorial choices of Mrs Tracy and her respondents, these strategies appear to have had an impact, perhaps most visibly in the case of the plumage trade, where the work of the Audubon Societies and the RSPB persuaded many women to stop wearing feathers, or at least to renounce the plumage of the egret and the bird of paradise in favour of the 'humanely' obtained plumage of the ostrich. Despite these successes, however, changing public attitudes and fashions was a slow and difficult process, and there was a danger that some species would be driven to extinction before consumers saw the error of their ways.

If giving up animal commodities was hard, two other options found favour with nineteenth-century conservationists: domestication and/or acclimatisation and the development of synthetic substitutes. The first of these, domestication, was a popular solution to perceived conservation crises and featured in many of the cases explored in this book. It was applied to African elephants, ostriches, South American vicuñas, egrets and even fur seals, and often entailed not merely domesticating wild species in their country of origin but relocating them to different (safer) parts of the globe – Australia in the case of alpacas, California in the case of ostriches and the Great Lakes in the case of the fur seal. It reflected, on the one hand, a desire to control and, if possible, improve upon the natural world and, on the other, a belief that there was a stark choice between domestication and extinction: either an animal surrendered to human management or it would be exterminated.

Touted as a panacea for species preservation, domestication had many proponents and was attempted on multiple occasions. In reality, however, domesticating previously wild creatures and naturalising them outside their native habitat proved trickier than its advocates anticipated, and few such experiments were successful. Of all the species subjected to domestication attempts in the nineteenth century, only ostriches were domesticated on a large enough scale to supply the market (civets were also successfully caged and farmed for their scent, but they were not truly domesticated as they did not generally breed in captivity). In the majority of other cases, domestication efforts fell foul of political opposition, mismanagement and ignorance, and the dream of farming elephants, egrets and vicuñas quickly faded. Even had such projects succeeded, more recent attempts at captive breeding – musk deer in China, vicuñas in Argentina, moon bears in Southeast Asia – have shown that farming endangered animals is not necessarily the key to their salvation; captive populations suffer from increased disease transmission and inbreeding, not to mention often appalling levels of welfare, while the removal of animals from the wild also takes away any incentive to protect their native habitat or to ensure the survival of remaining wild populations.[12] Naturalisation, meanwhile, even when it worked, could have tragic results, as illustrated by the propagation of the grey squirrel in Britain (at the expense of the native red squirrel), the success of ferrets and stoats in New Zealand (at the expense of the kiwi and the kakapo) and the proliferation of the cane toad in Australia (at the expense of native fauna).[13] Domestication was – and sometimes still is – a common answer to over-exploitation, but it did not always deliver the results its supporters hoped for.

The second solution, developing synthetic substitutes for animal products, was arguably more successful and did achieve some notable results.

---

[12] Michael Sas-Rolfes, 'Assessing CITES: Four Case Studies' in Jon Hutton and Barnabus Dickson (eds), *Endangered Species Threatened Convention: The Past, Present and Future of CITES* (London: Earthscan, 2000), pp. 82–4.

[13] See Philippa K. Wells, '"An Enemy of the Rabbit": The Social Context of Acclimatisation of an Immigrant Killer', *Environment and History* 12 (2006), pp. 297–324; Christopher Lever, *The Cane Toad: The History and Ecology of a Successful Colonist* (Otley: Westbury Academic Publishing, 2001), pp. 141–72. For a more positive assessment of invasive species and their ecological impact, see Fred Pearce, *The New Wild: Why Invasive Species Will Be Nature's Salvation* (London: Icon Books, 2016).

Celluloid, and later plastic, came to replace ivory in many manufacturing processes. '[M]anufactured seal cloth' and chiffon served as humane alternatives to sealskin (though chiffon was made of silk, obtained by killing silkworms).[14] Synthetic perfumes stood in for civet, musk and ambergris, while some women were persuaded to wear flowers or ostrich plumes on their bonnets instead of the carcasses of dead birds. Together, these new inventions helped curtail the demand for dwindling animal commodities and provided consumers with a comparatively painless alternative to the consumption of animal body parts – assuming, of course, that the alternative was functional and elegant enough. It is probably no exaggeration to claim that the invention of plastic saved the African elephant in the early twentieth century.

That said, even synthetic materials could have their downsides and their existence has not been an unmitigated triumph. As noted in Chapter 1, some supposedly man-made products were, in fact, the real thing, duping consumers into buying a product they might otherwise have renounced. In other cases, even the existence of viable (perhaps even indistinguishable) non-animal alternatives has not entirely eliminated demand for the genuine article – for instance, musk in perfumes or ivory piano keys. In yet others, the invention of new materials has simply replaced one environmental problem with another, an outcome most visible in the case of plastic, saviour of the elephant, which today poses a huge threat to marine life. Substitutes have thus played an important role in reducing demand for particular animal commodities but have sometimes come at an unexpected long-term cost.

Common Actors

To better understand the origin of some of these conservation proposals, we need to look at the key actors and institutions behind them. As the above examples demonstrate, nineteenth-century animal protection initiatives shared problems and tackled them through common strategies. They also shared common personnel and forms of expertise, pioneering approaches and methods that continue to shape conservation to this day.

First, science and technology played a critical role in almost all of our stories, both creating new threats to animals and facilitating their conservation. In the case of technology, the exploitation and popularity of all the animal products examined in this book rested in large part on the invention of new techniques for transporting and processing them. New

---

[14] 'Humane Dress', *The Animals' Friend* 1899, p. 43.

incubators enabled ostrich farming to operate on an industrial scale. New spinning machinery made it possible to process alpaca wool in larger quantities. Mechanised cutting and polishing equipment facilitated the use of ivory for piano keys, billiard balls and cutlery handles, while new techniques for depilating and dyeing sealskin transformed it into a popular feminine accoutrement. In coming up with these innovations, engineers and artisans were in part responsible for expanding the market in luxury animal products and generating demand for their feathers, fur and teeth. At the same time, however, some inventors contributed to conservation by devising synthetic alternatives to zoological materials, from celluloid combs and piano keys to chemical perfumes. Photography, another key technological advance of the nineteenth century, also played an important role in the humanitarian movement, furnishing seemingly irrefutable evidence of cruelty, providing an alternative to hunting with a gun and arousing public compassion for chained monkeys and starving egret chicks.[15] Technology thus generated demand for animal commodities but also offered ways of mitigating that demand.

While engineers and artisans devised new ways of using and viewing the natural world, naturalists played a central role in conserving it. As collectors in the field, students of natural history were among the first to appreciate the impact of commercial hunting on wild animal populations and among the first to draw attention to it. They provided crucial data to prove that animal populations were diminishing, used their zoological knowledge to demonstrate how the natural behaviour of particular species might render them especially vulnerable to over-exploitation or limit their capacity for recovery, and deployed their authority as accredited experts to discredit the lies peddled by the millinery industry, pelagic sealers or the purveyors of adulterated perfume. Some naturalists, indeed, took a broad approach to wild animal protection and featured in discussions about several animal commodities. British naturalist Frank Buckland, for instance, spoke out in favour of elephant domestication and alpaca acclimatisation, advocated conservation measures for seals (in this case, specifically North Sea seals, killed for their oil) and owned a range of exotic pets, including monkeys Jenny and Susey. American conservationist William Temple Hornaday condemned the killing of birds for their plumage, was a prominent advocate of bison conservation and proved instrumental in enacting the five-year moratorium on sealing proposed by Henry Elliott. London Zoo director Philip Lutley Sclater

[15] On the importance of photographic images in humanitarian campaigns, see J. Keri Cronin, *Art for Animals: Visual Culture and Animal Advocacy, 1870–1914* (University Park: Pennsylvania State University Press, 2018), p. 84.

exposed the non-existence of an egret farm in Tunis and repeatedly promoted the domestication of the African elephant, in this case in the British colonies. Of course, scientists did not always agree in their conclusions and the recommendations they made were sometimes contradictory. Henry Elliott and David Starr Jordan, for instance, disagreed over the best method of protecting fur seals, while their Canadian counterparts mobilised science to defend the killing of pinnipeds at sea. Despite such differences, however, naturalists were pivotal in the emerging conservation movement and among the most vocal proponents of conservation policies.

In terms of the institutions that contributed to animal use and/or conservation in the nineteenth century, zoological gardens merit a special mention, featuring in almost all of our case studies. Alpacas in Edward Cross's and Mrs Wombwell's menageries prompted calls for the species' acclimatisation. Captive elephants were adduced by writers such as William Yellowly and Philip Lutley Sclater as evidence that African elephants could (and should) be domesticated. Ostriches at the Crystal Palace and in Hagenbeck's Tierpark offered inspiration and information for the acclimatisation of the birds outside Africa, while the successful rearing of Alaskan fur seal pups Bismarck and Mamie at the Bureau of Fisheries in Washington DC persuaded contemporaries that the marine mammals might be successfully acclimatised in freshwater lakes.[16] Zoos functioned as both centres for research and repositories of expertise, providing venues for testing the natures and capabilities of wild animals. In this sense, they prefigured the contemporary role of zoos as (contested) sites of conservation that now operate breeding programmes to preserve endangered species and deploy their inmates as educational ambassadors for their counterparts in the wild. In 1998, for instance, Indianapolis Zoo celebrated the pregnancy of female African elephant Kubwa as the result of a new artificial insemination technique, trumpeting the achievement as a major boon for conservation.[17]

Finally, of course, women were important actors in the fight against animal cruelty, making a difference through their activism and consumer choices. Chastened by reports of animal suffering, many women elected not to buy fur or feathers and, like Mrs Tracy, actively sought substitutes for these controversial products. Others joined the RSPCA or the

---

[16] 'May Raise Fur Seals Here', *New York Sun*, 2 July 1910, cited in William T. Hornaday, *Scrapbook Collection on the History of Wild Life Protection and Extermination*, Vol. 5, Wildlife Conservation Society Archives Collection, 1007-04-05-000-a.

[17] 'African Elephant Pregnant from Artificial Insemination', *The Associated Press State & Local Wire*, 29 June 1998.

RSPB – the memberships of which were predominantly female – taking a stand against murderous millinery and seal slaughter and even intervening personally to prevent the abuse of living exotic animals. In 1881, Mrs Matilda Leckie 'of Merton Villa, Southsea' remonstrated in the street with Italian musician Octairo Pereschino after seeing him 'cruelly ill-treat a monkey by kicking it', receiving praise from the local magistrate.[18] While women took action in part out of love for animals, their activism can also be seen as a conscious assertion of female agency (in this case, in the form of consumer power) and an attempt to redeem themselves from charges of hypocrisy and wilful ignorance. In one fictional story, published in 1904, a girl named Cicely receives a sealskin coat from her rich uncle and, on learning how it was procured, pledges to tell 'every woman with whom I clasp hands' exactly where fur comes from, for currently 'Women do not know'.[19] Six years later, in a passionate speech on the role of women in the humanitarian movement, prominent anti-vivisectionist Louise Lind af Hageby urged women to educate themselves, arguing that it was their moral responsibility to do so: 'It is no excuse to say "we did not know"; it is a woman's business to know. It is a woman's business to teach her daughters as they grow up that these things should be despised.'[20] In defending animals, therefore, female humanitarians were defending their own reputations as morally upright, thinking and politically significant individuals – not to mention their competence as potential voters.[21]

Different Priorities

One last element that permeates all of our case studies is the complex relationship between conservation concerns and animal welfare issues. Most animal-based fashions entailed both the destruction of species and cruelty towards individual creatures, so it is not surprising to find both discussed by contemporary critics. Sometimes these concerns aligned and called for the same solution. On other occasions, however, they diverged, revealing the different motivations of their exponents.

For proto-conservationists, the key thing was preventing the extinction of a particular species and ensuring that its exploitation, if it continued,

---

[18] 'Portsmouth Police Court', *Portsmouth Evening News*, 28 May 1881.
[19] Ellen Hopkins, 'Cicely's Club', *The Animals' Friend*, 1904, pp. 19–21.
[20] 'Women as Humanitarians', *The Humanitarian*, June 1910, pp. 15–16.
[21] On the close connections between the female suffrage movement and humanitarianism – specifically the adoption of a vegetarian diet – see Leah Leneman, 'The Awakened Instinct: Vegetarianism and the Women's Suffrage Movement in Britain', *Women's History Review* 6:2 (1997), pp. 271–87.

did so at a sustainable level. Hence, it was acceptable to go on killing fur seals as long as the right type of seals (bachelors) were slaughtered in the right place (on the Pribilof hauling grounds) at the right time (outside the breeding period). Within the conservationist movement, however, there were significant differences of opinion and priorities, with two distinct approaches emerging clearly by the end of the nineteenth century. The first of these, which I described in Chapter 2 as a utilitarian approach, sought to preserve certain species primarily so that they could continue to be exploited for commercial purposes; this was often closely connected to the idea of improvement. The second approach, stewardship, focused on saving animals for posterity and for their own sake – because they had a right to exist and because it would be a tragedy for humanity, and for future generations, if they were to be lost. Calling for the preservation of another endangered species, the Indian rhinoceros, conservationist William Temple Hornaday characterised the animal as 'a gift handed down to us straight out of the Pleistocene age, a million years back' – a beautiful thing to be prized in its own right, and not for what it could do for man.[22] This was not equally true of all species, of course, and both breeds of conservationist adopted a selective attitude towards protection, focusing their conservation efforts on charismatic species that were either interesting or useful to humans (wolves, for example, received short shrift from almost all turn-of-the-century conservationists, as did lions, tigers and baboons). Until the 1920s, moreover, there was no real understanding of ecosystems and ecology (the word was coined by Ernst Haeckel in 1866, but the discipline itself did not come into existence until the 1920s), so few creatures were accorded protection that did not directly benefit people in some way.[23]

For proto-humanitarians, cruelty was the primary issue, whether a species was endangered or not. Thus, even sustainable killing was unacceptable and the mistreatment of a single monkey, tortoise or ostrich a cause for concern. Once again, however, there were different agendas at play, with some humanitarians calling for the mitigation and reduction of cruelty and others demanding its outright eradication. The pragmatic and gradualist RSPCA, at the more conservative end of the spectrum, accepted regulated vivisection, sustainably harvested ostrich feathers and humanely killed meat and moderated its criticism of elite big game

---

[22] William Temple Hornaday, *Our Vanishing Wild Life: Its Extermination and Preservation* (New York: Clark and Fritts, 1913), p. 189.

[23] On the advent of ecology, see Donald Worster, *Nature's Economy: A History of Ecological Ideas*, second edition (Cambridge: Cambridge University Press, 1994), pp. 191–204; Mark V. Barrow, *Nature's Ghosts: Confronting Extinction from the Age of Jefferson to the Age of Ecology* (Chicago: University of Chicago Press, 2009), pp. 200–33.

hunters, several of whom were to be found among its membership. The more radical Humanitarian League, on the other hand, promoted vegetarianism, opposed keeping wild animals in captivity and advocated a wide range of social reforms – including the abolition of the death penalty and female suffrage – a stance that posed a greater challenge to the established political order.[24] Both organisations disapproved of fur and feather fashions, but the RSPCA tended to treat them as isolated abuses while Humanitarians such as Henry Salt viewed them as part of a larger system of capitalist oppression.

In the nineteenth and early twentieth centuries, conservationism was the stronger force. This was certainly reflected in the framing of most conservation legislation, which banned, for instance, the importation of underweight (but not all) ivory and outlawed the killing of female (but not all) fur seals. More surprisingly, it was even prioritised to some degree by the RSPCA, which, in condemning the use of robins on women's hats, stated that 'we are not foolish enough to contend that robins may not be killed for man's use. We ask that they may not be exterminated – which they will speedily be if ladies do not discontinue wearing them on their bodies as wretched ornaments' – a pragmatic position, to be sure, but one that seemed to place preservation above suffering.[25] Despite this bias, however, animal welfare activism was a growing force during this period, and it seems likely that the descriptions of individual acts of barbarity – butchered seals or starving egret chicks – had a greater impact on consumers than scientific statistics about falling populations – just as they do today. Certainly, most of those women who explained their decision to abandon fur, feather or tortoiseshell did so for this reason, expressing concern about how quickly and painlessly an animal died. What we see here, therefore, are the beginnings of two distinct (though connected) approaches to animal protection, one of which privileges species survival while the other prioritises the welfare of individual animals.[26]

---

[24] On the priorities and strategies of the RSPCA, see Brian Harrison, 'Animals and the State in Nineteenth-Century England', *English Historical Review* 88 (1973), pp. 786–820. For a summary of the more radical agenda of the Humanitarian League, see Dan Weinbren, 'Against All Cruelty: The Humanitarian League, 1891–1919', *History Workshop Journal* 38 (1994), pp. 86–105.

[25] 'The Protection of Wild Birds', *The Animal World*, March 1876, p. 34.

[26] On the development of these two traditions in the twentieth and twenty-first centuries, see Lisa Kemmerer (ed.), *Animals and the Environment: Advocacy, Activism and the Quest for Common Ground* (Abingdon: Routledge, 2015); 'Mass Extinction and Mass Slaughter: Biodiversity, Violence and the Dangers of Domestication' in Ursula K. Heise, *Imagining Extinction: The Cultural Meanings of Endangered Species* (Chicago: University of Chicago Press, 2016), pp. 127–61.

As for individual consumers, they espoused a range of views and came to their own personal decisions as to what was and was not ethically acceptable – as reflected in the responses to Mrs Tracy's letter. *Animal's Friend* reader F.H., for instance, thought fur was acceptable as long as it was sourced humanely, urging women 'to examine the inside of skins purchased, to see if they contain shot-holes, in which case one may conclude that the animal died easily'.[27] *Animal World* reader M.S.H., whom we met in Chapter 2, banished all sealskin items from her wardrobe, so horrified was she at the measures used to procure it, but had no problem with seal oil, obtained just as brutally but seemingly put to less frivolous purposes (although naturalist Frank Buckland suggested that one of its uses was for treating jute, from which women's clothing was made).[28] *Animals' Friend* subscriber J.M.M. stated in a letter to the publication that she never wore 'fur or feathers' and that she would like to find a substitute for leather in her boots, though the latter seemed less reprehensible as 'certainly cows would not be slaughtered if their hides were all that was required'.[29] Fellow subscriber G.C.H., meanwhile, defended her continued use of ostrich feathers on the grounds that they were 'both pretty in themselves and becoming to the wearer' and that they were obtained more humanely than silk, which, 'I think, cannot be obtained without the death of the silk-worm'.[30] Individuals thus picked their own unique courses through the moral minefield of animal consumption and responded differently – and often inconsistently – to the promptings of humanitarian campaigners. Factors influencing their decisions included the aesthetic appeal and visibility of the animal victim; the use to which an animal product was put – specifically whether it was considered a luxury or a necessity; whether a commodity was obtained as a by-product of the meat industry or constituted the sole reason for the animal's death; and whether the animal was kept and/or killed in a humane fashion. These criteria continue to influence consumer choices into the twenty-first century.

---

[27] 'Humane Dress', *The Animals' Friend*, 1899, p. 44.
[28] 'Cruelty to Seals', *The Animal World*, April 1875, p. 62; 'Practical Uses of the Brighton Aquarium', *The Animal World*, July 1874, p. 102.
[29] 'Humane Dress', *The Animals' Friend*, 1899, p. 43.
[30] 'Ostrich Feathers', *The Animals' Friend*, 1905, p. 160.

Epilogue
Past, Present and Future

The nineteenth century witnessed the first major wave of concern for the survival and welfare of wild animals and the first stirrings of an international conservation movement. In the short term, these measures largely succeeded. None of the animals discussed in this book became extinct before 1914, despite fears for their survival, and several species were brought back from the brink by effective conservation legislation, the creation of viable substitutes and changing consumer habits. In the long term, the legacy of turn-of-the-century conservation measures has been more mixed. Some of our species have made a spectacular recovery, while others have succumbed to new, even more serious, threats, in many cases due to a growing consumer demand for their products in East Asia. To conclude this study, therefore, it seems appropriate to bring the stories of our animals up to date and ask what the long-term impact of exploitation and conservation has been. Are egrets, fur seals and African elephants doing better today than they were in 1900, or are they once again in trouble? To what extent do nineteenth-century precedents continue to influence contemporary conservation measures?

Egrets

The twentieth-century story of the egret is one of recovery and revival. In the USA, snowy egrets, great egrets and reddish egrets have all rebounded since the passing of the Migratory Birds Act in 1918, while the cattle egret, originally native to Africa, has become a regular fixture on the other side of the Atlantic.[1] Egrets are now frequently seen in the southern and mid-western states of the USA and are gradually recolonising their old heartland in Florida.[2]

Beyond the USA, egrets also appear to be making a comeback. In China, egrets are increasingly seen as valuable animals, on account of

[1] 'Cattle Egrets Spread', *The Ledger*, 2 December 1997.
[2] 'Decimated Egret Makes a Comeback', *St Petersburg Times*, 31 July 1995.

their role in destroying insects. Villagers in the Guangxi Zhuang region now actively protect the birds and have intervened twice on their behalf in recent years – once, in 2000, when they fed starving egrets during a severe drought, and a second time in 2014 when they rescued injured chicks after a typhoon.[3] In India, meanwhile, the authorities have taken measures to clamp down on the use of egret feathers in festivals to adorn traditional drums (*dhaks*), giving the species greater protection.[4] Habitat loss – particularly of wetland areas – remains a concern for the egret's future survival, but to some extent the species has compensated for this by extending its range northwards to countries such as Britain, where it is now seen frequently on the Somerset Levels.[5] The egret may therefore be an unwitting beneficiary of global warming; its story is certainly proof of a species' ability to recover once accorded proper legal protection from a deadly fashion.

Ostriches

The ostrich industry declined in the 1920s following changes in women's fashion but revived in the 1980s. Ostriches are now reared primarily for their hides, which are used to make handbags and boots, and for their meat, which is converted into steaks, sausages and fritters.[6] Farmers also sell ostrich feathers for dusters, ostrich eyelashes for paintbrush bristles, and ostrich tendons and corneas for use in human medical transplants.[7] South Africa still dominates the market for ostrich products, exporting around 3,000 ostrich skins per year, but it faces growing competition from the USA, Brazil, Australia and China.[8] Fledgling ostrich industries have also started to emerge in Britain, Japan, Nepal, Vietnam and Uzbekistan.[9]

---

[3] 'Chinese Villagers Help Egrets through Drought', *Xinhua General News Service*, 22 July 2000; 'Across China: Endangered Birds Protected after South China Typhoon', *Xinhua Economic News Service*, 31 July 2014.

[4] 'Egret Feathers of Pujadhaks Raise Hackles', *The Telegraph*, 28 September 2016.

[5] 'Nature Notes', *The Times*, 7 November 2017.

[6] 'Ostrich – Red Meat of the Future', *New Straits Times*, 13 July 1998.

[7] 'Ostrich Farming Then and Now', *World Poultry*, 1 March 2005.

[8] 'Cruel Price of Fashion's Ultimate Status Symbol', *Daily Mail*, 7 April 2016; 'A Tall Order', *St Petersburg Times*, 3 July 1994; 'Brazil Ostrich Breeders Revenue Seen at $26.56Mln', *SeeNews Latin America*, 5 October 2005; 'It May Be a Myth They Keep Burying Their Heads in the Sand but Ostrich Meat Wins by a Neck', *The Sunday Telegraph*, 16 March 1997; 'China Becomes Largest Ostrich Raiser in Asia', *Xinhua General News Service*, 3 April 2004.

[9] 'A Bird? Absurd', *Daily Mail*, 15 October 1993; 'Country's First Ostrich Farm on Expansion Spree', *My Republica*, 29 May 2013; 'Vietnam: Ostrich Business Booms in Quang Binh Province', *Thai News Service*, 11 October 2013; 'The Uzbek–British Joint

While the popularity of ostrich farming means that ostriches are not endangered (there are thought to be around 2 million ostriches globally), animal rights activists have expressed serious concerns about the welfare of farmed birds.[10] In 2017, a PETA (People for the Ethical Treatment of Animals) investigation exposed shocking levels of cruelty on several South African ostrich farms, where ostriches were reared in small barren feedlots, had their feathers plucked while still alive (now illegal in South Africa) and were killed by having their throats slit. Investigators also found that the toenails of ostrich chicks were routinely removed to prevent them from damaging their valuable hides – an operation that frequently resulted in the amputation of their whole toe.[11] Ostriches are thus thriving as a species but have become yet another victim of factory farming, processed on an industrial scale to meet a new market for luxury goods.

Fur Seals

The fur seal population climbed from around 200,000 in 1900 to nearly 1.7 million in 1983 – a year before hunting on the Pribilof Islands ended completely.[12] Since 1998, however, the number of fur seals has been steadily falling, and the species is currently classed as 'depleted'. Scientists are unsure of the precise reason for the decline, citing a decrease in fish stocks, increased predation, pollution and entanglement in discarded fishing tackle as possible causes.[13] Between 1998 and 2005, there were 795 sightings on St Paul Island of fur seals ensnared in debris, 282 of which were rescued.[14] In 2015, meanwhile, the Sausalito Marine Mammal Center reported record numbers of stranded seal pups washing up on the California coast, a phenomenon it attributed to the existence of a large area of unnaturally warm water in the Pacific Ocean (nicknamed 'the blob'), which may have caused fish to migrate beyond the seals' normal range.[15] Despite the positive impact of the 1911 Fur Seal Convention, therefore, the fur seal is again under threat and its future looks uncertain.

Venture "Straus Farm" in Fergana Region', *Uzbekistan National News Agency*, 15 January 2016.

[10] 'Raising Ostriches Takes Care, Cut-offs and Sometimes Cereal', *Lincoln Journal Star*, 29 October 2000.

[11] 'Cruel Price of Fashion's Ultimate Status Symbol', *Daily Mail*, 7 April 2016.

[12] 'Mystic Is Breeder of Fur Seals', *New York Times*, 31 July 1988.

[13] 'NOAA Reports Northern Fur Seal Pup Estimate Decline', *States News Service*, 15 January 2009.

[14] 'Seals' Island Is Dumping Ground for Marine Debris', *Associated Press Online*, 23 June 2008.

[15] 'Starved Seal Pups Washing Ashore', *Eureka Times-Standard*, 29 November 2015.

Once the victim of direct human exploitation for its coveted fur coat, the species is now a victim of more insidious environmental changes, which are, nonetheless, probably caused by man. The Pacific fur seal's marine relative, the harp seal, meanwhile, continues to be killed every spring on the ice floes of Canada in order to obtain its fur (used to trim coats) and its blubber (used in margarine, soap and cosmetics).[16]

Sealing is also implicated in wider debates over the wearing of fur. After a spike in popularity in the 1960s, fur became taboo in the 1970s and 1980s with the rise of animal rights activism and the anti-fur movement. Western consumers and designers have increasingly rejected fur in their clothing, and many are now opting for fake fur over the real thing. While the decline in fur wearing seems like a positive shift, there are signs that the battle against fur has not yet been won. On the one hand, although the trapping of endangered species remains illegal, the twenty-first century has witnessed a rise in demand for farmed furs and skins – such as mink, raccoon and alligator – fuelled primarily (though not exclusively) by a growing demand for furs among the rising Chinese and Russian middle classes.[17] On the other hand, and perhaps more worryingly, animal welfare organisations have expressed fears that even 'fake' fur may sometimes contain real animal products; a Humane Society International investigation in 2017 found that some products marketed as fake fur in fact contained dog, cat, mink, raccoon and rabbit fur – a modern-day incarnation of the artificial aigrette.[18] While fur no longer constitutes a major conservation risk, it thus continues to raise serious welfare issues and poses some awkward ethical questions for consumers. Is fur farming any more reprehensible than factory farming animals for meat or dairy? Is it possible to distinguish between humanely raised fur and inhumanely raised fur? And by buying fake fur, are purchasers actively supporting animal welfare or unwittingly promoting cruelty?

African Elephants

The African elephant has had a rather chequered career since the early twentieth century. The species made an impressive recovery once the demand for ivory waned in the West, and populations rebounded across Africa. Unfortunately, poaching then resumed in the 1970s and 1980s to meet a growing demand for ivory in the Far East – initially from Japan

---

[16] 'Axe About to Fall Again on Fur Seal', *Washington Post*, 7 March 1971.
[17] 'Why Fur Is Back in Fashion', *National Geographic*, September 2016.
[18] 'Fashion Faux-pas', *The Times*, 11 April 2017.

and Hong Kong, now, increasingly, from China. As a result, the elephant population has plummeted from around 2 million in 1973 to just 415,000 in 2017.[19] Recent data suggest that around ninety-six elephants perish in Africa every day at the hands of poachers, which equates to an elephant being killed roughly every fifteen minutes.[20] This means that elephants are dying faster than they are reproducing, putting the species' survival at risk.

In a bid to halt the decline, wildlife organisations and governments have reprised some of the measures first devised at the start of the twentieth century. First, to protect elephants on the ground, many African states have established wildlife reserves, employing rangers to deter and apprehend poachers. As in the nineteenth century, moreover, the latest technology is being applied to the problem of elephant conservation and is helping scientists to better understand and manage the surviving elephant population. In 1989, the ZSL used battery-powered transmitters to track six elephants in Kenya's Rift Valley, having first trialled the apparatus on a captive elephant at Whipsnade Zoo.[21] From 2014 to 2016, scientists conducted an elephant census to determine the number, distribution and range of elephants across Africa, using aerial surveys and counts of elephant dung.[22] Together, these efforts have achieved some positive results and are enabling conservationists to better understand elephant behaviours and social interactions.

A second strand of conservation has focused on curbing the international trade in ivory by preventing or limiting its importation into non-African countries. This has proven contentious, since stakeholders disagree as to whether a whole or a partial ban on ivory would be more effective in preventing poaching. East and Central African countries, which have suffered the most extensive slaughter of their elephants, want a blanket ban on ivory trading, on the assumption that any legal ivory on the market will stimulate demand and act as a cover for illegally traded ivory. Southern African countries such as South Africa, Zimbabwe, Botswana and Namibia, on the other hand, have well-managed and sustainable elephant populations and have argued that they should be permitted to sell ivory commercially in order to generate funds for conservation. In 1989, the African elephant was placed on Appendix I of CITES, making any trade in ivory illegal. Since then, however, two

---

[19] 'Poaching behind Worst African Elephant Losses in 25 Years', *Eturbo News*, 25 September 2016.
[20] 'Elephants Wiped out for Ivory', *Pretoria News*, 17 August 2017.
[21] 'London Zoo Will "Track" Elephants by Satellite', *The Times*, 12 September 1989.
[22] 'African Elephant Population Dropped 30 Percent in 7 Years', *New York Times*, 1 September 2016.

sales of ivory stockpiles have been sanctioned and elephants in states with healthy populations have been downgraded to Appendix II.[23] Conservationists believe that the latter move has encouraged poaching, since there is no reliable way of distinguishing legal from illegal ivory and nothing to stop elephants themselves from migrating between states.

As in the nineteenth century, a few more eccentric individuals have proposed domestication as a way of saving the African elephant. In 1981, Dr Stewart Eltringham, a lecturer in applied biology at the University of Cambridge, noted that 'Asian elephants had been used for centuries as working animals, and the African species could be used in the same way, perhaps to carry tourists of game viewing safaris'.[24] In 1998, American animal trainers Randall Moore and Michael Lorentz tamed fifteen young elephants orphaned from a cull and trained them to carry tourists around Botswana's Okavango Delta, using Sri Lankan mahouts.[25] Advocates of domestication have presented it as a way of preserving elephants and reducing human–animal conflict in Africa. Conservationists are sceptical, however, arguing that domesticated elephants would require too much food to be sustained on a large scale.

Ultimately, reducing demand for ivory is probably the only reliable way of preserving the African elephant from extinction, together with tighter controls on the importation and sale of ivory. Animal protection organisations have focused increasing attention on this goal, targeting Chinese consumers in particular and attempting to persuade them that ivory is not a status symbol. Steps are also being taken to suppress the trade in so-called antique (pre-1975) ivory, which has often served as a cover for freshly killed tusks; the USA and China both banned the domestic trade in antique ivory in 2016, and the UK followed suit in 2018.[26] Changing consumer behaviour probably represents the best way of ending the ivory trade, for, as in the nineteenth century, it is the demand that fuels the supply. Bringing about this change is likely to be a slow process, however, and it may be decades before the demand for ivory ceases – too long to sustain current levels of poaching.

---

[23] 'Southern Africa Seeks CITES Approval for Sale of Ivory Stocks', *Africa News*, 18 March 2010.

[24] 'Applied Biology: Elephants Face Risk of Extinction', *The Times*, 4 September 1981.

[25] 'Exchanging Wild and Free for a Place in the Workforce', *The Ottawa Citizen*, 7 January 1998.

[26] 'US Adopts Near Total Ban on African Elephant Ivory', *Associated Press International*, 2 June 2016; 'IFAW Applauds China Extending Bans on African Elephant Ivory Imports', *Targeted News Service*, 23 March 2016; 'UK to Tighten Laws on "Abhorrent" Ivory Trade', *The Guardian*, 3 April 2018.

Alpacas and Vicuñas

Alpacas must qualify as a twenty-first-century success story. Ledger's efforts to acclimatise the animals in Australia may have failed, but since the 1990s alpacas have become an increasingly common sight 'Down Under' and have also colonised Britain, Canada, the USA and much of continental Europe. Farmed primarily for their wool, alpacas are popular hobby farm animals and now feature regularly at livestock shows. They have also expanded their repertoire to include working as therapy animals and guarding poultry. In Berkshire, farmer Tom Copas employs ten alpacas to protect his 2,400 free-range turkeys from foxes.[27] In Japan, you can even hire an alpaca to attend your wedding, specially groomed for the occasion and decked out in a snazzy red bow tie![28]

While alpacas have greatly extended their global range, Peru remains the world's largest exporter of alpaca wool. Of the 3.5 million alpacas thought to live in the world today, 3 million live in Peru, with a further 500,000 in Bolivia. According to the International Alpaca Association, Peru exports an average of 4,000 tons of alpaca fibre every year, mostly to China, Italy and the United Kingdom. It also exports a range of finished alpaca goods, predominantly to the USA. The government of Peru is very conscious of the value of alpaca wool to the country's economy and has instituted breeding programmes to improve the quality of the fibre, sometimes with the assistance of international organisations. In 1984, the International Alpaca Association was founded in Arequipa to 'protect the image of the alpaca fiber and its derivatives, as well as [to promote] international consumption and ensure the quality of their products'.[29] In 2003, the Peruvian Government promulgated law number 28041 to promote 'the rearing, production, commercialisation and consumption of domestic South American camelids', establishing genealogical registers for both alpacas and llamas so that the best examples of each could be bred and maintained.[30] These initiatives are intended to raise the value of wool exports and to counteract competition from foreign alpaca producers in countries such as Australia, where alpacas can be farmed on a larger scale and their breeding more closely monitored.

The alpaca's wild relative the vicuña has had a more traumatic path into the twentieth century, although it now appears to be on the road to

---

[27] Elizabeth Hotson, 'The Alpacas Protecting 24,000 Christmas Turkeys', *BBC News*, 1 December 2016.
[28] 'New Craze in Japan', *The Daily Mail*, 16 July 2015.
[29] 'About Us', International Alpaca Association, aia.org.pe/.
[30] 'Ley que promueve la Crianza, Producción, Comercialización y Consumo de los Camélidos Sudamericanos Domésticos Alpaca y Llama, FAOLEX'.

recovery. By the 1960s, overhunting had left the vicuña critically endangered and on the verge of extinction, with a mere 6,000 in the wild in 1974. Since then, however, a ban on international trade and the establishment of special vicuña reserves at Pampa Galeras and Salinas y Aguada Blanca have enabled the species to make an impressive recovery, to the extent that its fleece can once again be shorn using the traditional Inca *chakku* technique. By 2015, it was estimated that there were 340,000 vicuñas in the Andean highlands of Peru, Bolivia, Argentina and Chile, with the vast majority concentrated in Peru.[31] The species has also been successfully acclimatised in Ecuador, where the first official vicuña shearing took place in 2017.[32] The vicuña thus appears to offer a tangible example of what can be achieved when sufficient energy and resources are put into conservation. It is sometimes cited as a successful example of the sustainable use paradigm and a possible model for the management of other species.

Despite these encouraging signs, however, there is no room for complacency. On the one hand, conservationists have expressed fears about the possible stress caused to vicuñas by catch and release shearing, which may impact on their health, fertility and physical well-being. Vicuñas that have been shorn may be unable to cope with temperature extremes high in the Andes; human interference during the *chakku* may lead to the separation of *crias* from mothers; injuries may be sustained during shearing; shearing may cause stress to pregnant females, resulting in miscarriage; and muscle damage caused during the *chakku* may prevent the animals from escaping from predators and impair their ability to find food or shelter.[33] Criticisms have also been made of recent attempts to subject vicuñas to captive management, which may trigger changes in behaviour and increase the animals' vulnerability to disease.[34] On the other hand, poaching continues to present a problem, fuelled in part by the legalisation of the wool trade. With trade in legally obtained vicuña wool permitted since 1997, it is harder to identify the origin of fleeces and easier for poachers to smuggle illegal skins onto the market. This, coupled with a rise in demand for vicuña wool, has led to an increase in

[31] 'Legalizing Rhino Horn Trade Won't Save Species', *National Geographic*, 8 January 2015.
[32] 'Luego de 29 años, se Esquila a las Vicuñas', *El Comercio*, 7 September 2017.
[33] Cristian Bonacic, Jessica Gimpel and Pete Goddard, 'Animal Welfare and Sustainable Use of the Vicuña' in Iain Gordon (ed.), *The Vicuña: The Theory and Practice of Community-Based Wildlife Management* (New York: Springer, 2009), pp. 49–62.
[34] 'Legalizing Rhino Horn Trade Won't Save Species', *National Geographic*, 8 January 2015.

illegal hunting. The transition from conservation to sustainable use will thus need to be managed carefully if the vicuña is to survive in the long term.

Musk Deer

The musk deer is still heavily exploited for its musk, which continues to be used in perfumes (despite the existence of synthetic substitutes) and now increasingly in Chinese medicine, as both a sedative and an aphrodisiac. It is severely endangered throughout its range, with alarming drops in population reported in India, China and Russia. Male deer produce around 25 grams of musk each, so at least forty have to die to obtain just one kilogram of musk.[35] Killing, moreover, is indiscriminate, with many females and juveniles perishing in poachers' snares. A report by TRAFFIC Europe in 1999 claimed that 160 deer were killed to obtain just two pounds of musk.[36] The musk deer was placed on CITES Appendix I in 1976 and is considered at risk of extinction.[37]

Major efforts are being made to preserve the musk deer, many of which reprise nineteenth-century tactics. First, hunting bans and trade bans have been implemented in several Asian countries and reserves created for the protection of the deer. China upgraded the status of the musk deer to State Protection Scale Grade 1 in 2002, giving it the same level of protection accorded to the giant panda.[38] India established the Kedarnath Musk Deer Sanctuary in 1972 in the Chamoli district of Uttarakhand and later created two more reserves in Pithoragarh in 2008.[39] Nepal mobilised its army in 2010 to arrest poachers and smugglers and destroy any musk deer traps they discovered (1,500 were found in just a few months).[40] Despite heightened awareness, the musk deer population continues to fall, with conservation efforts marred by severe underfunding and lack of appropriately trained personnel. In Mongolia, where hunting musk deer has been illegal since 1953, an average of 2,000 male musk deer were poached annually between 1996 and 2001.[41] In India, the number of musk deer in the Kedarnath sanctuary fell from 600–1,000 in 1986 to under 100 in 2016, a decline blamed on

---

[35] 'Fading Scent of the Musk Deer', *Down to Earth*, 30 September 2016.
[36] 'Perfume Trade Decimates Musk Deer', *United Press International*, 6 July 1999.
[37] 'Scent of the End for Tigers and Deer', *South China Morning Post*, 17 December 1997.
[38] 'Musk Deer Joins Giant Panda on China's Most Endangered Species List', *Xinhua General News Service*, 5 November 2002.
[39] 'Musk Deer to Get Two Reserves in Pithoragarh', *The Pioneer*, 28 June 2008.
[40] 'Nepali Army in Operation to Save Musk Deer', *Himalayan Times*, 3 June 2010.
[41] 'Musk Deer under Threat in Russia, Mongolia', *US Newswire*, 15 July 2004.

insufficient funding and a lack of forest guards – only six to police a park spanning 2,390 square kilometres.[42]

As well as protecting musk deer in the wild, several attempts were made in the second half of the twentieth century to breed musk deer in captivity and rear them on farms for their musk. These have had mixed results. In India, a captive breeding programme was initiated at Kedarnath Musk Deer Sanctuary in 1982, but this ended in failure; only one deer, a female named Pallavi, remained at the sanctuary in 2010, the rest succumbing to stomach complaints, pneumonia and snake bites.[43] In China, breeding programmes have been more successful, with an experimental centre for musk deer domestication established in Shaanxi province in 1958 and a scheme begun in 1982 to resettle musk deer on the islands of the Zhoushan archipelago off the coast of Shanghai, where they could be better monitored and protected.[44] Despite these apparent successes, however, domestication has not been without problems and its long-term future remains in doubt. Inbreeding, stress and poor nutrition are serious issues among farmed populations, leading many captive deer to suffer diarrhoea, pneumonia and even birth defects, while the focus on captive animals does not help protect wild populations and may lead to a loss of genetic diversity.[45] This has prompted one researcher, Dr Mike Green, to propose a switch from farming musk deer to catching them in the wild, extracting their musk and releasing them again – essentially a Chinese version of the Peruvian *chakku de vicuñas*.[46] Domestication is thus no panacea, and, as in the nineteenth century, has proven inadequate as a means of wildlife conservation.

Civets

Civets are still farmed in Africa for their perfume, which is used by brands such as Chanel, Lancôme and Cartier. Around 200 civet farms existed in Ethiopia in 1999, exporting 1,000 kilograms of civet musk per year, mostly to France.[47] Captured civets are kept in small cages and,

[42] 'Fading Scent of the Musk Deer', *Down to Earth*, 30 September 2016.
[43] 'Mini-Zoo to Replace Musk Deer Centre at Kanchula Khark', *The Pioneer*, 26 August 2011.
[44] 'China – Environment: Musk Deer Make a Comeback in Hot Shanghai', *Inter Press Service*, 9 May 1996.
[45] Lan He et al., 'Welfare of Farmed Musk Deer: Changes in the Biological Characteristics of Musk Deer in Farming Environments', *Applied Animal Behaviour Science* 156 (2014), pp. 1–5.
[46] 'Scent of the End for Tigers and Deer', *South China Morning Post*, 17 December 1997.
[47] 'Scent from Hell', *The People*, 4 July 1999.

according to one observer, 'fed on a diet of cornmeal porridge ... raw and smoked beef' and the occasional chicken egg.[48] Civet is removed by pinning the animals down to expose their perineal glands, squeezing the gland to release the scent and scraping the latter off with a spatula. Animal rights campaigners claim that the extraction process is cruel and that nearly 40 per cent of trapped civets die from shock in the first few days after their capture.[49]

In Asia, the perfume industry has largely ceased – at least with regard to exports to the West – but civets continue to be exploited in other ways. In China, the animals are now farmed extensively for their meat, and were held responsible for the transmission of the SARS virus in the early 2000s.[50] In India, civets are poached for their nails, meat, oil and skin, while in Southeast Asia they have become victims of the exotic pet trade; in 2016 there were an estimated 10,000 civets kept as pets across Indonesia alone.[51]

More recently, civets have been enlisted into the coffee production process in the Philippines, Indonesia and Vietnam, following the 'discovery' that coffee beans digested and excreted by the cats possess a superior, chocolatey flavour. Initially, so-called civet-poo coffee – *kopi luwak* – was collected from the droppings of wild civets, who were believed to select only the choicest berries in their scavenging.[52] Quickly, however, the industry expanded, and the majority of beans are now obtained from farmed civets, kept in tiny cages and forced to ingest an unnatural quantity of coffee beans.[53] The future of the civet, therefore, looks rather bleak, although there are hopes that consumer action, educational campaigns and captive breeding may gradually improve the lot of this persecuted species. In 2013, for instance, UK retailers Harrods and Selfridges stopped stocking civet coffee in response to campaigns by animal welfare organisations.[54] Whether this is enough to save the civet from abuse remains to be seen; for now, the civet continues to suffer in the service of human luxury.

---

[48] 'Perfume: A Defense of Using the Civet', *New York Times*, 13 May 1973.
[49] 'Scent from Hell', *The People*, 4 July 1999.
[50] 'From Gourmet to Health Risk', *Sydney Morning Herald*, 21 June 2003.
[51] 'Indonesian Activists Against Exotic Pets', *The Straits Times*, 16 May 2016.
[52] 'Coffee Connoisseurs' Demands Set off a Gold Rush', *International Herald Tribune*, 19 April 2010.
[53] 'WSPA International: The Cruelty behind the World's Most Expensive Coffee', *Business Wire*, 13 September 2013.
[54] 'Selfridges Stops Selling Luxury Cat Dung after Campaign over Treatment of Animals in Rainforests of Indonesia', *Mail Online*, 21 January 2014.

Parrots

Parrots are still victims of the exotic pet trade, which has shifted increasingly from Europe and the USA to the Middle East and Asia. Thousands of birds are captured every year to meet consumer demand, and between 50 and 75 per cent of these perish in transit from stress, disease, overcrowding, asphyxiation and dehydration.[55] In consequence, parrot numbers in the wild are plummeting and many species are critically endangered. According to a recent scientific study, 111 of the 398 known parrot species are threatened with extinction, while the most popular pet species of the nineteenth century, the African grey, has already become locally extinct in areas of the Democratic Republic of the Congo (now the main supplier of the species), Kenya, Rwanda, Gabon and Uganda.[56] Captive parrots, deprived of company and stimulation, often develop neuroses and engage in self-mutilation.[57] The screaming, feather-plucking parrot of Victorian advice columns is still very much present in the modern world.

Wildlife organisations and governments have taken action to try to protect these exploited birds, seeking ways to counter habitat loss and the appalling casualties of the international pet trade. In 1992, the USA took an important step when it prohibited the importation of wild-caught birds.[58] In 2007, the European Union also instituted an import ban – a measure supplemented in 2016 by the upgrading of the African grey parrot to Appendix I of CITES, making all trade in this popular species illegal.[59] The difficulty of enforcing import bans, however, makes the success of these measures uncertain, while some fear that the export ban will simply drive up prices and push the trade underground. The future of African parrots hangs in the balance.

Monkeys

Like parrots, monkeys remain popular exotic pets and a thriving market exists to supply them. Since the USA banned monkey imports in 1975,

[55] 'New Report: US Demand Fuelling Illegal Capture and Trade of Certain Endangered Mexican Parrots', *US Fed News*, 14 February 2007.
[56] George Olah, Stuart H. M. Butchart, Andy Symes, Iliana Medina Guzmán, Ross Cunningham, Donald J. Brightsmith and Robert Heinsohn, 'Ecological and Socio-Economic Factors Affecting Extinction Risk in Parrots', *Biodiversity and Conservation* 25 (2016), pp. 205–23; 'Total Ban in Trade in Wild African Grey Parrots', WWF, 2 October 2016, wwf.panda.org/wwf_news/?279870/African%2DGrey%2DParrots.
[57] 'Who's a Pretty Bird Brain?', *Vancouver Sun*, 30 January 2004.
[58] 'Worth Squawking About: Wild Bird Import Ban Hailed', *The West Briton*, 25 January 2007.
[59] 'African Grey Parrot Has Global Summit to Thank for Protected Status', *The Guardian*, 2 October 2016.

captive breeding facilities have emerged in the country to meet the demand for young primates – especially capuchins, squirrel monkeys and marmosets – with babies frequently being removed from their mother's care prematurely and given to human owners. Alongside this ethically questionable but legal trade, a huge illegal trade in monkeys persists across Africa, Asia and South America, which sees animals trapped in the wild and smuggled across national frontiers into North America, Europe and, increasingly, the Middle and Far East. In 2017, an undercover BBC investigation exposed the shocking trade in baby chimpanzees in Senegal, showing how easily export permits could be forged to allow the removal of endangered species from their native countries.[60] This booming illegal traffic in primates not only threatens the survival of multiple primates but also raises concerns over the transmission of human diseases such as tuberculosis, Herpes B, salmonella and Ebola, all of which can be carried by monkeys.[61]

The keeping of primates as pets also has major welfare implications. Monkeys are not domestic animals and they cannot be tamed like dogs or cats; they destroy furniture, urinate to mark their territories and almost always end up biting their owners unless – as often happens – their canine teeth are removed. Most pet monkeys lack the space, diet and social stimulation they need to develop into healthy adults, and, as a consequence, they suffer from illnesses such as diabetes (caused by excessive sugar in their diet), rickets (caused by lack of vitamin D) and depression, which often expresses itself through continual rocking or self-mutilation. In 2013, the RSPCA rescued a marmoset named Milo who had been kept in a dirty bird cage without light or toys, fed only cherries and grapes and taken to the pub by his owner; in 2015, the Society seized three capuchin monkeys who had lived for years in a cold outbuilding and were suffering from arthritis and diabetes.[62] With an estimated 5,000 pet monkeys in the UK, cases like these probably represent the tip of the iceberg, and animal welfare organisations are calling on the government to prohibit the keeping of primates as pets – a practice already illegal in fifteen EU countries, over twenty

---

[60] David Shukman and Sam Piranty, 'The Secret Trade in Baby Chimps', *BBC News*, 30 January 2017.
[61] 'Cruel Owner Took Monkey down the Pub', *Daily Mirror*, 9 December 2013; 'Calls on UK Government to Ban People Keeping Monkeys as Pets', *The Express*, 25 September 2015.
[62] 'RSPCA Seize Marmoset Monkey', *Daily Mail*, 27 April 2017.

US states, India, Brazil, Israel, Mexico and Singapore, but (at the time of writing) still permitted in Britain.[63] Monkeys thus continue to be commodified in the modern world, while ostriches, elephants, civets and parrots all suffer in the production of human luxuries. The age of fashion victims is far from over.

[63] 'Monkey Business', *Express Online*, 10 December 2015.

References

Archival Sources

Archivo Digital de la Legislación del Perú

Decreto Estableciendo la Prohibición de Extraer del Territorio Peruano las Alpacas, Vicuñas y Animales que Proceden del Cruzamiento de Ambas Razas, 8 October 1868

Archivo General de Indias, Seville

AGI Lima 651
AGI Lima 652
AGI Indiferente 1549

British Library

'The Bird of Paradise', RSPB Pamphlet Number 20, 1895
'The Trade in Birds' Feathers', RSPB Pamphlet Number 28, 1895
'Moulted Plumes', RSPB Pamphlet Number 60, 1908
'How the Osprey Feathers Are Procured', RSPB Pamphlet Number 61, 1909

Chethams's Library, Manchester

Animal Care Journal, Belle Vue Gardens, Jennison Collection, F.5.04

London Metropolitan Archives

LMA 4605/02/002

Parliamentary Papers

Great Britain Parliament, *An Act to Consolidate and Amend the Several Laws Relating to the Cruel and Improper Treatment of Animals, and the*

Mischiefs Arising from the Driving of Cattle, and to Make Other Provisions in Regard Thereto, 1835: 5 & 6 William 4 c.59

Great Britain Parliament, *An Act for the More Effectual Prevention of Cruelty to Animals*, 1849: 12 & 13 Victoria c.92

Great Britain Parliament, *An Act for the Preservation of Sea Birds*, 1869: 32 & 33 Victoria c.17

Treaty Series, 1912. Convention between the United Kingdom, the United States, Japan and Russia, Respecting Measures for the Preservation and Protection of the Fur Seals of the North Pacific Ocean. Signed at Washington, July 7, 1911, House of Commons, Command Papers, Cd 6034 (London: His Majesty's Stationery Office, 1912)

Royal Pharmaceutical Society Museum

Collection Reference MA3
Collection Reference YBC3

State Library of New South Wales

Annotated Watercolour Sketches by Santiago Savage, 1857–1858, Being a Record of Charles Ledger's Journeys in Peru and Chile, MLMSS 630/1

Letter from Thomas Holt to Sir Henry Parkes, 23 May 1873, *Sir Henry Parkes Papers*, 1833–96

Victoria and Albert Museum

Hummingbird earrings, c.1865, M.11:1, 2-2003

R. Norman, 'Grimaldi and the Alpaca in the Popular Pantomime of the RED DWARF', 11 January 1813, H. Beard Print Collection

Wildlife Conservation Society Archives

William T. Hornaday, *Scrapbook Collection on the History of Wild Life Protection and Extermination*, Vol. 5, Wildlife Conservation Society Archives Collection, 1007-04-05-000-a

Newspapers and Periodicals

Aberdeen Journal
Adelaide Observer
The Advocate of Peace
Africa News
The Animals' Friend
The Animal World
The Anti-Slavery Reporter

Associated Press International
Associated Press Online
The Associated Press State & Local Wire
The Bazaar, the Exchange and Mart and Journal of the Household
Belfast News-Letter
Berrow's Worcester Journal
Bird Lore
Birmingham Daily Post
Blackwood's Edinburgh Magazine
Bradford Observer
Brighton Patriot
Bristol Mercury
Business Wire
Caledonian Mercury
Canadian Record of Science
Chambers's Journal of Popular Literature, Science, and Art
Chelmsford Chronicle
Chester Observer
Collection for Improvement of Husbandry and Trade
El Comercio
The Contemporary Review
Cornish Guardian
The Cornishman
Courier and Argus
Daily Mail
Daily Mirror
Daily News
Derby Mercury
Devon and Exeter Gazette
Devon Journal
Down to Earth
Dundee Courier and Argus
The Edinburgh Magazine
The Era
Essex Standard
Eturbo News
Eureka Times-Standard
The Evening News
Evening Telegraph and Star and Sheffield Daily Times
The Express
Express Online
Freeman's Journal
The Friendly Companion
Gazette de Charleroi
The Gentleman's Magazine
Glasgow Herald
The Graphic

The Guardian
Hampshire Advertiser
Hampshire Telegraph
Himalayan Times
Huddersfield Daily Chronicle
Hull Daily Mail
Hull Packet and East Riding Times
Humanity: The Journal of the Humanitarian League (renamed *The Humanitarian* in 1902)
The Illustrated London News
The International Herald Tribune
Inter Press Service
Jackson's Oxford Journal
Journal of the Society for the Preservation of the Wild Fauna of the Empire
Journal of the Society of Arts
The Ladies' Cabinet of Fashion, Music and Romance
The Ledger
Leeds Mercury
The Leisure Hour
Lincoln Journal Star
Liverpool Mercury
Living Age
Lloyd's Illustrated Newspaper
Lloyd's Weekly Newspaper
The London Reader
Mail Online
Maitland Mercury
Manchester Courier
The Manchester Guardian
Manchester Times
Manchester Times and Gazette
Melbourne Argus
The Morning Chronicle
The Morning Post
Music Trade Review
My Republica
National Geographic
The National Review
The Newcastle Courant
New Straits Times
New York Daily Tribune
New York Sun
New York Times
New York Tribune
The North American Review
Northampton Mercury
North-Eastern Daily Gazette

Northern Echo
Northern Star
North Wales Chronicle
The Nottinghamshire Guardian
Once a Week
The Ottawa Citizen
Our Young Folk's Weekly Budget
Pall Mall Gazette
Penny Illustrated Paper
Penny London Post
Penny Magazine of the Society for the Diffusion of Useful Knowledge
The People
El Peruano
Philosophical Transactions
The Pioneer
Portsmouth Evening News
Post Boy
Preston Chronicle
Pretoria News
Proceedings of the Royal Geographical Society and Monthly Record of Geography
Punch
The Review of Reviews
Reynold's Miscellany
Royal Cornwall Gazette
The Scientific Monthly
SeeNews Latin America
Semanario de Agricultura, Industria y Comercio
Semanario de Agricultura y Artes Dirigido a los Parrócos
Sheffield and Rotherham Independent
Sheffield Daily Telegraph
Sheffield Independent
South Australian Register
Southampton Herald
South China Morning Post
The Spectator
The Standard
The Star
States News Service
St Petersburg Times
The Straits Times
The Sunday Telegraph
Sunderland Daily Echo
Suplemento al Peruano
Sydney Morning Herald
Tamworth Herald
Targeted News Service

The Telegraph
Thai News Service
The Times
The Times of India
Trewman's Exeter Flying Post
United Press International
US Fed News
US Newswire
Uzbekistan National News Agency
Vancouver Sun
Washington Post
Weekly Standard and Express
The West Briton
Western Times
World Poultry
Xinhua Economic News Service
Xinhua General News Service
Yorkshire Evening Post
York Herald

Printed Primary Sources

Alexander, T., *Across the Great Craterland to the Congo* (New York: Alfred A. Knopf, 1924)

Anon., *Documentos para la Historia del Río de la Plata*, Vol. III (Buenos Aires: Compañía Sud-Americana de Billetes de Banco, 1913)

 List of Animals in the Liverpool Zoological Gardens (Liverpool: Ross and Nightingale, 1839)

 The Most Severe Case of Mr Thomas Chapman who First Discovered the Means of Making the Fur of the Seal Available (London: C. Cox, 1818)

 The Parrot Keeper's Guide (London: Thomas Dean and Son, 1857)

 Report of the Council and Auditors of the Zoological Gardens of London (London: Taylor and Francis, 1853)

 Report of the Council and Auditors of the Zoological Gardens of London (London: Taylor and Francis, 1857)

Bennett, George, *The Third Annual Report of the Acclimatisation Society of New South Wales* (Sydney: Joseph Cook, 1864)

Bingley, William, *Useful Knowledge: or A Familiar Account of the Various Productions of Nature, Mineral, Vegetable and Animal, which Are Employed for the Use of Man* (London: Baldwin, Craddock and Joy, 1821)

Caldas, Francisco José de, 'Memoria sobre la Importancia de Connaturalizar en el Reino la Vicuña del Perú y Chile' in Francisco José de Caldas, *Obras Completas de Francisco José de Caldas* (Bogotá: Imprenta Nacional, 1966)

Cobo, Bernabé, *Historia del Nuevo Mundo* (Seville: Imprenta de E. Rasco, 1895)

Collinson, Joseph, *How Sealskins Are Obtained* (London: Animal's Friend Society, 1910)

Cornish, Charles John, *Life at the Zoo: Notes and Traditions of the Regent's Park Gardens* (London: Seeley, 1895)

Critchell, James Troubridge and Raymond, Joseph, *A History of the Frozen Meat Trade* (London: Constable, 1912)

Cumberland, C., *The Guinea Pig, or Domestic Cavy, for Food, Fur and Fancy* (London: L. Upcott Gill, 1886)

Danson, William, *Alpaca, the Original Peruvian Sheep before the Spaniards Invaded South America, for Naturalisation in Other Countries* (Liverpool: M. Rourke, 1852)

De la Vega, Garcilaso, *Primera Parte de los Comentarios Reales* (Madrid: Imprenta de Doña Catalina Piñuela, 1829)

Deville, M. E. *Considérations sur les Avantages de la Naturalisation en France de l'Alpaca* (Paris: Imprimerie de L. Martinet, 1851)

Downham, C. F., *The Feather Trade: The Case for the Defence* (London: London Chamber of Commerce, 1911)

Elliott, Henry W., *Report on the Condition of the Fur-Seal Fisheries of the Pribylov Islands in 1890* (Paris: Chamerat et Renouard, 1893)

Flattery, M. Douglas, *The Truth about the Fur-Seal Question* (Danville: Edward Fox, 1897)

Hagenbeck, Carl, *Beasts and Men*, translated by Hugh S. R. Elliot and A. G. Thacker (London: Longmans, 1912)

Heath, Harold, *Special Investigation of the Alaska Fur-Seal Rookeries, 1910* (Washington: Government Printing Office, 1911)

Holder, Charles Frederick, *The Ivory King: A Popular History of the Elephant and Its Allies* (New York: Charles Scribner's Sons, 1902)

Hornaday, William Temple, *The Last Fight for the Persecuted Fur Seal* (New York: Office on Game Protection and Preserves, 1912)
  Our Vanishing Wild Life: Its Extermination and Preservation (New York: Clark and Fritts, 1913)

Jordan, David Starr, *Observations on the Fur Seals of the Pribilof Islands, Preliminary Report* (Washington: Government Printing Office, 1896)

Lankester, Edwin, *The Uses of Animals in Relation to the Industry of Man* (London: Robert Hardwicke, 1860)

Ledger, George, *The Alpaca: Its Introduction into Australia and the Probabilities of its Acclimatisation There. A Paper Read before the Society of Arts, London. Republished by the Acclimatisation Society of Victoria* (Melbourne: Mason and Firth, 1861)

Markham, Clements R. (ed. and trans.), *The Travels of Pedro Cieza de León, A.D. 1532–50, Contained in the First Part of his Chronicle of Peru* (London: Hakluyt Society, 1864)

Markham, Clements, *Travels in Peru while Superintending the Collection of Cinchona Plants and Seeds in South America, and Their Introduction into India* (London: John Murray, 1862)

Markham, Colonel Frederick, *Shooting in the Himalayas: A Journal of Sporting Adventures and Travel* (London: Richard Bentley, 1854)

Miller, John, *The Memoirs of General Miller* (New York: AMS Press, 1973)

Molina, Juan Ignacio, *Compendio de la Historia Geográfica, Natural y Civil del Reyno de Chile* (Madrid: Sancha, 1788)

Moore, E. D., *Ivory Scourge of Africa* (New York and London: Harper and Brothers Publishers, 1931)

Mosenthal, Julius de and Harting, James Edmund, *Ostriches and Ostrich Farming* (London: Trübner and Co., 1877)

O'Leary, Daniel Florencio, *Memorias del General O'Leary* (Caracas: Imprenta de El Monitor, 1883)

Patterson, Arthur, *Notes on Pet Monkeys and How to Manage Them* (London: L. Upcott Gill, 1888)

Piesse, Charles H., *Piesse's Art of Perfumery*, fifth edition (London: Piesse and Lubin, 1895)

Rennie, James, *The Art of Preserving the Hair on Popular Principles: Including an Account of the Diseases to which it Is Liable* (London: Septimus Prowett, 1826)

Rimmel, Eugene, *The Book of Perfumes* (London: Chapman and Hall, 1865)

Ross, Alexander, *A Treatise on Bear's Grease* (London: Printed for the Author, 1795)

RSPB, *Feathers and Facts: Statement by the Royal Society for the Protection of Birds* (London: Royal Society for the Protection of Birds (RSPB), 1911)

Ruíz, Hipólito, *The Journals of Hipólito Ruíz, Spanish Botanist in Peru and Chile 1777–1788*, translated by Richard Evans Schultes and María José Nemry von Thenen de Jaramillo-Arango (Portland: Timber Press, 1998)

Salt, Henry, *Animal Rights Considered in Relation to Social Progress* (New York: Macmillan, 1894)

Sawer, J. C., *Odorographia: A Natural History of the Raw Materials and Drugs Used in the Perfume Industry* (London: Gurney and Jackson, 1894)

Sims, Edwin W., *Report on the Alaskan Fur-Seal Fisheries, 31 August 1906* (Washington: Government Printing Office, 1906)

SPB, *Annual Reports* (London: J. Davy and Sons, 1891–1900)

Tavernier, Jean-Baptiste, *The Six Voyages of John Baptista Tavernier, Baron of Aubonne through Turky, into Persia and the East-Indies, for the Space of Forty Years*, translated by Daniel Cox (London: William Godbid, Robert Littlebury and Moses Pitt, 1677)

Tschudi, Johann von, *Travels in Peru, during the Years 1838–1842*, translated by Thomasina Ross (London: David Bogue, 1857)

Walker Scott, Alexander, *Mammalia Recent and Extinct* (Sydney: Thomas Richards, 1873)

Walton, William, *The Alpaca: Its Naturalisation in the British Isles Considered as a National Benefit, and as an Object of Immediate Utility to the Farmer and Manufacturer* (New York: Office of the New York Farmer and Mechanic, 1845)

  A Memoir Addressed to Proprietors of Mountain and Other Waste Lands and Agriculturalists of the United Kingdom, on the Naturalisation of the Alpaca (London: Smith, Elder and Co., 1841)

Wood, J. G., *Illustrated Natural History* (London: Routledge, 1876)

Secondary Sources

Abreyava Stein, Sarah, *Plumes: Ostrich Feathers, Jews and a Lost World of Global Commerce* (New Haven: Yale University Press, 2008)

Actman, Jani, 'Woolly Mammoth Ivory Is Legal, and That's a Problem for Elephants', *National Geographic*, 23 August 2016

Aguirre, Robert, *Informal Empire: Mexico and Central America in Victorian Culture* (Minneapolis: University of Minnesota Press, 2005)

Albritton Jonson, Frederik, *Enlightenment's Frontier: The Scottish Highlands and the Origins of Environmentalism* (New Haven: Yale University Press, 2013)

Alpers, Edward, 'The Ivory Trade in Africa: An Historical Overview' in Doran H. Ross (ed.), *Elephant: The Animal and Its Ivory in African Culture* (Los Angeles: Fowler Museum of Cultural History, 1992), pp. 367–81

Amato, Sarah, *Beastly Possessions: Animals in Victorian Consumer Culture* (Toronto: University of Toronto Press, 2015)

Baratay, Elisabeth and Hardouin-Fugier, Eric, *Zoo: A History of Zoological Gardens in the West* (London: Reaktion Books, 2002)

Barrell, John, *The Spirit of Despotism: Invasions of Privacy in the 1790s* (Oxford: Oxford University Press, 2006)

Barrow, Mark V. *Nature's Ghosts: Confronting Extinction from the Age of Jefferson to the Age of Ecology* (Chicago: University of Chicago Press, 2009)

Beckert, Sven, *Empire of Cotton: A New History of Global Capitalism* (London: Penguin, 2014)

Beetham, Margaret and Boardman, Kay (eds), *Victorian Women's Magazines: An Anthology* (Manchester: Manchester University Press, 2001)

Beinart, William, *The Rise of Conservation in South Africa: Settlers, Livestock and the Environment, 1770–1950* (Oxford: Oxford University Press, 2003)

Beinart, William and Hughes, Lottie, *Environment and Empire* (Oxford: Oxford University Press, 2007)

Berg, Maxine, *Luxury and Pleasure in Eighteenth-Century Britain* (Oxford: Oxford University Press, 2005)

Bonacic, Cristian, Gimpel, Jessica and Goddard, Pete, 'Animal Welfare and Sustainable Use of the Vicuña' in Iain Gordon (ed.), *The Vicuña: The Theory and Practice of Community-Based Wildlife Management* (New York: Springer, 2009), pp. 49–62

Boomgaard, Peter, 'Oriental Nature, Its Friends and Its Enemies: Conservation of Nature in Late-Colonial Indonesia, 1889–1949', *Environment and History* 5 (1999), pp. 257–92

Brockway, Lucille, *Science and Colonial Expansion: The Role of the British Royal Botanic Gardens* (New Haven: Yale University Press, 2002)

Brown, Matthew, *Adventuring through Spanish Colonies: Simón Bolívar, Foreign Mercenaries and the Birth of New Nations* (Liverpool: Liverpool University Press, 2006)

Bulliet, Richard, *Hunters, Herders and Hamburgers* (New York: Columbia University Press, 2005)

Camerini, Jane, 'Wallace in the Field', *Osiris* [2nd series] 11 (1996), pp. 44–65

Clark, Fiona, *Hats* (London: Batsford, 1982)

Collins, E. J. T., 'Food Adulteration and Food Safety in Britain in the 19th and Early 20th Centuries', *Food Policy* 18:2 (1993), pp. 95–109

Connor, Neil, 'Booming Trade in Mammoth Ivory Fuels Fears over Elephants', *The Telegraph*, 2 May 2017

Cooper Busch, Briton, *The War against the Seals: A History of the North American Seal Fishery* (Montreal: McGill-Queen's University Press, 1985)

Cowie, Helen, *Conquering Nature in Spain and Its Empire 1750–1850* (Manchester: Manchester University Press, 2011)

Exhibiting Animals in Nineteenth-Century Britain: Empathy, Education, Entertainment (Basingstoke: Palgrave Macmillan, 2014)

Llama (London: Reaktion Books, 2017)

Crawford, Matthew, *The Andean Wonder Drug: Cinchona Bark and Imperial Science in the Spanish Atlantic, 1630–1800* (Pittsburgh: University of Pittsburgh Press, 2016)

Crespy, Daniel, Bozonnet, Marianne and Meier, Martin, '100 Years of Bakelite, the Material of 1,000 Uses', *History of Science* 47 (2008), pp. 3322–8

Cronin, J. Keri, *Art for Animals: Visual Culture and Animal Advocacy, 1870–1914* (University Park: Pennsylvania State University Press, 2018)

Crosby, Alfred, *The Columbian Exchange: Biological and Cultural Consequences of 1492* (Westport: Greenwood Press, 1972)

Cushman, Gregory T., *Guano and the Opening of the Pacific World: A Global Ecological History* (Cambridge: Cambridge University Press, 2013)

Custred, Glynn, 'Hunting Technologies in Andean Cultures', *Journal de la Société des Américanistes* LXVI (1979), pp. 7–19

Donald, Diana, *Women against Cruelty: Protection of Animals in Nineteenth-Century Britain* (Manchester: Manchester University Press, 2019)

Dorsey, Kurk, 'Putting a Ceiling on Sealing: Conservation and Cooperation in the International Arena, 1909–1911', *Environmental History Review* 15:3 (1991), pp. 27–45

Whales and Nations: Environmental Diplomacy on the High Seas (Seattle: University of Washington Press, 2013)

Doughty, Robin, *Feather Fashions and Bird Preservation: A Study in Nature Protection* (Berkeley: University of California Press, 1975)

Drayton, Richard, *Nature's Government: Science, Imperial Britain and the 'Improvement' of the World* (New Haven: Yale University Press, 2000)

Earle, Rebecca, *The Return of the Native: Indians and Myth-Making in Spanish America, 1810–1930* (Durham: Duke University Press, 2007)

Fan, Fa-ti, 'Victorian Naturalists in China: Science and Informal Empire', *British Journal for the History of Science* 36:1 (2003), pp. 1–26

Flindell-Klarén, Peter, *Peru: Society and Nationhood in the Andes* (New York: Oxford University Press, 2000)

Flores Ochoa, Jorge, *Pastoralists of the Andes: The Alpaca Herds of Paratía* (Philadelphia: Institute for the Study of Human Issues, 1979)

Gates, Barbara T., *Kindred Nature: Victorian and Edwardian Women Embrace the Living World* (Chicago: University of Chicago Press, 1998)

Gentry, Roger L., *Behavior and Ecology of the Northern Fur Seal* (Princeton: Princeton University Press, 1998)

Gillbank, Linden, 'A Paradox of Purposes: Acclimatization Origins of the Melbourne Zoo' in R. J. Hoage and William A. Deiss (eds), *New Worlds, New Animals: From Menagerie to Zoological Park in the Nineteenth Century* (Baltimore: Johns Hopkins University Press, 1996), pp.76-9

Gissibl, Bernhard, *The Nature of German Imperialism: Conservation and the Politics of Wildlife in Colonial East Africa* (New York: Berghahn, 2016)

Gómez-Centurión Jiménez, Carlos, *Alhajas para Soberanos: Los Animales Reales en el Siglo XVIII* (Madrid: Junta de Castilla y León, 2011)

Goody, Jack, 'Industrial Food' in Carole Counihan and Penny Van Esterik (eds), *Food and Culture: A Reader* (London: Routledge, 1997), pp. 338–56

Grier, Katherine C., *Pets in America: A History* (Chapel Hill: University of North Carolina Press, 2006)

Grove, Richard, *Green Imperialism: Colonial Expansion, Tropical Island Edens and the Origins of Environmentalism, 1600–1860* (Cambridge: Cambridge University Press, 1995)

Guerrini, Anita, *Experimenting with Humans and Animals: From Galen to Animal Rights* (Baltimore: Johns Hopkins University Press, 2003)

Hanson, Elizabeth, *Animal Attractions: Nature on Display in American Zoos* (Princeton: Princeton University Press, 2002)

Harrison, Brian, 'Animals and the State in Nineteenth-Century England', *English Historical Review* 88 (1973), pp. 786–820

He, Lan et al., 'Welfare of Farmed Musk Deer: Changes in the Biological Characteristics of Musk Deer in Farming Environments', *Applied Animal Behaviour Science* 156 (2014), pp. 1–5

Heise, Ursula K., *Imagining Extinction: The Cultural Meanings of Endangered Species* (Chicago: University of Chicago Press, 2016)

Hochadel, Oliver, 'Watching Exotic Animals Next Door: "Scientific" Observations at the Zoo (ca.1870–1910)', *Science in Context* 24:2 (2011), pp. 183–214

Hochschild, Adam, *King Leopold's Ghost: A Story of Greed, Terror and Heroism in Colonial Africa* (Basingstoke: Pan Macmillan, 2006)

Hotson, Elizabeth, 'The Alpacas Protecting 24,000 Christmas Turkeys', *BBC News*, 1 December 2016, www.bbc.co.uk/news/business-38133658

Howell, Philip, 'Animals, Agency and History' in Philip Howell and Hilda Kean, *Handbook for Animal–Human History* (London: Routledge, 2018), pp.197-221
    At Home and Astray: The Domestic Dog in Victorian Britain (Charlottesville and London: University of Virginia Press, 2015)

Irwin, Robert, 'Canada, Aboriginal Sealing, and the North Pacific Fur Seal Convention', *Environmental History* 20 (2015), pp. 57–82

Isenberg, Andrew, *The Destruction of the American Bison* (Cambridge: Cambridge University Press, 2001)

Kadwell, Miranda, Wheeler, Jane, et al., 'Genetic Analysis Reveals the Wild Ancestors of the Llama and the Alpaca', *Proceedings of the Royal Society, London* 268 (2001), pp. 2575–85

Kean, Hilda, *Animal Rights: Political and Social Change in Britain since 1800* (London: Reaktion Books, 1998)

Kemmerer, Lisa (ed.), *Animals and the Environment: Advocacy, Activism and the Quest for Common Ground* (Abingdon: Routledge, 2015)

Kemp, Christopher, *A Natural (& Unnatural) History of Ambergris* (Chicago: University of Chicago Press, 2012)

Kennedy, Dane, *The Last Blank Places: Exploring Africa and Australia* (Cambridge: Harvard University Press, 2013)

Kete, Kathleen, *The Beast in the Boudoir: Petkeeping in Nineteenth-Century Paris* (Berkeley: University of California Press 1994)

King, Anya H., *Scent from the Garden of Paradise: Musk and the Medieval Islamic World* (Leiden: Brill, 2017)

Kolbert, Elizabeth, *The Sixth Extinction: An Unnatural History* (London: Bloomsbury, 2014)

Kranzer, Jonas, 'Tickling and Clicking the Ivories: The Metamorphosis of a Global Commodity in the Nineteenth Century' in Bernd-Stefan Grewe and Karin Hofmeester (eds), *Luxury in Global Perspective: Objects and Practices 1600–2000* (Cambridge: Cambridge University Press, 2016), pp. 242–62

Krech III, Shepherd, *The Ecological Indian* (New York and London: Norton, 1999)

Lansbury, Coral, *The Old Brown Dog: Women, Workers and Vivisection in Edwardian England* (Madison: University of Wisconsin Press, 1985)

Leal, Claudia, *Landscapes of Freedom: Building a Postemancipation Society in the Rainforests of Western Colombia* (Tucson: University of Arizona Press, 2018)

Leneman, Leah, 'The Awakened Instinct: Vegetarianism and the Women's Suffrage Movement in Britain', *Women's History Review* 6:2 (1997), pp. 271–87

Lever, Christopher, *The Cane Toad: The History and Ecology of a Successful Colonist* (Otley: Westbury Academic Publishing, 2001)

Livingstone, David, *Putting Science in Its Place: Geographies of Scientific Knowledge* (Chicago: University of Chicago Press, 2003)

MacKenzie, John, *The Empire of Nature: Hunting, Conservation and British Imperialism* (Manchester: Manchester University Press, 1988)

Mallapur, Avanti and Choudhury, B. C., 'Behavioural Abnormalities in Captive Nonhuman Primates', *Journal of Applied Animal Welfare Science* 6:4 (2003), pp. 275–84

Mandala, Viajaya Ramadas, 'The Raj and the Paradoxes of Wildlife Conservation: British Attitudes and Experiences', *Historical Journal* 58:1 (2015), pp. 75–110

Martin, Janet, *Treasure of the Land of Darkness: The Fur Trade and Its Significance for Medieval Russia* (Cambridge: Cambridge University Press, 1986)

Matthews David, Alison, *Fashion Victims: The Dangers of Dress Past and Present* (London: Bloomsbury, 2015)

McCook, Stuart, '"It May Be Truth, but It Is not Evidence": Paul du Chaillu and the Legitimation of Evidence in the Field', *Osiris* [2nd series] 11 (1996), pp. 177–97

McIver, Stuart B., *Death in the Everglades: The Murder of Guy Bradley, America's First Martyr to Environmentalism* (Gainesville: University Press of Florida, 2003)

Mengoni Goñalons, Luis, 'Camelids in Ancient Andean Societies: A Review of the Zooarcheological Evidence', *Quaternary International* 185 (2008), pp. 59–68

Miller, Rory, 'The Wool Trade in Southern Peru, 1850–1915', *Ibero-Amerikanisches Archiv* 8:3 (1982), pp. 297–311

Moss, Cynthia, *Elephant Memories* (Chicago: University of Chicago Press, 2000)

Nance, Susan, *Animal Modernity: Jumbo the Elephant and the Human Dilemma* (New York: Palgrave Macmillan, 2015)

Nance, Susan (ed.), *The Historical Animal* (Syracuse: Syracuse University Press, 2015)

Neuman, Roderick P., *Imposing Wilderness: Struggles over Livelihood and Nature Preservation in Africa* (Berkeley: University of California Press, 1992)

Olah, George, Butchart, Stuart H. M., Symes, Andy, Guzmán, Iliana Medina, Cunningham, Ross, Brightsmith, Donald J. and Heinsohn, Robert, 'Ecological and Socio-Economic Factors Affecting Extinction Risk in Parrots', *Biodiversity and Conservation* 25 (2016), pp. 205–23

Osborne, Michael, *Nature, the Exotic and the Science of French Colonialism* (Bloomington: Indiana University Press, 1994)

Paddle, Robert, *The Last Tasmanian Tiger: The History and Extinction of the Thylacine* (Cambridge: Cambridge University Press, 2000)

Pearce, Fred, *The New Wild: Why Invasive Species Will Be Nature's Salvation* (London: Icon Books, 2016)

Pemberton, Neil, Strange, Julie-Marie and Worboys, Michael, *The Invention of the Modern Dog: Breed and Blood in Victorian Britain* (Baltimore: Johns Hopkins University Press, 2018)

Plumb, Christopher, *The Georgian Menagerie* (London: I. B. Tauris, 2015)

Pouillard, Violette, *Histoire des Zoos par les Animaux: Imperialisme, Contrôle, Conservation* (Ceyzérieu: Champ Vallon, 2019)

Quintero Toro, Camilo, *Birds of Empire, Birds of Nation* (Bogotá: Universidad de los Andes, 2012)

Raj, Kapil, *Relocating Modern Science: Circulation and the Construction of Knowledge in South Asia and Europe, 1650–1900* (Basingstoke: Palgrave Macmillan, 2007)

Rappaport, Erika Diane, *Shopping for Pleasure: Women and the Making of London's West End* (Princeton: Princeton University Press, 2000)

Reinarz, Jonathan, *Past Scents: Historical Perspectives on Smell* (Urbana: University of Illinois Press, 2014)

Ritvo, Harriet, *The Animal Estate: The English and Other Creatures in the Victorian Age* (Cambridge: Harvard University Press, 1987)

   Noble Cows and Hybrid Zebras: Essays on Animals and History (Charlottesville: University of Virginia Press, 2010)

   The Platypus and the Mermaid and Other Figments of the Classifying Imagination (Cambridge: Harvard University Press, 1997)

Robbins, Louise, *Elephant Slaves and Pampered Parrots: Exotic Animals in Eighteenth-Century Paris* (Baltimore: Johns Hopkins University Press, 2002)

Rothfels, Nigel, *Savages and Beasts: The Birth of the Modern Zoo* (Baltimore: Johns Hopkins University Press, 2002)

Sas-Rolfes, Michael, 'Assessing CITES: Four Case Studies' in Jon Hutton and Barnabus Dickson (eds), *Endangered Species Threatened Convention: The Past, Present and Future of CITES* (London: Earthscan, 2000), pp. 71–84

Schaffer, Simon, Roberts, Lissa, Raj, Kapil and Delbourgo, James (eds), *The Brokered World: Go-Betweens and Global Intelligence, 1770–1820* (Sagamore Beach: Watson Publishing International, 2009)

Schell, Patience, *The Sociable Sciences: Darwin and his Contemporaries in Chile* (Basingstoke: Palgrave Macmillan, 2013)

Schiebinger, Londa, *Plants and Empire: Colonial Bio-Prospecting in the Atlantic World* (Cambridge: Harvard University Press, 2004)

Shayt, David A., 'The Material Culture of Ivory Outside Africa' in Doran H. Ross (ed.), *Elephant: The Animal and Its Ivory in African Culture* (Los Angeles: Fowler Museum of Cultural History, 1992), pp. 367–81

Sheriff, Abdul, *Slaves, Spices and Ivory in Zanzibar* (Athens: Ohio University Press, 1987)

Shukman, David and Piranty, Sam, 'The Secret Trade in Baby Chimps', *BBC News*, 30 January 2017, www.bbc.co.uk/news/resources/idt-5e8c4bac-c236-4cd9-bacc-db96d733f6cf

Sivasundaram, Sujit, 'Trading Knowledge: The East India Company's Elephants in India and Britain', *Historical Journal* 48:1 (2005), pp. 27–63

Soluri, John, 'On the Edge: Fur Seals and Hunters along the Patagonian Littoral, 1860–1930' in Martha Few and Zeb Tortorici (eds), *Centering Animals in Latin American History* (Durham: Duke University Press, 2013), pp. 243–69

Spary, Emma, *Utopia's Garden: French Natural History from Old Regime to Revolution* (Chicago: University of Chicago Press, 2000)

Stearns, Peter N., *The Industrial Turn in World History* (London: Routledge, 2017)

Stephenson, Marcia, 'From Marvelous Antidote to the Poison of Idolatry: The Transatlantic Role of Andean Bezoar Stones during the Late Sixteenth and Early Seventeenth Centuries', *Hispanic American Historical Review* 90:1 (2010), pp. 3–39

Tague, Ingrid, *Animal Companions: Pets and Social Change in Eighteenth-Century Britain* (Philadelphia: Penn State University Press, 2015)

Torres, Hernán (ed.), *South American Camelids: An Action Plan for Their Conservation* (Gland: IUCN Species Survival Commission, 1992)

Trautmann, Thomas R., *Elephants and Kings: An Environmental History* (Chicago: University of Chicago Press, 2015)

Tucker Jones, Ryan, *Empire of Extinction* (Oxford: Oxford University Press, 2014)
    'Running into Whales: The History of the North Pacific from Below the Waves', *American Historical Review* 118:2 (2013), pp. 349–77

Vincent, Susan J., *Hair: An Illustrated History* (London: Bloomsbury, 2018)

Walkid, Emily, 'Saving the Vicuña: The Political, Biophysical and Cultural History of Wild Animal Conservation in Peru, 1964–2000', *American Historical Review* 125:1 (2020), pp. 54–88

Warsh, Molly, *American Baroque: Pearls and the Nature of Empire, 1492–1700* (Chapel Hill: University of North Carolina Press, 2018)

Weinbren, Dan, 'Against All Cruelty: The Humanitarian League, 1891–1919', *History Workshop Journal* 38 (1994), pp. 86–105

Wells, Philippa K., '"An Enemy of the Rabbit": The Social Context of Acclimatisation of an Immigrant Killer', *Environment and History* 12 (2006), pp. 297–324

Whitfield, Susan, *Life Along the Silk Road* (Berkeley: University of California Press, 2015)

Wilson, David, 'Racial Prejudice and the Performing Animals Controversy in Early Twentieth-Century Britain', *Society and Animals* 17 (2009), pp. 149–65

Woods, Rebecca J., 'From Colonial Animal to Imperial Edible: Building an Empire of Sheep in New Zealand, ca.1880–1900', *Comparative Studies of South Asia, Africa and the Middle East* 35:1 (2015), pp. 117–36

The Herds Shot around the World: Native Breeds and the British Empire, 1800–1900 (Chapel Hill: University of North Carolina Press, 2017)

Worster, Donald, *Nature's Economy: A History of Ecological Ideas*, second edition (Cambridge: Cambridge University Press, 1994)

Wylie, Dan, *Elephant* (London: Reaktion Books, 2008)

Yacobaccio, Hugo, 'The Historical Relationship between People and the Vicuña' in Iain Gordon (ed.), *The Vicuña: The Theory and Practice of Community-Based Wildlife Management* (New York: Springer, 2009), pp. 7–20

Websites

collections.vam.ac.uk/item/O86513/earrings-emanuel-harry/

perfumesociety.org/dudley-zoos-big-cats-go-crazy-salome/

'About Us', International Alpaca Association, aia.org.pe/

'Total Ban in Trade in Wild African Grey Parrots', WWF, 2 October 2016, wwf.panda.org/wwf_news/?279870/African%2DGrey%2DParrots

Index